Applied Theory of Functional Differential Equations

Mathematics and Its Applications (*Soviet Series*)

Volume 85

Applied Theory of Functional Differential Equations

by

V. Kolmanovskii

and

A. Myshkis

Department of Cybernetics,
Moscow Institute of Electronic Machinery,
Moscow, Russia, C.I.S.

KLUWER ACADEMIC PUBLISHERS

DORDRECHT / BOSTON / LONDON

Library of Congress Cataloging-in-Publication Data

Kolmanovskiĭ, Vladimir Borisovich.
Applied theory of functional differential equations /
V. Kolmanovskii and A. Myshkis.
p. cm. -- (Mathematics and its applications. Soviet series ;
85)
Includes bibliographical references.

1. Functional differential equations. I. Myshkis, A. D.
(Anatoliĭ Dmitrievich) II. Title. III. Series: Mathematics and its
applications (Kluwer Academic Publishers). Soviet series ; 85.
QA372.K77 1992
515'.35--dc20 92-35413

ISBN 978-90-481-4215-6

Published by Kluwer Academic Publishers,
P.O. Box 17, 3300 AA Dordrecht, The Netherlands.

Kluwer Academic Publishers incorporates
the publishing programmes of
D. Reidel, Martinus Nijhoff, Dr W. Junk and MTP Press.

Sold and distributed in the U.S.A. and Canada
by Kluwer Academic Publishers,
101 Philip Drive, Norwell, MA 02061, U.S.A.

In all other countries, sold and distributed
by Kluwer Academic Publishers Group,
P.O. Box 322, 3300 AH Dordrecht, The Netherlands.

Printed on acid-free paper

SERIES EDITOR'S PREFACE

'Et moi, ..., si j'avait su comment en revenir, je n'y serais point allé.'

Jules Verne

The series is divergent; therefore we may be able to do something with it.

O. Heaviside

One service mathematics has rendered the human race. It has put common sense back where it belongs, on the topmost shelf next to the dusty canister labelled 'discarded non-sense'.

Eric T. Bell

Mathematics is a tool for thought. A highly necessary tool in a world where both feedback and non-linearities abound. Similarly, all kinds of parts of mathematics serve as tools for other parts and for other sciences.

Applying a simple rewriting rule to the quote on the right above one finds such statements as: 'One service topology has rendered mathematical physics ...'; 'One service logic has rendered computer science ...'; 'One service category theory has rendered mathematics ...'. All arguably true. And all statements obtainable this way form part of the raison d'être of this series.

This series, *Mathematics and Its Applications*, started in 1977. Now that over one hundred volumes have appeared it seems opportune to reexamine its scope. At the time I wrote

> "Growing specialization and diversification have brought a host of monographs and textbooks on increasingly specialized topics. However, the 'tree' of knowledge of mathematics and related fields does not grow only by putting forth new branches. It also happens, quite often in fact, that branches which were thought to be completely disparate are suddenly seen to be related. Further, the kind and level of sophistication of mathematics applied in various sciences has changed drastically in recent years: measure theory is used (non-trivially) in regional and theoretical economics; algebraic geometry interacts with physics; the Minkowsky lemma, coding theory and the structure of water meet one another in packing and covering theory; quantum fields, crystal defects and mathematical programming profit from homotopy theory; Lie algebras are relevant to filtering; and prediction and electrical engineering can use Stein spaces. And in addition to this there are such new emerging subdisciplines as 'experimental mathematics', 'CFD', 'completely integrable systems', 'chaos, synergetics and large-scale order', which are almost impossible to fit into the existing classification schemes. They draw upon widely different sections of mathematics."

By and large, all this still applies today. It is still true that at first sight mathematics seems rather fragmented and that to find, see, and exploit the deeper underlying interrelations more effort is needed and so are books that can help mathematicians and scientists do so. Accordingly MIA will continue to try to make such books available.

If anything, the description I gave in 1977 is now an understatement. To the examples of interaction areas one should add string theory where Riemann surfaces, algebraic geometry, modular functions, knots, quantum field theory, Kac-Moody algebras, monstrous moonshine (and more) all come together. And to the examples of things which can be usefully applied let me add the topic 'finite geometry'; a combination of words which sounds like it might not even exist, let alone be applicable. And yet it is being applied: to statistics via designs, to radar/sonar detection arrays (via finite projective planes), and to bus connections of VLSI chips (via difference sets). There seems to be no part of (so-called pure) mathematics that is not in immediate danger of being applied. And, accordingly, the applied mathematician needs to be aware of much more. Besides analysis and numerics, the traditional workhorses, he may need all kinds of combinatorics, algebra, probability, and so on.

In addition, the applied scientist needs to cope increasingly with the nonlinear world and the extra mathematical sophistication that this requires. For that is where the rewards are. Linear models are honest and a bit sad and depressing: proportional efforts and results. It is in the nonlinear world that infinitesimal inputs may result in macroscopic outputs (or vice versa). To appreciate what I am hinting at: if electronics were linear we would have no fun with transistors and computers; we would have no TV; in fact you would not be reading these lines.

There is also no safety in ignoring such outlandish things as nonstandard analysis, superspace and anticommuting integration, p-adic and ultrametric space. All three have applications in both electrical engineering and physics. Once, complex numbers were equally outlandish, but they frequently proved the shortest path between 'real' results. Similarly, the first two topics named have already provided a number of 'wormhole' paths. There is no telling where all this is leading - fortunately.

Thus the original scope of the series, which for various (sound) reasons now comprises five subseries: white (Japan), yellow (China), red (USSR), blue (Eastern Europe), and green (everything else), still applies. It has been enlarged a bit to include books treating of the tools from one subdiscipline which are used in others. Thus the series still aims at books dealing with:

- a central concept which plays an important role in several different mathematical and/or scientific specialization areas;
- new applications of the results and ideas from one area of scientific endeavour into another;
- influences which the results, problems and concepts of one field of enquiry have, and have had, on the development of another.

The shortest path between two truths in the real domain passes through the complex domain.

J. Hadamard

La physique ne nous donne pas seulement l'occasion de résoudre des problèmes ... elle nous fait pressentir la solution.

H. Poincaré

Never lend books, for no one ever returns them; the only books I have in my library are books that other folk have lent me.

Anatole France

The function of an expert is not to be more right than other people, but to be wrong for more sophisticated reasons.

David Butler

Bussum, March 1992

Michiel Hazewinkel

Contents

Preface

The purpose of this book is to consider in sufficient detail problems and methods connected with the application of systems with memory, also called hereditary systems, that can be described by functional differential equations (FDEs).

A distinguishing feature of the FDEs under consideration is that the evolution rate of the processes described by such equations depends on the past history.

During the last two decades, the theory and application of FDEs has developed and spread to an extent never experienced earlier. In this book we study mathematical models described by FDEs which are applicable to phenomena of quite different natures. Appropriate applications include: immunology, nuclear power generation, heat transfer, track signal processing, regulation systems, medicine, economy, etc.

This book gives an introduction to the theory of FDEs. We have attempted to give the reader an insight into the wide environment in which this theory is embedded.

We have made every attempt to include the most important methods and techniques used in applications. The first chapter plays a special role. Its function is largely motivational, and it also serves to show 'where FDEs come from', and what kind of problems typically arise in applications.

This part of the book is also concerned with modeling. The basic view on scientific modeling is that a *model* is any 'simplified description of a system (etc.) that assists calculations and predictions' (Oxford English Dictionary). The main reason to model something is to provide for an efficient organization of information and experiences in order to enhance understanding and enable (wise) decision making.

Modeling is ubiquitous in human activities, because a model is a condensed representation of available information. Models are used to describe aspects of experience, and to predict, influence and regulate future developments.

We hope that this chapter is useful for mathematicians who want to find applications of their theories, and also for specialists who want to create new models.

Some model problems discussed in the first chapter will be taken up again in subsequent chapters, to test the available theory.

Numerous investigations have shown that temporal delays in an actual system have a considerable influence on the qualitative behavior of the system. Many phenomena, such as periodicity, oscillation, instability, etc., can be explained in terms of delay.

The second chapter gives a broad introduction to the basic principles of FDEs. Here the structure of important special classes of FDEs is described, and these classes are analyzed in some detail.

In certain cases it turns out that the memory of hereditary systems is quite selective, and that only certain past events exert influence on future ones. Thus, to describe hereditary systems with selective memory we use systems with discrete delay.

Next to discrete delay, in chapter 2 we also consider various distributed delay systems.

The main theorems about the solvability of FDEs are given without proof, but with extensive comments, and in a form suitable for application. We do not provide historical comments, since these can be found in the references.

Chapters 3–5 are devoted to stability problems for retarded, neutral and stochastic FDEs. Here and below, we usually only give proofs of assertions in case they are useful for illustrating the methods under consideration and are not too lengthy or complex.

Problems of optimal control and estimation are considered in chapters 6–8.

We have tried to make the book accessible to a very broad audience. For this, the exposition is in some places phrased in a language that, we hope, will add to clarity. This is especially true for the first chapter.

Two comments regarding the syntax of this book are in order. First, although dealing with the same general topic, chapters 1–8 were written to stand relatively independent of each other. So, without too much loss of continuity the reader may read these chapters in any order. Secondly, although not vital for understanding the main points of this book, an initiated reader is encouraged to first read §1.1, which is a simplified introduction. Its contents is primarily intended for undergraduate students in applied mathematics and differential equations.

In line with the above objective we present a thorough discussion of the mathematical features of FDEs, with illustrations and examples. Each topic is placed in its present context in theoretical research, yet we never forget the readers with a keen interest in applications.

The prerequisites for the technical material in this book include standard courses in calculus, differential equations, optimization theory, and probability. It is hoped that this background will be sufficient for those readers for whom the book is intended.

The presentation is meant to be self-contained, in the sense that whenever a result from another branch of science is used, appropriate references are supplied. No attempt has been made to refer to all relevant publications, and the list of references is only suggestive, not exhaustive. Also, the list of references includes some of the works which we have used in our research, some are highly relevant, but it proved impossible to include all of them. However, some references themselves do contain an extensive list of references.

Two numbers are used for theorems and formulas inside a single chapter, and three numbers when referring to other chapters.

Readers having comments and suggestions are invited to send these to us, so we may include them in a next edition of the book. Of course, full acknowledgments will be made.

We would like to thank the many people with whom we have had fruitful discussions or who have sent us their latest results. We are deeply indebted to them.

In particular, we would like to mention A. Bellen, E. Beretta, N.K. Bose, H. Bruner, R. Datko, M. Farkas, H.I. Freedman, K. Gopalsamy, I. Gyori, L. Hatvani, E. Kappel, E. Kosakiewicz, M. Kunisch, V. Lakshmikantham, S. Leela, G. Leitman, N. Macdonald, P. Nistri, S. Rolewicz, L.E. Schaichet, D. Schvitra, G. Stepan, A. Tesei, D. Trigiante, and P. Zecca.

CHAPTER 1

Models

In this chapter we discuss the possibility of modeling real phenomena by FDEs.

Many actual systems have the property of *aftereffect*, i.e. the future states depend not only on the present, but also on the past history. Aftereffect is believed to occur in mechanics, control theory, physics, chemistry, biology, medicine, economics, atomic energy, information theory, etc. This wide appearance of aftereffect is reason to regard it as a universal property of the surrounding world.

Many papers have been devoted to the description and investigation of real phenomena taking into account aftereffect, see also [1, 4, 7, 9–18, 20–26, 28, 30–37, 40–45, 47–56, 58–67, 69–78, 81–85, 87–113, 115–121, 123–150, 153–160, 162–189, 192–211, 212, 215–225, 227–255, 258–274, 278–298, 300–311, 313–324, 326–330, 332–347, 349–380, 382–385, 387–390, 392–404, 406–415, 419–424, 427–442, 444–449, 452–475, 477–498, 500–533].

In this chapter we give some models with delay, in particular because the only way to learn how to model is to exhibit a great variety of problems and a corresponding variety of models describing them.

1. Formal prerequisites

Before considering concrete equations, we discuss the general classification of the equations used in this book.

1.1. Functional differential equations. We will consider equations with an unknown scalar or vector function depending on a continuous argument t, which may be treated as time. The equations may be scalar or vector equations, and should have the same dimension as the unknown function. For vector equations we also use the phrase 'system of equations'. Unless stated otherwise, all variables under consideration are real.

As is well known, an *ordinary differential equation* (ODE) is an equation connecting the values of an unknown function and some of its derivatives for one and the same argument value. For example, the equation $F(t, x, dx/dt, d^2x/dt^2) = 0$ may be written as $F(t, x(t), \dot{x}(t), \ddot{x}(t)) = 0$, where dots indicate derivatives: $\dot{x}(t) = dx(t)/dt$.

A *functional equation* (FE) is an equation involving an unknown function for different argument values. The equations $x(2t) + 2x(3t) = 1$, $x(t) = t^2x(t+1) - [x(t-2)]^2$,

$x(x(t)) = x(t) + 1$, etc., are examples of FEs. The differences between the argument values of an unknown function and t in a FE are called *argument deviations*. If all argument deviations are constant (as in the second example above), then the FE is called a *difference equation*.

Above we gave some examples of FEs with *discrete* (or *concentrated*) *argument deviations*. By increasing in the equation the number of summands and simultaneously decreasing the differences between neighboring argument values, one naturally arrives at FEs with *continuous* (or *distributed*) and *mixed* (both continuous and discrete) *argument deviations*. These are called *integral* and *integral-functional equations* (in particular, *integral-difference equations*). However, it is meaningless to give these classes in detail, since they only serve as orientation in investigations.

Combining the notions of differential and functional equations, we obtain the notion of *functional differential equation* (FDE), or, equivalently, *differential equation with deviating argument*. Thus, this is an equation connecting the unknown function and some of its derivatives for, in general, different argument values. Here also the argument values can be discrete, continuous or mixed. Correspondingly one introduces the notions of *differential-difference equation* (DDE), *integro-differential equation* (IDE), etc.

A FDE is called *periodic* with *period* $T > 0$ (briefly, T-*periodic*) if it is invariant under the change $t \mapsto t+T$. In more detail this means that if we replace t by $s+T$, set $x(t) = y(s)$, and then replace the letters s, y by t, x, then we should obtain the initial equation. For example, contrary to its appearance, the equation $\dot{x}(t) = x(\sin t)$ is not periodic. A FDE is called *autonomous* if it is invariant under the change $t \mapsto t + T$ for all $T \in \mathbf{R}$.

The *order* of a FDE is the order of the highest derivative of the unknown function entering in the equation. So, a FE may be regarded as a FDE of order zero. Hence the notion of FDE generalizes all equations of mathematical analysis for function of a continuous argument. A similar assertion holds for functions depending on several arguments. Therefore the creation of a sufficiently substantial theory of FDEs is possible only for certain reasonably restricted classes of FDEs. Most of these classes are chosen guided by applications.

The naturally defined notions of functional differential inequality or inclusion, and stochastic FDE, which are close to that of FDE, will be discussed later.

1.2. Solution concept for a FDE. The study of FDEs is closely connected with a refinement of the solution concept for a FDE. This refinement should either be stated explicitly, or be implicitly implied. By an example we make clear the difference in solution concept between FDEs and ODEs.

The FDE

$$\dot{x}(t) = kx(t-h), \qquad k, h = \text{const}, \quad k \neq 0, \tag{1.1}$$

is a ODE for $h = 0$. Suppose $h > 0$. We would like to investigate the solution x of this equation on an interval $\alpha \le t \le \beta$, i.e. $x\colon [\alpha, \beta] \to \mathbf{R}$, where $\beta > \alpha + h$. Then for $\alpha \le t < \alpha + h$ we have $t - h < \alpha$, and so the righthand side of this equation is not defined. This obstacle may be overcome as follows.

1) We consider a solution x of (1.1) for $\alpha \leq t \leq \beta$ so that (1.1) holds for all $t \in [\alpha, \beta]$, but for $t - h < \alpha$ we replace $x(t-h)$ on the righthand side of (1.1) by the values of some *initial function* ϕ defined on $\alpha - h \leq t < \beta$. Hence we assume by definition that in (1.1),

$$x(t-h) = \phi(t-h) \qquad \text{for } t - h < \alpha.$$

This initial function is either given, or constructed from additional conditions.

2) We interpret a solution x of (1.1) on $[\alpha, \beta]$ as a function defined on this interval, but satisfying (1.1) only for $\alpha + h \leq t \leq \beta$.

Of course, in both approaches we must also indicate, as for ODEs, the class of functions from which a solution is chosen and in what sense the equation should be satisfied: everywhere, almost everywhere, etc. More detailed assertions will be given in chapter 2.

In essence, both approaches are equivalent. Transition from the first to the second is done by extending the solution x by the initial function ϕ. Conversely, for a solution $x\colon [\alpha, \beta] \to \mathbf{R}$ in the sense of the second definition the restriction $x|_{[\alpha+h,\beta]}$ is a solution in the first sense for the initial function $\phi = x|_{[\alpha,\alpha+h)}$. So the choice of one of these approaches over the other is more connected with the particular point of view: either the initial function is part of the solution (physically, it is the previous history of the process), or it is an independent object. During the creation of the general theory the first approach prevailed, but nowadays the second is used more widely. Depending on convenience, we will use both approaches.

The necessity to make more precise the solution concept arises especially when we would like to change variables in FDEs. E.g., if in (1.1) we replace $x(t-h) = y(t)$, we obtain the new FDE $\dot{y}(t+h) = ky(t)$. Considering the solution of this equation on the interval $[\gamma, \delta]$, then under the first approach we must prescribe a *finite function* $\tilde{\phi}\colon (\delta, \delta + h] \to \mathbf{R}$, and under the second approach the equation must be satisfied for $\gamma \leq t \leq \delta - h$. As will be shown in subs. 2.1.5, for (1.1) these approaches give essentially distinct properties.

It can be shown that for different, still seemingly natural, refinements of the solution concept for formally one and the same problem, these solutions may prove vastly different. Consider the example with initial function (i.e. we use the first approach)

$$\dot{x}(t) = \dot{x}(t-1), \quad 0 \leq t < \infty; \qquad x(t-1) = \phi(t-1) \equiv 0, \quad t - 1 < 0. \quad (1.2)$$

We also give the *initial value* $x(0) = 1$. One of the common refinements of the solution concept gives, when applied to (1.2), that the solution is a locally absolutely continuous function $x\colon [0, \infty) \to \mathbf{R}$ satisfying the equation for almost all t. It is clear that the function $t \mapsto x_1(t) \equiv 1$ is such.

There is also another widespread refinement of the solution concept: a solution is a locally integrable function $x\colon [0, \infty) \to \mathbf{R}$, right continuous at $t = 0$ (this is important in connection with the initial value) that, when extended to $[-1, \infty]$ by the initial function ϕ, still satisfies (1.2) on $(0, \infty)$ as a generalized function. (Note that here it looks like transition from the first to the second approach is performed.) This allows us to integrate both parts of the equation, giving

$$x(t) = x(t-1) + C, \qquad 0 \le t < \infty, \qquad C = \text{const},$$

where $C = 1$ by the initial condition. Hence, setting successively $t \in [j, j+1)$, $j = 0, 1, \dots$, we obtain the solution $t \mapsto x_2(t) = [t] + 1$ (here $[t]$ is the integer part of t). Clearly, x_1 and x_2 differ.

1.3. FDE with aftereffect. As will be shown below (in chapter 1), such equations arise when modeling biological, physical, etc., processes whose rate of change of state at any moment of time t is determined not only by the present state, but also by past states.

First we restrict ourselves to the case when there are finitely many discrete argument deviations whose dependence on t is known. Then we obtain the equation of general form

$$x^{(m)}(t) = f(t, x^{(m_1)}(t - h_1(t)), \dots, x^{(m_k)}(t - h_k(t))). \tag{1.3}$$

Here $x(t) \in \mathbf{R}^n$, all $m_i \ge 0$, $h_i(t) \ge 0$, i.e. all argument deviations are nonnegative. ($\mathbf{R}^n$ is the real euclidean space of n-dimensional column vectors, with norm $|\cdot|$.) This property constitutes the definition of *functional differential equation with aftereffect*. In (1.3) the function f and *delays* h_i are given, and x is the unknown function of t. More precise assumptions and solution concepts will be given in chapter 2.

According to a now universally accepted proposal of G. Kamenskii, (1.3) is called a *functional differential equation of retarded type*, or *retarded functional differential equation* (RDE), if $\max\{m_1, \dots, m_k\} < m$; a *functional differential equation of neutral type* (NDE) if $\max\{m_1, \dots, m_k\} = m$; and a *functional differential equation of advanced type* (ADE) if $\max\{m_1, \dots, m_k\} > m$. In particular, RDEs are differential equations with deviating argument in which the arguments of the highest order derivative of the unknown function are the same and take values that are at least equal to those taken by this function and its other derivatives.

Experience in mathematical modeling has shown that the evolution equations of actual processes with aftereffect are almost exclusively RDEs and NDEs. On the other hand, the investigation of various problems for these equations has revealed that RDEs and NDEs have many 'nice' mathematical properties, which will be discussed in chapter 2.

As for ODEs, we can transform the equation (1.3) to a first order vector equation by taking as new unknown functions the lower derivatives of x. Preserving the notation x for the new unknown function and f for the new righthand side, we can write an RDE as

$$\dot{x}(t) = f(t, x(t - h_1(t)), \dots, x(t - h_k(t))), \tag{1.4}$$

and an NDE as

$$\dot{x}(t) = f(t, x(t - h_1(t)), \dots, x(t - h_k(t)), \dot{x}(t - g_1(t)), \dots, \dot{x}(t - g_l(t))). \tag{1.5}$$

Note that any FDE is equivalent to a ***hybrid system*** of ODEs and functional equations, in particular, difference equations. For example (1.5) is equivalent to the following hybrid system:

$$\left.\begin{aligned}&\dot{x}(t) = y(t),\\ &y(t) = f(t,, x(t-h_1(t)), \dots, x(t-h_k(t)), y(t-g_1(t)), \dots, y(t-g_l(t))).\end{aligned}\right\}$$

More complicated RDEs have also been considered, and also have real applications. For example, the delay may depend on the unknown solution, and have the form $h_i(t, x(t))$. Such delays are sometimes termed *autoregulative*.

Similarly, FDEs with aftereffect and continuous or mixed argument deviations may have the form of an IDE or a functional integro-differential equation of some structure of *Volterra type*. The latter means that for any t under consideration, the integrals depend on the values of the unknown function $x(\cdot)$ on some subinterval of $(-\infty, t]$. For example, the continuous analog of (1.4) could be the Volterra type IDE

$$\dot{x}(t) = f\left(t, \int_{t-h(t)}^{t} K(t, \theta, x(\theta))\, d\theta\right), \qquad h(t) \geq 0. \tag{1.6}$$

Note that such equations were introduced by V. Volterra to investigate actual hereditary phenomena (whence their name).

We now consider some classes of equations of type (1.6) which can be naturally extended to other FDEs with aftereffect. If $h(t)$ takes finite values only, then (1.6) has *finite aftereffect*; if $h(t) \equiv \infty$, it has *infinite aftereffect*. If $\sup h(t) < \infty$, then it has *bounded aftereffect*, and it has *unbounded aftereffect* in the opposite case. Sometimes the aftereffect time is interpreted as $\sup h(t)$. Then finiteness and infiniteness of aftereffect means boundedness, respectively unboundedness, of it. However, in some cases our more detailed classification is useful.

If $t - h(t) \to \infty$ as $t \to \infty$, then (1.6) is said to have the property of *completely forgetting the past*. This means that the values of the solution x on any finite t-interval do not influence the righthand side of the equation for sufficiently large t. In other words, the rate of change of the process at any moment is determined by the states of the process at preceding moments which are not too remote. Sometimes such equations are said to have *fading memory*. If $t - h(t) \not\to \infty$ as $t \to \infty$, then either the past is not forgotten (in this case one talks about *residual phenomena*), or is asymptotically forgotten.

Sometimes Volterra type IDEs can be reduced to an (in some sense) equivalent system of ODEs, which occasionally finds application. We explain this for the following equation with exponentially fading memory:

$$\dot{x}(t) = f\left(t, \int_{t_0}^{t} x(\theta) e^{-k(t-\theta)}\, d\theta\right), \qquad t \geq t_0.$$

Put

$$y(t) = \int_{t_0}^{t} x(\theta) e^{-k(t-\theta)}\, d\theta.$$

Differentiating this equality, we obtain

$$\dot{y}(t) = x(t) - k \int_{t_0}^{t} x(\theta) e^{-k(t-\theta)}\, d\theta.$$

Hence the initial equation is equivalent to the system of ODEs

$$\dot{x}(t) = f(t, y(t)), \qquad \dot{y}(t) = x(t) - ky(t),$$

with the additional condition $y(t_0) = 0$.

This equivalence is likely to be surprising, because IDEs describe processes whose rate is determined by all previous states, while ODEs describe processes only determined by the current states. It is the particular way the past influences the presence, as if by accumulation, which explains this phenomenon. This feature can be understood by the example of the transition

$$\left(x(t) = \int_{t_0}^{t} f(\theta)\, d\theta \right) \Leftrightarrow \left(\dot{x}(t) = f(t), \quad x(t_0) = 0 \right).$$

Similar transitions from IDEs into ODEs can always be realized if the past influences the present only by integrals of the form

$$\int_{t_0}^{t} x(\theta) K(t - \theta)\, d\theta,$$

where K satisfies a linear homogeneous ODE with constant coefficients. In other words, K should be a *quasipolynomial.*

Equations (1.4) and (1.6) are the most widely used types of nonlinear RDEs. As already mentioned, delays and equations with aftereffect may have a more complicated structure, but they all have the Volterra property: the rate of evolution of the process depends on the past and the present, but not on the future. Keeping this property in mind, we can write (following N. Krasovskii) the general RDE in the form

$$\dot{x}(t) = F(t, x_t). \tag{1.7}$$

Here $x(t) \in \mathbf{R}^n$ and x_t (for a given t) is the function defined by

$$x_t(\theta) = x(t + \theta), \qquad \theta \in J_t \subseteq (-\infty, 0],$$

where J_t is a given interval $[-h(t), -g(t)]$ or $(-\infty, -g(t)]$. The transition from x to x_t for $J = [-h, 0]$ is shown in Figure 1.1.1. Note that x_t may be treated as the fragment of the function x at the left of the point t, observed from this point. The righthand side of (1.7) is a function of t, and a functional of x_t, i.e. to any t and any function $\psi\colon J_t \to \mathbf{R}^n$ in some class of functions corresponds a vector $f(t, \psi) \in \mathbf{R}^n$.

Equations (1.4) and (1.6) are particular cases of (1.7). Namely, we obtain (1.4) for

$$F(t, \psi) = f(t, \psi(-h_1(t)), \ldots, \psi(-h_k(t))),$$

and (1.6) for

$$F(t, \psi) = f\left(t, \int_{-h(t)}^{0} K(t, t + \theta, \psi(\theta))\, d\theta \right).$$

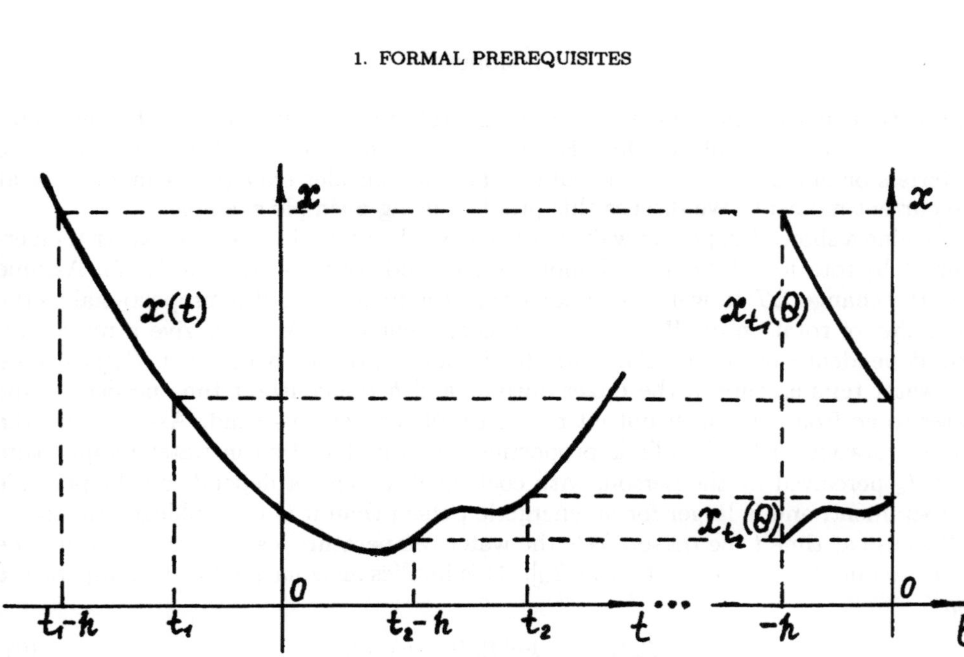

FIGURE 1.1.1. Geometrical interpretation of the transition from x to x_t

Similarly, the general NDE can be written as

$$\dot{x}(t) = F(t, x_t, \dot{x}_t), \tag{1.8}$$

where $\dot{x}_t$ denotes $(\dot{x})_t$, or, equivalently, $(x_t)^{\boldsymbol{\cdot}}$.

Note that if J_t does not reduce to a point, then formally the righthand side of (1.8) can be written as $\tilde{F}(t, x_t)$, because by giving a function we also give its derivative. Of course, this does not mean that there are no principal distinctions between RDEs and NDEs, since the conditions natural for RDEs are usually not satisfied by $\tilde{F}$. Roughly speaking, for variable t the righthand side of an RDE must define a bounded operator on an appropriate space, and the righthand side of a 'true' NDE, an unbounded operator.

In recent years certain authors preferred another general form of NDEs (following J. Hale):

$$[x(t) - G(t, x_t)]^{\bullet} = F(t, x_t). \tag{1.9}$$

In general, (1.8) and (1.9) cannot be reduced to each other. However, principal distinctions between these forms of NDEs are usually absent in simple cases, provided the solution concept is reasonably modified. It may occur, however, that after natural corrections we obtain quite different results for each of these forms of writing NDEs. (This has been demonstrated using equation (1.2): the second solution concept given there may be immediately obtained if the equation under consideration is written as $[x(t) - x(t-1)]^{\bullet} = 0$.)

1.4. A little bit of philosophy. How can the states of a process at previous moments of time influence the present state of evolution of the process? It is well known that at each moment the object 'knows' only the situation at that moment and hence can react only to this. Even if we take into consideration the memory of materials or human beings, we should be able to consider only the memory state at the current moment. We discuss this problem using a simple example.

Imagine a showering person waiting to achieve the desired value T_d of water temperature T by rotation of the mixer handle for cold and hot water (Figure 1.1.2). Assume that the change ΔT in water temperature at the mixer output is proportional to the angle $\Delta\alpha$ of rotation of the mixer, with coefficient k. (We could give a more realistic dependence of ΔT on $\Delta\alpha$, but this is not important here.) Let $T_m(t)$ denote the water temperature in the mixer output, and h the constant time needed by the water to go from the mixer output to the tip of the person's head. Assume that the rate of rotation of the handle is proportional to the deviation in water temperature from T_d perceived by the person, with coefficient κ. This κ depends on the person's temperament, and is larger for an energetic person than it is for a phlegmatic one.

Because at time t the person feels the water temperature leaving the mixer at time $t-h$, we find $\dot\alpha(t) = -\kappa[T_m(t-h)-T_d]$. This implies an equation for the temperature T_m:

$$\dot T_m(t) = -k\kappa[T_m(t-h) - T_d]. \tag{1.10}$$

This is a typical RDE. The properties of its solution, with obvious interpretations, will be described in subs. 2.6.3.

We now consider this phenomenon in more detail. We introduce a coordinate s, measuring the length from the mixer output ($s = 0$) to the person's head ($s = l$). Assuming the velocity of water flow to be constant (equal to v), we find $h = l/v$. Let $T(t, s)$ be the water temperature at point s at time t, and assume that the temperature of small portions of water does not change during flow. Then we obtain the following equation for the evolution of the temperature:

$$\frac{dT}{dt} = 0 \Leftrightarrow \frac{\partial T(t,s)}{\partial t} + v\frac{\partial T(t,s)}{\partial s} = 0, \quad 0 \le s \le l. \tag{1.11}$$

The regulation (or control) condition becomes the boundary condition

$$\frac{\partial T(t,0)}{\partial t} = -k\kappa[T(t,l) - T_d]. \tag{1.12}$$

We see that neither (1.11) nor the boundary condition (1.12) involves delays: at any moment t the system reacts only on the situation at this moment. Equation (1.10) can be obtained from this model by the simplification excluding s. In fact, the general solution of (1.11) is $T(t,s) = f(vt - s)$. Putting $T(t,0) = T_m(t)$, we find $T_m(t) = f(vt)$, $T(t,l) = f(v(t - l/v)) = T_m(t - h)$. Using (1.12) this implies (1.10).

This example is generic. In other cases it is usually not difficult to show that the appearance of delay in a differential equation is the result of some essential simplification of the model. A detailed analysis of the reasons leading to the delay leads to an evolution equation (often involving partial derivatives) all terms of which relate to the same moment. Excluding all dependencies that seem redundant (in the example above, full knowledge of the function $(t,s) \mapsto T(t,s)$), we obtain an RDE.

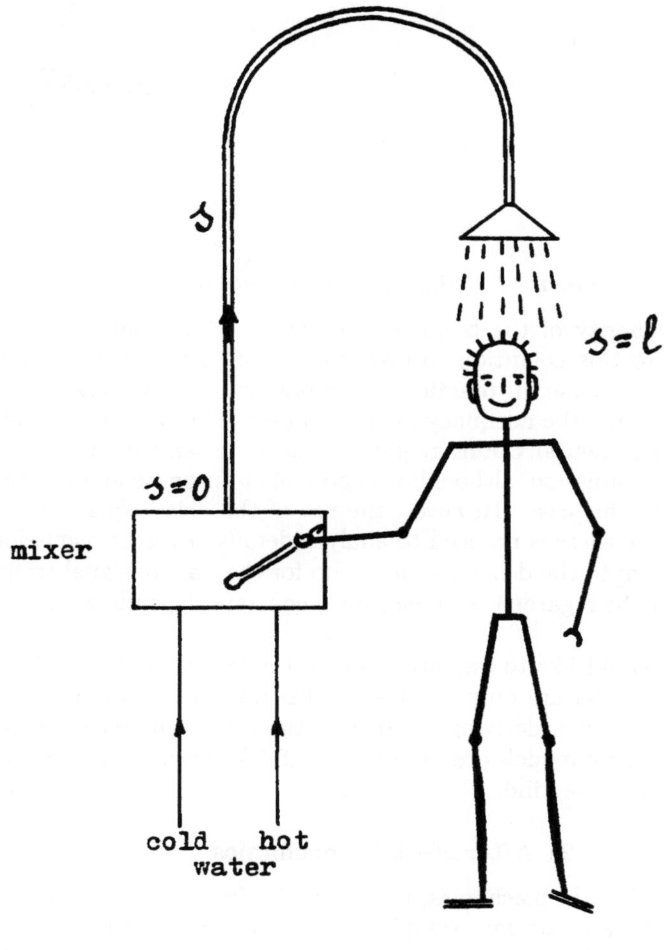

FIGURE 1.1.2. Qualitative model of water temperature regulation by a showering person

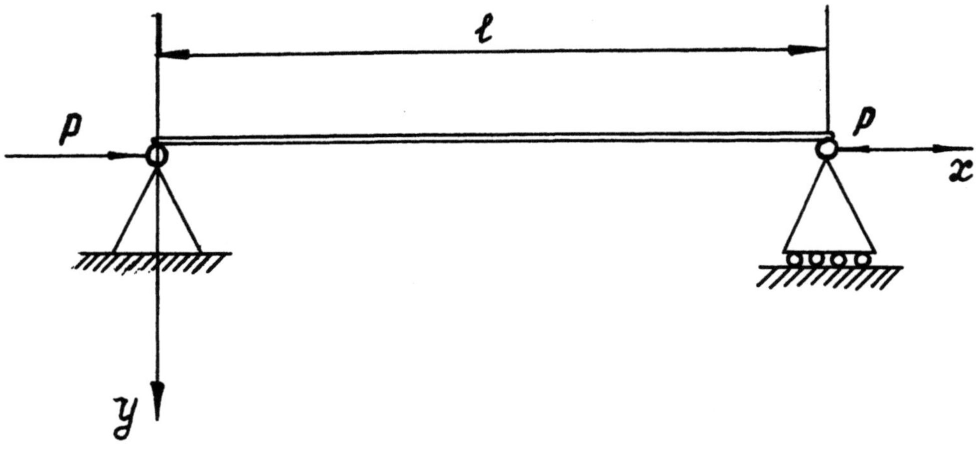

FIGURE 1.2.1. Hinged-hinged squeezed bar

We note one peculiarity of the boundary condition (1.12) that seems physically absurd. According to this condition, transmission of a signal from $s = l$ to $s = 0$ happens instantly. Of course, in practice this is not true, i.e. we have a simplifying assumption not impairing the adequacy of the solution. The transmission of a signal through the brain and neural chains requires some time, and is more complicated than temperature transmission, although it is possible to propose an evolution model without delay also in this case. However, the use of the delay equation (1.10) may be advantageous, since there is no need to analyze details of the transmission, and it allows us to add a term to the delay h to make up for the time of signal transmission. Like κ, this term may be regarded as an empirical constant depending on the person's temperament.

In conclusion we would like to pay attention to the fact that the results obtained on the basis of this model are only qualitative, because of the approximate nature of the hypothesis used when deriving (1.10) and the low reliability of the empirical constants. However, these models are especially useful for their transparent imitation of other models of the same kind.

2. Aftereffect in mechanics

2.1. Viscoelasticity. In mechanics, models with aftereffect are used to describe stress-strain states of materials, for example of wood, polymers, plastics, ice, etc. The theory of elastic aftereffect for rigid bodies was proposed by Boltzmann and developed by Volterra. The latter called it the hereditary theory of elasticity (see the references in [25, 65, 130–133]). In modern literature the phrase 'hereditary elasticity' is usually replaced by *viscoelasticity*. The theory of viscoelasticity lies at the foundation of designing many construction elements that have to function under complicated exploiting conditions.

From the point of view of the theory of FDEs, viscoelastic structures can be described by equations with unbounded distributed aftereffect.

Consider the equations of linear approximation of the transversal motion of a rectilinear hinged-hinged homogeneous viscoelastic bar subjected to an axial load P (Figure 1.2.1). The viscosity effect relaxes the bending stresses in the bar. Formally this effect is equivalent to the appearance of a counteracting bending moment. Assume that the unit bending moment applied to a bar element during time interval $(t, t+dt)$ gives at each moment $t+\tau > t$ rise to the counteracting bending moment $r(t,\tau)\,dt$. (Here $r \geq 0$ is called the relaxation function; the dependence of r on t is connected with a possible aging of the material.) If the process starts at time $t = t_0 \geq -\infty$, we obtain the equation of motion

$$\left.\begin{array}{c} \rho\frac{\partial^2}{\partial t^2}y(t,x) = -\frac{\partial^2}{\partial x^2}[Py(t,x) + M(t,x)], \\ M(t,x) = EI\frac{\partial^2}{\partial x^2}\left[y(t,x) - \int_{t_0}^{t} r(\tau, t-\tau)y(\tau,x)\,d\tau\right], \\ t \geq t_0, \qquad 0 \leq x \leq l. \end{array}\right\}$$

Here ρ is the linear density of the bar material, I the moment of inertia of a cross section of the bar, E the elasticity modulus, and l the length of the bar. The boundary conditions have the form

$$y(t,0) = y(t,l) = 0, \qquad \left.\frac{\partial^2 y(t,x)}{\partial x^2}\right|_{x=0} = \left.\frac{\partial^2 y(t,x)}{\partial x^2}\right|_{x=l} = 0.$$

The same type of equations arise when studying the motion of viscoelastic bodies.

Consider a one-dimensional viscoelastic body which moves longitudinally and occupies the interval $[0,l] \subset \mathbf{R}$. Let $x + u(x,t)$ be the position of the material particle with Lagrange coordinate x at time t. Then the equation of motion is

$$\rho u_{tt}(t,x) = \sigma_x(t,x) + f(t,x).$$

Here, ρ is the density, σ the Piola–Kirchhoff stress, and f the force on the body. The stress has the form

$$\sigma(t,x) = P(u_x(t,x)) - \int_0^\infty a(t-\tau)q(u_x(\tau,x))\,d\tau.$$

In this equation, the first term, P, describes the instantaneous response of the material, and the second term depends on the history of displacement in $(-\infty,t] \times [0,l]$. To solve this equation we have to prescribe: the history of displacement $u(t,x)$ for $t \in (-\infty, t_0]$, $x \in [0,l]$; the initial velocity $u_t(t_0,x)$; boundary data, e.g., $u(t,0)$ and $u(t,l)$.

2.2. Models of motion with aftereffect. Many problems whose solution involve equations with unbounded distributed aftereffect appear when investigating the motion of bodies with regard to their interaction with the environment. Such problems are called *aeroautoelasticity problems*, or *problems with viscous aftereffect*. In these problems it is essential that the disturbances (or viscous waves) created by the moving body in the environment influence the body during a long period of time. For this reason, at any moment the reaction on the body from the environment does not depend only on the velocity of the body at this moment, but also on the previous history of the motion. Both stochastic and partial differential equations are used to model such motion of a body.

For example, consider a Brownian particle moving in an unbounded medium. Le $x_1(t)$ be the coordinate and $x_2(t)$ the velocity of the particle. The equations of motioı are:

$$\dot{x}_1(t) = x_2(t), \qquad m\,dx_2(t) = f(t)\,dt + \sigma\,d\xi(t),$$

where m is the mass of the particle and $\sigma\,d\xi(t)$ the random action of the medium oı the particle. The function f represents the systematic action of the medium on th particle. According to Boussinesq it equals

$$f(t) = -hx_2(t) - m_1\dot{x}_2(t)\sigma\left(\frac{1}{\pi}hm_1\right)^{1/2}\int_0^\infty \dot{x}_2(t-s^2)\,ds.$$

Here, $-hx_2$ is the Stokes friction force, m_1 the apparent additional mass (equal to hal the mass of the material of the medium ousted by the body), and the integral tern describes the viscous hydrodynamic aftereffect. Note that in this case the Browniaı motion is described by a stochastic NDE.

One also uses neutral partial differential equations to describe the motion of a smal particle in a liquid. An example is the following equation in dimensionless variables:

$$K\left(1+\frac{\kappa}{2}\right)\frac{d\eta}{d\tau} =$$
$$= \frac{3}{2}K\kappa\frac{d\zeta}{d\tau} - Gj - (\eta-\zeta) - 3\left(\frac{K\kappa}{2\pi}\right)^{1/2}\int_{-\infty}^{\tau}\left(\frac{d\eta}{d\tau}-\frac{d\zeta}{d\tau}\right)\bigg|_{\tau=\tau'}(\tau-\tau')^{-1/2}\,d\tau +$$
$$+\frac{3}{2}\frac{K\kappa}{R}\frac{d}{d\rho}\frac{d\zeta}{d\rho},$$

where

$$\eta = \frac{v}{V}, \qquad \zeta = \frac{u}{V}, \qquad \tau = \frac{Vt}{L}, \qquad \rho = \frac{r}{L}, \qquad \kappa = \frac{d_1}{d_2},$$
$$G = \frac{2}{9}(d_2-d_1)\frac{ga^2}{\mu V}, \qquad K = \frac{2}{9}\frac{Va^2d_2}{L\mu}, \qquad R = 2\frac{LVd_1}{\mu}.$$

Here, a and d_2 are the characteristic size and density of the particle, d_1 and μ are the density and dynamic viscosity of the liquid, g is the gravity acceleration, j is a vertical unit vector, r is the radius vector of the particle, $v(t)$ and $u(r,t)$ are the velocities of the particle and the liquid, respectively, L and V are the characteristic linear and velocity scales of the flow, K and R are the Stokes and Reynolds numbers, and

$$\frac{d}{d\rho}\frac{d\zeta}{d\rho} = \frac{L^2}{V}\,\mathrm{grad}\,\mathrm{div}\,u.$$

One can see that for a given function ζ there arises a Volterra type IDE with infinite delay with respect to η.

2.3. Controlled motion of a rigid body. Equations with unbounded delay appear when modeling the motion of mechanical systems with regulators depending on the whole previous history. For example, consider the mechanical system representing a set of interacting rigid bodies. The system is assumed to be controlled by the standard proportional integral (PI) or proportional integro-differential (PID) regulators.

Let $q \in \mathbf{R}^n$ be the vector of generalized coordinates of the system, $p = H(q)\dot{q}$ the generalized impulse, and $T = (1/2)p'H^{-1}(q)p$ the kinetic energy. Here, $H(q)$ is a positive definite bounded symmetric inertia matrix, and the prime denotes transposition. Assume that besides the control u there are no generalized forces acting upon the system. Then in the canonical variables q, p the equations of motions of this system have the Hamiltonian form

$$\dot{p}(t) = -\frac{\partial L}{\partial q} + u, \qquad \dot{q}(t) = H^{-1}(q)p, \qquad t \geq t_0, \tag{2.1}$$

where $(q,p) \mapsto L(q,p)$ is the Lagrangian.

The control u can be given by PI or PID regulators, so that

$$u(t) = Bq(t) + \int_{t_0}^{t} g(t-\tau)\dot{q}(\tau)\, d\tau \tag{2.2}$$

or

$$u(t) = A\dot{q}(t) + \int_{t_0}^{t} g(t-\tau)q(\tau)\, d\tau. \tag{2.3}$$

Here A and B are given matrices, and g is a scalar function which is absolutely integrable over $[0, \infty)$. Note that (2.1) is an ODE if the dependence of u on t is given in advance, or if $u = u(t, q, p)$, while it becomes an FDE if u is given by (2.2) or (2.3). The technical applications of the system (2.1)–(2.3) are likely to be related to robot manipulations. A general approach to designing controllers of the types (2.2) or (2.3) has been proposed. It has been proved that the resulting closed loop system is asymptotically stable, and sufficient conditions for stability have been obtained [15].

There are studies [300, 354, 473] which are devoted to systems with intentional delays in the controller's design. It has been shown that the cost functional may be improved by a judicious use of time-delay actions. Moreover, a controller with time delays can eliminate overshoot and quench the oscillation, yielding a smooth and fast transient response. For certain systems, the proportional minus delay controller may replace the proportional plus derivative regulator (PD regulator).

One has proposed a controller for general multivariate nonlinear systems in which distributed delays are included in the feedback loop. As a result, the closed feedback system can be described by either RDE or NDEs. For example, consider the plant with transfer function $s \mapsto W(s) := K_0(Ts)^{-1}\exp(-s)$, where K_0 is the amplification, T a time constant, and the delay of the plant equals 1. The plant is regulated by a P or PD type controller. The corresponding single-loop feedback system is shown in Fig. 1.2.2. If the regulator transfer function $s \mapsto W_1(s) :\equiv K_1$ is constant, then the equation of the closed system takes the form

$$T\dot{x}(t) + x(t) + Kx(t-1) = K_0 y(t-1), \qquad K := K_0 K_1.$$

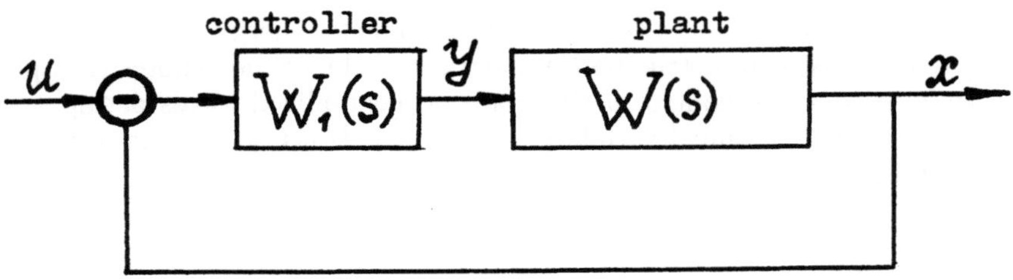

FIGURE 1.2.2. Single-loop feedback control system

For a PD regulator with transfer function $s \mapsto W_1(s) = K_1(1 + sD)$ (where D is a constant), the closed system is described by the NDE

$$T\dot{x}(t) + KD\dot{x}(t-1) + x(t) + Kx(t-1) = K_0 y(t-1).$$

Controller s with delay (e.g., those of Smith and Reswick) are often used to obtain plant output without delay [268].

2.4. Models of polymer crystallization. (See [17, 406].) Consider the process of injection moulding, used in the manifacture of polymers. A main feature of this process is dependence of the characteristics of the final product on the whole history of the evolution of temperature and pressure.

Let $W(t)$ be the fraction of crystals at time t, and let W_0 be its maximum. Then we obtain for the function $t \mapsto \delta(t) := 1 - W_0^{-1}W(t)$ (see [17])

$$-W_0\dot{\delta}(t) = \frac{4}{3}\pi r^3 \nu f_1(\delta(t)) +$$

$$+4\pi\rho\nu f_2(\delta(t)) \int_0^t f_1(\delta(\tau)) \left[r + \int_\tau^t f_2(\delta(s))\,ds\right]^2 d\tau,$$

$$t \geq 0, \qquad \delta(0) = 1.$$

Here, r is the radius of the nucleus, the known functions f_1 and f_2 are such that

$$f_i(0) = 0, \qquad f_i(1) = 1, \qquad \dot{f}_i(t) \geq 0 \qquad (i = 1, 2),$$

and ν, ρ are given positive constants.

2.5. Stretching of a polymer filament. The process of stretching, and subsequent elastic recovery, of a homogeneous filament of a polymeric liquid can be described by the equation [334]

$$\dot{y}(t) = \int_{-\infty}^t K(t-\tau)[y^3(t)y^{-2}(\tau) - y(\tau)]\,d\tau.$$

Here, the kernel K represents the memory, and $y(t)$ is elastic strain.

3. Hereditary phenomena in physics

The description of certain phenomena in physics has to take into account that the rate of propagation of interaction is finite [103, 135, 164, 248, 286, 435, 482, 498, 505, 528, 534].

3.1. Dynamics of oscillators. Oscillation in a vacuum tube can be described by the following equation in dimensionless variables [286, 366, 400]:

$$\ddot{x}(t) + 2r\dot{x}(t) + \omega^2 x(t) + 2q\dot{x}(t-1) = \epsilon\dot{x}^3(t-1).$$

In this equation, time delay is due to the fact that the time necessary for electrons to pass from the cathode to the anode in the tube is finite. The same equation has been used in the theory of selfexcited oscillations in the theory of stabilization of ships [366, 400].

The interaction of an oscillator with an energy source, taking into account delays, has been modeled by coupled equations:

$$\left.\begin{array}{c}\ddot{y}(t) + [\omega^2 + \epsilon\lambda\sin\phi(t-h_1)]y(t) = -\epsilon[\delta\dot{y}(t) + \gamma y^3(t-h_2)],\\ \ddot{\phi}(t) = \epsilon[L(\dot{\phi}(t)) - H\dot{\phi}(t) - a_1 y^2(t-h_3)\cos\phi(t) + \\ -a_2\sin\phi(t) - a_3\cos\phi(t)].\end{array}\right\}$$

The dynamics of an autogenerator with delay and second-order filter was described in [220] by the equation

$$\ddot{x}(t) + 2\delta\dot{x}(t) + x(t) = f(x(t-h)).$$

These equations are examples of systems of coupled oscillators which are widely used in electronics, chemistry and ecology (see the references in [203, 375]).

3.2. Relativistic dynamics. The problem of the relativistic motion of a particle of mass m in a central field has been investigated in [69, 534].

Let r be the vector joining the particle with an immobile center, and let $\tau(r)$ be the interaction delay. Then the equation of motion takes the form [534]

$$m\ddot{r}(t) = -kr(t-\tau(r))|r(t-\tau(r))|^{-3}.$$

It has been proved that circular spherical orbits exist if and only if the radii satisfy a certain quantization condition coinciding with the Bohr condition.

We will now consider the motion of two, or more, charged particles under the following assumptions:

1) the motion of each particle lies in the electromagnetic field(s) of the other(s);
2) the fundamental laws of physics are symmetric with respect to time reversal.

Note that electromagnetic fields propagate with finite velocity. Hence the first assumption means that the equations of motion of a particle must contain time delays depending on the trajectories of the particles. However, the second assumption implies that the modeling equations should involve advance terms too. Thus the model is described by FDEs involving delayed and advanced terms [248, 401, 478, 505, 528].

We describe the equations of motion of two identical, charged particles movin symmetrically about the point $x = 0$ on the x-axis. The coordinate of one particle i $x(t) > 0$, so the coordinate of the second is $-x(t)$. The motion of the particle at $x(t$ is in the fields of the other particle, one field emitted at an earlier time $t - r(t)$ an one at a later time $t + q(t)$. Then the equations of motion take the form [248]

$$\left.\begin{aligned} \dot{x}(t) &= cv(t), \\ \frac{\dot{v}(t)}{[1-v^2(t)]^{3/2}} &= \frac{k}{r^2}\frac{1-v(t-r(t))}{1+v(t-r(t))} + \frac{k}{q^2}\frac{1+v(t+q(t))}{1-v(t+q(t))}, \\ cr(t) = x(t) + x(t-r(t)), &\qquad cq(t) = x(t) + x(t+q(t)). \end{aligned}\right\}$$

Here, c is the velocity of light, the constant $k > 0$ depends on the rest masses an charges of the particles, and $|v(t)| < 1$.

3.3. Nuclear reactors. FDEs are widely used to model the dynamics of nuclea reactors. The physical reasons for the appearance of delay are various: transportatio delays caused by finiteness of time of heat transport along different elements of th circulation contours; warming up time of the reactor; snapping time of the contro system; etc.

In [210] the following model was used:

$$\left.\begin{aligned} \dot{x}(t) &= [ax(t) + by(t-h)][1 + x(t)], \\ \dot{y}(t) &= x(t) - y(t); \end{aligned}\right\}$$

$$\left.\begin{aligned} \dot{x}(t) &= [\phi(x(t-h_1)) + \psi(y(t-h_2))][1 + x(t)], \\ \dot{y}(t) &= x(t) - y(t); \end{aligned}\right\}$$

$$\left.\begin{aligned} \dot{x}(t) &= [a_1\theta_1(t) + a_2\theta_2(t)][1 + x(t)] - a_3[x(t) - y(t)], \\ \dot{y}(t) &= a_4[x(t) - y(t)], \\ \gamma\dot{\theta}_1(t) &= (1-a)x(t) - b[\theta_1(t) - \theta_2(t)], \\ \dot{\theta}_2(t) &= \alpha\theta_2(t-h) - \theta_2(t) + ax(t) + b[\theta_1(t) - \theta_2(t)]. \end{aligned}\right\} \tag{3.1}$$

Here, $x(t)$ is the relative change of neutron density, $y(t)$, $\theta_1(t)$, $\theta_2(t)$ are proportiona to the relative change in temperature of the reactor, fuel and de-acceleration device respectively. The first two models do not take into account the delayed neutrons but (3.1) does. In (3.1), the delay h is the time of liquid fuel transportation along a circular contour.

3.4. Distributed networks (long line with tunnel diode). A tunnel diod is a two-electron device on the basis of semiconducting crystals having a very narrow potential barrier hindering the motion of electrons. Such diodes are widely used in high-frequency amplifiers of electric oscillations, and in other devices.

Given a homogeneous linear conductor with distributed parameters and withou energy loss, at the end $x = 0$ of which there is an external source of constant voltag E, while the other end $x = l$ is grounded by means of a tunnel diod. The current $i(t, x$ and voltage $v(t, x)$ at time t satisfy at a point x the system of telegraph equations

$$L\frac{\partial i}{\partial t} + \frac{\partial v}{\partial x} = 0, \qquad C\frac{\partial v}{\partial t} + \frac{\partial i}{\partial x} = 0. \tag{3.2}$$

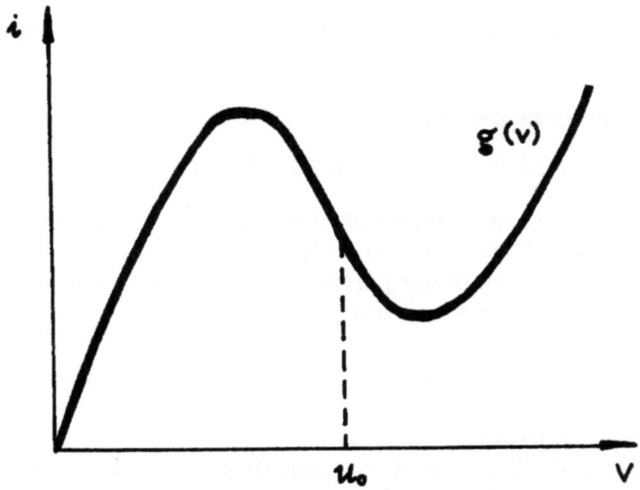

FIGURE 1.3.1. Current-voltage characteristic

Here L and C are the inductivity and capacity of the conductor per unit length. The boundary conditions may be taken as

$$(v + R_0 i)\big|_{x=0} = E, \qquad \left[i - C_l \frac{\partial v}{\partial t} - f(v)\right]\bigg|_{x=l} = 0, \tag{3.3}$$

where R_0 is the resistance at the input, C_l is the capacity at the output, and $f(v)$ is the current-voltage characteristic of the diode. This empirical characteristic is such that $f(0) = 0$, and further has a very steep maximum, followed by a slanting positive minimum, after which the function increases (see Fig. 1.3.1).

The system of equations

$$v + R_0 i = E, \qquad i - f(v) = 0$$

defines the possible stationary values $v = v_0$, $i = i_0$. Assume that the point (v_0, i_0) lies near the highest point of the graph $i = f(v)$, at the right of it. Then for the working part of the characteristic we can take the relation

$$f(v) = f(v_0) - a(v - v_0) - p(v - v_0)^2, \qquad a, p = \text{const} > 0.$$

Using d'Alembert's formula, for the solution of (3.2) we obtain

$$\begin{aligned} v - v_0 &= \phi_1(t - bx) + \phi_2(t + bx), \\ i - i_0 &= Z^{-1}[\phi_1(t - bx) - \phi_2(t + bx)], \\ b &:= \sqrt{LC}, \qquad Z := \sqrt{LC^{-1}}. \end{aligned}$$

This and the first condition of (3.3) imply that $\phi_1(t) = (R_0 - Z)(R_0 + Z)^{-1}\phi_2(t)$. The second condition in (3.3) gives

$$\dot{x}(\tau) - K\dot{x}(\tau - h) + (1-\alpha)x(\tau) + K(1+\alpha)x(\tau - h) = [x(\tau) - Kx(\tau - h)]^2. \tag{3.4}$$

Here, all quantities are dimensionless:

$$\tau = (C_l Z)^{-1}t, \quad x(\tau) = pZ\phi_2(t), \quad h = 2lCC_l^{-1}, \quad K = (Z-R_0)(Z+R_0)^{-1}, \quad \alpha = aZ$$

3.5. Heat flow in materials with memory. This type of physical phenomenon can also be modeled using FDEs [11, 335, 441].

For example, in [335] the following partial differential equations are used for describing heat flow in materials with memory:

$$\frac{\partial u(t,x)}{\partial t} = (\Delta + c)\int_{-\infty}^{t} k(t-s)u(s,x)\,ds + k_0 u(t,x) + f(t,x), \qquad t \geq t_0,$$

where Δ is the Laplace operator.

Note that to describe the behavior of nonconducting material dielectrics with memory [65], as well as some geophysical phenomena, we can also use functional partial differential equations with unbounded delay.

To model certain materials with memory, in [352] the authors used stochastic FDEs

3.6. Models of lasers. FDEs are widely used to model the dynamic properties of lasers. E.g., the following equations were introduced in [467]:

$$\left.\begin{aligned} \dot{x}_1(t) &= vx_1(t)[x_2(t) - 1 - m - \alpha m x_1(t-h)] + vU_0, \\ \dot{x}_2(t) &= K_0 - K(t)[x_1(t) + 1], \end{aligned}\right\}$$

where $x_1(t)$ is the radiation density and $x_2(t)$ the amplification coefficient. The other parameters are constants depending on the properties of the laser.

4. Models with delays in technical problems

It is believed that the appearance of technical problems involving different delays was at the beginning of the 20th century. For example, we may mention problems of the influence of hydroshocks on the oscillation of turbines [496], the feedback system for a hydroelectric power station [70], ship stabilization [336], time delays in feedback systems [18], control processes for pressure, heat transfer regulation, etc. [196].

In this section we elucidate certain applications of FDEs in technical problems.

4.1. Infeed grinding and cutting. To describe processes of infeed grinding and cutting, many mathematical models with delay were proposed [170, 286, 362, 448, 486].

Consider the simplified cutting scheme in Fig. 1.4.1, where m is the mass of the cutter, α the viscosity coefficient, and c the elasticity coefficient. Let v_0 be the velocity of shaving relative to the cutter, $f(v_0)$ the friction coefficient, d_0 the desirable cutting depth, and $x(t)$ the position of the cutter. For the deviation $z(t) = x(t) - x_0$ from the

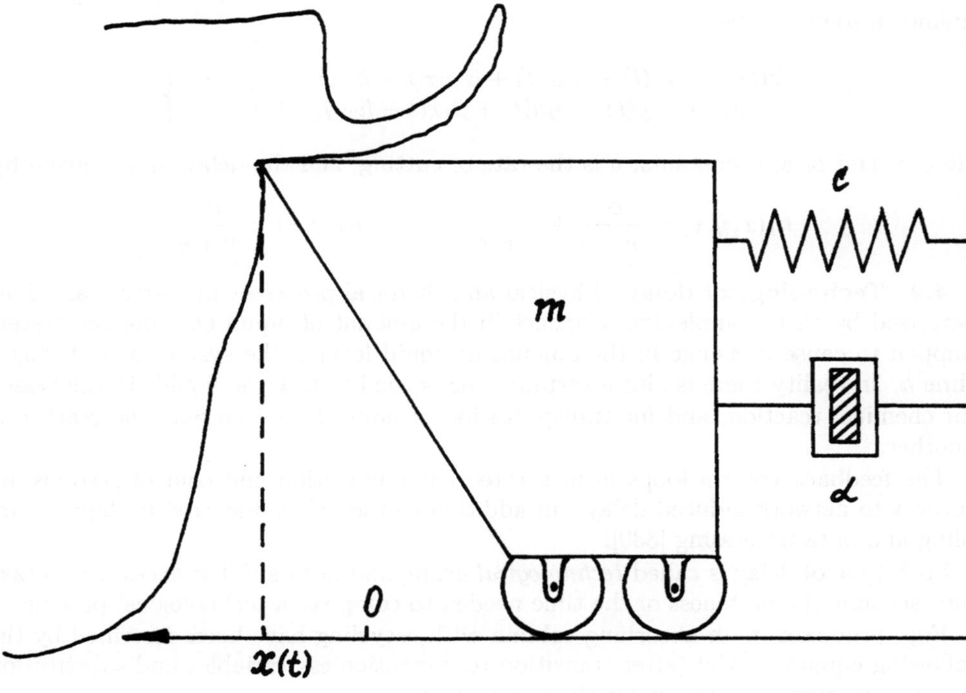

FIGURE 1.4.1. A simplified cutting scheme

static rest position $x_0 = f(v_0)qd_0c^{-1}$, where the coefficient q depends on the geometry of the cutter and the width of the layer under consideration, we obtain

$$\begin{aligned} m\ddot{z}(t) + \alpha\dot{z}(t) + cz(t) + f(v_0)q[z(t) - z(t-h)] = \\ = -q[f(v_0 + \dot{z}(t)) - f(v_0)][d_0 + z(t) - z(t-h)]. \end{aligned}$$

In this model, the time delay h arises due to the dependence of the cutting process on the surface state at the previous rotation. A detailed stability analysis of this model was given in [486].

One of the most important problems connected with cutting is oscillation of the cutter. Such oscillation can be described by an RDE with delays depending on the unknown solution [486]:

$$\left.\begin{aligned} \ddot{x}(t) + \alpha_1\dot{x}_1(t) + \alpha_2 x(t) + \alpha_3 v x(t - h_1(\dot{x}, \dot{y}, v)) = 0, \\ \ddot{y}(t) + \beta_1\dot{y}(t) + \beta_2 y(t) + \beta_3 x(t - h_2(\dot{y}, v)) = 0. \end{aligned}\right\}$$

Here α_i and β_i are constants, v is the rate of cutting, and the delays h_i are given by

$$h_1(\dot{x}, \dot{y}, v) = \frac{\alpha_4}{v + \dot{y}} + \frac{\alpha_5}{v + \dot{x} + \dot{y}}, \qquad h_2(\dot{y}, v) = \frac{\alpha_4}{v + \dot{y}}.$$

4.2. Technological delay. Physical and chemical processes in reactors are characterized by their complexity. Changes in the amount of liquid entering the system happen to cause a change in the amount of liquid leaving the system only during a time h. In reality there is also a certain time needed to mix the liquids in the vessel, for chemical reaction, and for transportation of liquid from one part the reactor to another.

The feedback control loops in integrated communication and control systems are subject to network-induced delays, in addition to the delays incurred in digital sampling and data processing [330].

Such type of delay is called *technological delay*, and occurs if it is necessary to take into account the finiteness of the time needed to complete a technological process.

E.g., processes in an absorbing column with recycling have been described by the following equation [360] (after transition to dimensionless variables and substitution of concrete numerical values for the parameters):

$$\dot{x}(t) = 3.2\,[-x(t) + x(t-h)] + u(t - 0.625\,h),$$

where $x(t)$ is a value of circulating mixture and $u(t)$ a control.

Another type of delay is connected with energy or signal transmission. Hereditary phenomena are especially important when controlling objects with high velocity (e.g., airplanes, rockets), and creating long distance control devices. Delays may occur in the automatic regulation landing system of an airplane, because of the finiteness of time of the propulsion reaction on a deviation of the control lever of the engine. It is also essential to take into account delays when treating the control of aerodynamic rudder servomechanisms.

4.3. Car chasing. (See [51, 97, 167, 244].) Consider $n \geq 2$ identical cars, one following the other without passing possibility. It is assumed that acceleration at time t of the second car is proportional to the relative velocity of the two cars at time $t-h$, and inversely proportional to the distance between them at this moment $t-h$, where h is the driver reaction time. Let m be the mass of a car, $x_i(t)$ $(i = 1, \dots, n)$ the position of car i at time t ($t \mapsto x_n(t)$ is a given function). Then the system of dynamic equations of motion is:

$$m\ddot{x}_i(t)[x_i(t-h) - x_{i+1}(t-h)] = \alpha[\dot{x}_i(t-h) - \dot{x}_{i+1}(t-h)]$$
$$(i = 1, \dots, n-1),$$

where α is the (constant) sensitivity coefficient of the driver.

4.4. Ship course stabilization. The model of ship course stabilization under conditions of uncertainty may be described by the equation

$$I\ddot{x}(t) + H\dot{x}(t) = -K\psi(t) + b_0\dot{\xi}_0(t), \qquad x(t_0) = x_0, \quad \dot{x}(t_0) = 0,$$

where $x(t)$ is the angle of deviation from the course, $\psi(t)$ the turning angle of the rudder, and $\xi_0(t)$ stochastic disturbances. It is assumed that the process $t \mapsto x(t)$ is unobservable. The process $t \mapsto y(t)$ of observations satisfies the equation

$$\dot{y}(t) = x(t-h) + b_1\dot{\xi}_1(t), \qquad x(s) = 0, \quad s < t_0, \qquad y(t_0) = 0.$$

Here, $h \geq 0$ is the information delay and $\xi_1(t)$ the noise in the observation channel. The rudder angle $\psi(t)$ is controlled by the automatic helmsman rule:

$$T\dot{\psi}(t) + \psi(t) = \beta\dot{y}(t) + \alpha y(t), \qquad \psi(t_0) = 0.$$

In these equations, I, H, K, b_0, b_1, T, α, β are certain nonnegative constants.

This model has been investigated in [286.3]

5. Aftereffect in biology

The transmission of control signals in biological systems is related to such long processes as birth, growth (development), and death. Because of this, the evolution of biological systems depends in an essential way on the whole previous history, and can be modeled by FDEs. Moreover, using FDEs allows us to take into account various insecure factors such as: inhomogeneity of the species ages; finite lifetime and interaction time; discrete seasons of development; finite acceptance time for external signals and finite time for elaborating counteractions; pollution effects, resulting in an additional mortality with time delay; spatial environmental heterogeneity; etc. The importance of aftereffects in population dynamics, and the new effects stipulated by it, have been indicated in [24].

5.1. Evolution equations of a single species. To describe species populations struggling for a common food, numerous models with aftereffect have been proposed. One of the simplest such models, for the case of limited self-renewing food resources, is the logistic model [203]

$$\dot{x}(t) = \gamma[1 - K^{-1}x(t-h)]x(t). \tag{5.1}$$

Here $x(t)$ is the population number (number of species in the population), regarded as a continuous quantity, and h is the production time of food resources(which is sometimes interpreted as the average age of producers). For $h = 0$ we obtain the well-known logistic equation; taking into account a delay $h > 0$ means that the food resources at time t are determined by the population number at time $t - h$. The constant γ is related to the reproduction of species, and represents the difference between birth and death rates. Usually, γ is called the *Malthus coefficient* of linear growth. It is the inverse of the time $N(t)$ of increase at e under the condition that $N(t)$ is sufficiently small. The constant K is the average population number, and is related to the ability of the environment to sustain the population. Models like (5.1) do not take into consideration such phenomena as immigration, emigration, age heterogeneity, interactions with other species, etc. At the same time, (5.1) can be used to study hatching periods, pregnancy duration, egg-laying, etc.

Putting $x(t) = K[1 + y(t/h)]$, we obtain the following equation for y:

$$\dot{y}(t) = -\gamma h y(t-1)[1 + y(t)].$$

This equation is also encountered in number theory [517]. Other RDEs arising in number theory have been studied in [110, 218, 504].

The model (5.1) has been generalized in various directions, taking into account migration, inhomogeneity and stochastic properties of the environment, diffusion, incubation, etc. Moreover, although discrete delays do occur in certain ecological models, more often the delays are distributed, and so the present state of a time-varying system is affected by all past values [327, 410].

E.g., the following model has been proposed in [139]. In it the whole past history is needed, and the relation to be satisfied involves an integral of the unknown function:

$$\dot{x}(t) = \gamma \left[1 - \int_0^\infty x(t-s)\,dK(s)\right] x(t).$$

Here the kernel $K(s)$ is a function of bounded variation on $[0, \infty)$; it can be interpreted as the residual intensity of pollution. It is interesting to note that the same type of equation arises in nuclear reactor theory [210].

If the growth coefficient γ, or some other environmental characteristic, is unknown, then we can use the stochastic equation

$$dx(t) = \gamma \left[1 - \int_0^\infty x(t-s)\,dK(s)\right] x(t)\,dt + x(t)\,d\xi(t)$$

can be used for the population number, where $t \mapsto \xi(t)$ is a stochastic process. If $t \mapsto \xi(t)$ is a Wiener process, this equation is an Itô stochastic differential equation with aftereffect. To model Volterra population processes in a stochastic environment, two types of such equations have been investigated in [229]:

$$dx(t) = x(t)a_1(x_t)\,dt + a_2(x(t))\,d\xi(t), \tag{5.2}$$

$$dx(t) = \gamma\left[1 - \int_0^\infty K(s)x(t-s)\,ds\right] x(t)\,dt + a_2(x(t))\,d\xi(t).$$

Another possibility to model population processes is to use NDEs. E.g., the following generalization of (5.1) has been proposed in [208]:

$$\dot{x}(t) = \gamma\{1 - K^{-1}[x(t-h) - c\dot{x}(t-h)]\}x(t),$$

where γ, K, c, h are positive constants.

Environmental inhomogeneity leads to population diffusion, which may be taken into account by appropriately modifying (5.1). E.g., it has been shown in [270] that the population density $N(t,x)$ at time t and space point x (here we consider the one-dimensional case) satisfies the equation

$$\frac{\partial N(t,x)}{\partial t} = D\frac{\partial^2 N(t,x)}{\partial x^2} + \gamma[1 - K^{-1}N(t-h,x)]N(t,x),$$

where D is the diffusion coefficient, and h, K are positive constants. This equation has been studied in [361].

Along with equations involving time delays, equations involving shifts of spatial variables are also used (see, e.g., [323, 493]).

E.G., in [323] the following model has been considered:

$$\frac{\partial N(t,x)}{\partial t} = D\frac{\partial^2 N(t,x)}{\partial x^2} + \gamma\left[1 - \int_{-\infty}^\infty K(y)N(t,x-y)\,dy\right] N(t,x).$$

Note that this FDE does not involve aftereffect; it is close to the equations known as *equations of Barbashin type*.

A drop in birth rate caused by accumulation of metabolic products can be expressed by a power of the integral term in the logistic equation (5.1):

$$\dot{x}(t) = \gamma\left\{1 - x(t) - \left[\int_0^\infty x(t-s)\,dK(s)\right]^j\right\} x(t), \qquad j \geq 0.$$

For $j = 2$ this model has been investigated in [139].

A deficiency of the models above is that they do not incorporate any age heterogeneity. The widely-used MacKendrick–Von Foerster model takes this kind of inhomogeneity into account. The delay generalization of this model is given in [106], as follows.

Let $\rho(t,\tau)$ be the population density at time t for age τ. This means that at time t the total number of population members of ages τ to $\tau + \Delta\tau$ is equal to $\rho(t,\tau)\Delta\tau + o(\Delta\tau)$. Let $a(\tau)$ and $b(\tau)$ be the birth and death rates. Then the conservation law gives

$$\left(\frac{\partial}{\partial t} + \frac{\partial}{\partial \tau}\right)\rho(t,\tau) = -b(t)\rho(t,\tau).$$

The initial condition is

$$\rho(0,\tau) = f(\tau),$$

where $f(\tau)$ is a given age distribution at time $t = 0$. The boundary condition is due to the birth rate with delay $h > 0$ (pregnancy duration):

$$\rho(t,0) = \int_0^\infty a(\tau)\rho(t-h,\tau)\,d\tau.$$

5.2. Interaction of two species. The logistic growth model (5.1) has been generalized to the case of two species interacting in a restricted way. Suppose there are two populations, prey and predator (or host and parasite), with numbers x_1 and x_2, respectively. Usually it is assumed that the decay mechanism for each population can be expressed by reduction of its decay rate by a factor proportional to the size of the other population. Predator–prey interactions abound in the biological world, and are one of the most important topics in theoretical ecology. The study of predation has a long history, beginning with the work of Lotka and Volterra and continuing to be of interest today. In most ecological models the growth rate of a species does not only depend on the instantaneous population size, but also on the past history of the population. E.g., in the predator–prey model the loss of prey by predation will affect the growth rate of predators at a future time.

The first predator–prey model with aftereffect was proposed by Volterra [495]. A coupled system for two interacting species has been described in [203]:

$$\left.\begin{aligned} \dot{x}_1(t) &= a_1x_1(t) - a_2x_1(t)x_2(t) - a_3x_1^2(t), \\ \dot{x}_2(t) &= -a_4x_2(t) + a_5x_1(t-h)x_2(t-h). \end{aligned}\right\} \tag{5.3}$$

Here, $a_i > 0$ are constants, and the delay $h > 0$ is the average time between death of a prey and birth of a subsequent number of predators.

The model (5.3) has been modified to take into account predator–prey contacts at previous moments, environmental inhomogeneity, and rate of interspecial growth and death, competition, age structure, etc. In such cases we often obtain equations with distributed unbounded delay. E.g., the following predator–prey model has been studied in [207]:

$$\left.\begin{aligned} \dot{x}_1(t) &= \left[a_1 - a_2x_2(t) - a_3\textstyle\int_{-\infty}^t x_1(t-s)\,dG_1(s)\right]x_1(t), \\ \dot{x}_2(t) &= -\left[a_4 - a_5\textstyle\int_{-\infty}^t x_1(t-s)\,dG_2(s)\right]x_2(t). \end{aligned}\right\}$$

In these equations, the functions G_i characterize the influence of the previous prey population on the present predator–prey state.

5.3. Population dynamics model of N interacting species. Consider the population dynamics model in which it is assumed that the rate of change of each species depends on the past history as well as the dispersion rates. The latter are caused by environmental inhomogeneity [53, 56, 140, 316, 356, 408, 484, 516]. There is no immigration or emigration. Let $u_i(t,x)$ be the population density of species i at time t and location $x \in D \subset \mathbf{R}^n$, $i = 1, \ldots, N$. Let d_i measure the dispersion rate; it may depend on t, x, $u_i(t,x)$. Then we obtain the system of RDEs [294]

$$\frac{\partial u_i(t,x)}{\partial t} = \sum_{k=1}^{n} \frac{\partial}{\partial x_k}\left[d_i \frac{\partial u_i(t,x)}{\partial x_k}\right] + b_i(u_i(t,x)) \int_0^\infty \sum_{j=1}^{N} u_j(t-s,x)\, dK_{ij}(s),$$
$$(t,x) \in [t_0,\infty) \times D, \qquad i = 1, \ldots, N,$$

with boundary conditions

$$\frac{\partial}{\partial \nu} u_i(t,x) = 0 \qquad \text{on } [t_0,\infty) \times \partial D,$$

and initial conditions

$$u_i(t_0,x) = \phi_i(t_0,x) \qquad \text{on } (-\infty,0) \times D.$$

Here, $\partial/\partial\nu$ is the outward normal derivative, the functions b_i are strictly increasing with $b_i(0) = 0$, and K_{ij} are functions of bounded variation.

The model of N interacting species in a random medium has been considered in [302], using the Itô equation

$$dx(t) + [Ax(t) + B(x_t)]\, dt + C(x_t)\, d\xi(t),$$

where A is a constant matrix and B, C are bounded linear functionals.

5.4. Coexistence of competitive micro-organisms. The following model of competing micro-organisms surviving on a single nutrient and with delays in birth and death processes has been described in [187]:

$$\left.\begin{aligned} \dot{x}_0(t) &= 1 - x_0(t) - x_1(t)f_1(x_0(t)) - x_2(t)f_2(x_0(t)), \\ \dot{x}_1(t) &= [f_1(x_0(t-h_1)) - 1]x_1(t), \\ \dot{x}_2(t) &= [f_2(x_0(t-h_2)) - 1]x_2(t). \end{aligned}\right\}$$

Here x_0 is the nutrient concentration, x_1, x_2 are the concentrations of competing micro-organisms, $h_i > 0$ are (constant) delays, and $f_i(0) = 0$, $\dot{f}_i(x) > 0$ for $x > 0$.

The stability of a periodic solution of this model has been studied in [187].

5.5. Control problems in ecology. In each of the above mentioned models we can introduce a control $u(t)$, taking into account some purposeful action of various factors on the system. For example, consider the action of pesticides, and assume these act only on preys. Using the model (5.3), this action can be described by the equations

$$\begin{aligned} \dot{x}_1(t) &= [a_1 - a_2x_2(t) - a_3x_1(t) - b_1(t)u(t)]x_1(t), \\ \dot{x}_2(t) &= -a_4x_2(t) + a_5x_1(t-h)x_2(t-h), \qquad b_1(t) \geq 0. \end{aligned}$$

If pesticides act only on predators, then a control appears at the second equation in (5.3):

$$\begin{aligned} \dot{x}_1(t) &= [a_1 - a_2x_2(t) - a_3x_1(t)]x_1(t), \\ \dot{x}_2(t) &= -a_4x_2(t) + a_5x_1(t-h)x_2(t-h) - b_2(t)x_2(t)u(t), \qquad b_2(t) \geq 0. \end{aligned}$$

If pesticides act upon both species, then a control appears in both equations in (5.3).

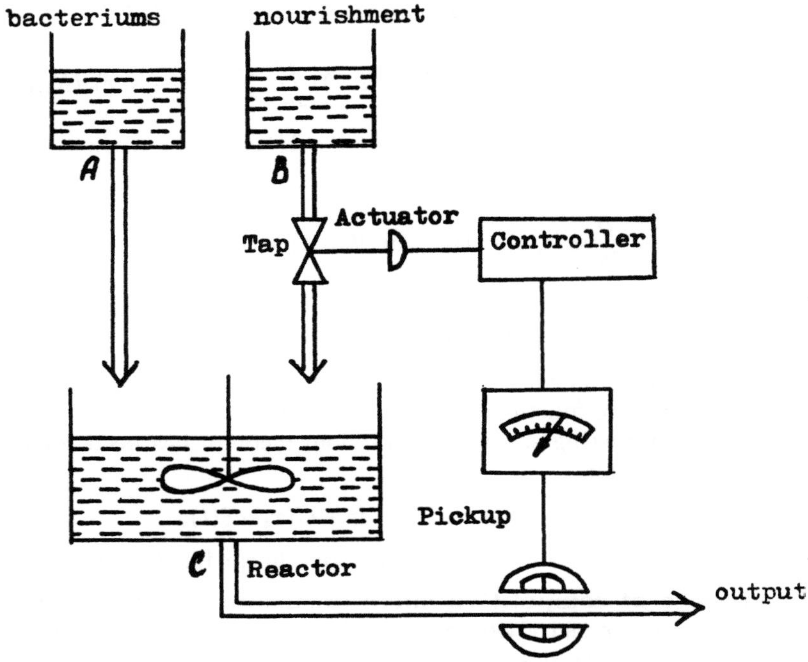

FIGURE 1.5.1. Biological reactor

Finally, it is also possible that pesticides have, next to an instantaneous effect, a prolonged effect by means of a residuum $R(t)$. Then the behavior of the predator–prey system can be modeled by the equations

$$\left.\begin{aligned} \dot{x}_1(t) &= [a_1 - a_2x_2(t) - a_3x_1(t) - b_1(t)u(t) - a_6R(t)]x_1(t), \\ \dot{x}_2(t) &= [-a_4 - b_2(t)u(t) - a_7R(t)]x_2(t) + a_5x_1(t-h)x_2(t-h), \\ \dot{R}(t) &= -a_8R(t) + u(t), \qquad a_i \geq 0, \qquad b_i(t) \geq 0. \end{aligned}\right\}$$

For the model of the production of white blood cells, optimal control problems for a Lotka–Volterra system have been considered in [283] and [456].

5.6. Control problems in microbiology. Certain delay models are used to control processes of microbiological growth of cells and production of a useful product. We will consider one of them, describing the continuous reproduction of micro-organisms, the production of ferments, the degradation of wastes, etc. The process here discussed is as follows (Fig. 1.5.1).

Bacteria are introduced into a vessel with an entrance for nourishing substances and an entrance for extraction of resulting products. The bacteria consume the nourishing substances, reproduct, and produce at a certain moment of time some quantity of the resulting product.

This process can be described by the bilinear delay model [283]

$$\left.\begin{aligned}\dot{x}_1(t) &= \gamma(t)x_1(t) - u(t)x_1(t) - \beta x_1(t-h),\\ \dot{x}_2(t) &= \gamma(t)\alpha^{-1}x_1(t) - u(t)x_2(t) + bu(t).\end{aligned}\right\} \tag{5.4}$$

The first equation is the balance equation of biological substrate, the second equation characterizes the production of resulting mass by the bacteria. Here,

$x_1(t)$ is the volume of microbiological substrate;
$x_2(t)$ is the volume of the resulting product;
$u(t)$ is the volume of nourishing environment in the vessel;
$\gamma(t)$ is the rate of biological growth;
$x_1(t-h)$ accounts for the loss of bacteria of great vitality during a finite time h;
β and b are constants in the model;
α is the rate of growth of the useful product.

Models like (5.4) can be used in the process of biological clearance of sewage, when dirty water goes into a vessel with active substances that come to react with the contaminator. The rates of supply of dirty water and active substances are controlled. As a result of the reaction we obtain clean water, and a residue of biopolymers. The values of the delays in this model depends on the rate of mixing, temperature, density, etc.

5.7. Nicholson blowflies model. To describe the dynamics of Nicholson's blowflies, the following equation has been used in [228]:

$$\dot{x}(t) = -ax(t) + bx(t-h)e^{-\gamma x(t-h)}, \qquad t \geq t_0.$$

Here, $x(t)$ is the size of the population at time t, b is the maximum daily production rate of eggs per capita, γ^{-1} is the size at which the production rate of the population is maximal, a is the daily adult death rate per capita, and h is the lifetime.

5.8. Helical movement of tips of growing plants. The tip of a growing plant has been modeled in [103, 255, 297, 333, 465] by the equation

$$\ddot{x}(t) + \alpha\dot{x}(t) + \beta f(x(t-h)) = 0.$$

Here α, β, h are positive constants, and $xf(x) > 0$, $x \neq 0$.

6. Aftereffect in medicine

6.1. Mathematical models of the sugar quantity in blood. FDEs can be efficiently used to describe various processes in living organisms. Various hereditary models have been proposed in [456] to describe the functioning of thyroid gland, the system of maintaining the sugar level in blood, and blood production. Certain parameters in these models can be regulated (temperature, diet, drugs, etc.). E.g., the control model for the sugar level in blood has the form [456]

$$\left.\begin{aligned}\dot{x}_1(t) &= a_1\{a_2x_4(t) + a_3[a_2x_4(t) - a_4x_2(t)] - a_5x_1(t-h)\}x_1(t),\\ \dot{x}_2(t) &= a_6\{a_2x_4(t) + b_1u(t) - a_7[a_2x_4(t) - a_5x_1(t)] - a_4x_2(t)\}x_2(t),\\ \dot{x}_3(t) &= a_8\{a_5x_1(t) + b_2u(t) + a_9[a_5x_1(t) - a_4x_2(t)] - a_{10}x_3(t)\}x_3(t),\\ \dot{x}_4(t) &= a_{11}\{1 + u(t) + a_{12}[1 - a_4x_2(t)] - a_2x_4(t)\}x_4(t).\end{aligned}\right\}$$

Here, $x_1(t)$ is the amount of insulin produced by the pancreas, $x_2(t)$ is the amount of active insulin in the blood, $x_3(t)$ is the total amount of insulin in the blood, $x_4(t)$ is the amount of sugar in the blood (all at time t); a_2, a_4, a_5, a_{10} are the averages of these amounts; the delay h characterizes the finite time needed for production of insulin, and a_1 is the rate of insulin production; a_6, a_8, a_{11} reflect the increase of insulin, total amount of insulin and sugar in the blood; finally, $b_1 \geq 0$, $b_2 \geq 0$, a_3, a_7, a_9, a_{12} are feedback coefficients. The control $u(t)$ is fulfilled by choice of a diet, and may affect the amount of sugar in the blood.

We also mention the Mackay–Glass model of blood production:

$$\dot{x}(t) = \frac{ax(t-h)}{1 - x^2(t-h)} - bx(t).$$

A numerical treatment of this equation has been given in [176], in which also the appearance of chaotic attractors has been shown. In [222], an analysis of the blood control mechanism is based on the equation

$$\dot{x}(t) = \frac{kx^m(t-h)}{1 + k_0x^n(t-h)} - k_1x(t).$$

6.2. Model of arterial blood pressure regulation. In [202], FDEs have been used to describe arterial pressure control and mechanisms for the origin of hypertension. In it, for the first time these phenomena were treated as resulting from the stress-strained vascular state, measured by numerous receptor systems. It proved possible to completely re-evaluate well-known experimental and clinical observations, and to design an orderly concept of them. The cardiac vacular system is presented in Fig. 1.6.1, where A and B are arterial vessels connected with each other, R is the peripheral resistance, Q is the rate of blood flow from A to B, and Q_h is the productivity of the heart. The rate of the incoming liquid is Q_i, that of the outgoing Q_o.

Let $V_a(t)$ and $V_b(t)$ be the stressed arterial and venous volume. Then

$$\begin{aligned}\dot{V}_a(t) &= k_1k_bV_b(t) - k_4k_aV_a(t) +\\ &+ f(t) - [k_aV_a(t) - k_bV_b(t)] \cdot [k_2 - k_3V_a(t-h)]^{-1},\end{aligned} \tag{6.1}$$

$$\dot{V}_b(t) = Q_i + [k_aV_a(t) - k_bV_b(t)] \cdot [k_2 - k_3V_a(t-h)]^{-1} - k_1k_bV_b(t). \tag{6.2}$$

The constants k_i depend on the characteristics of the cardiac vascular system, and h is the delay of baroreceptors. The function f in (6.1) is equal to $t \mapsto V_a(t-h_1)$ (where h_1 is the delay in the aldosterone circuit) if the aldosterone circuit is taken into account, and $f = 0$ otherwise.

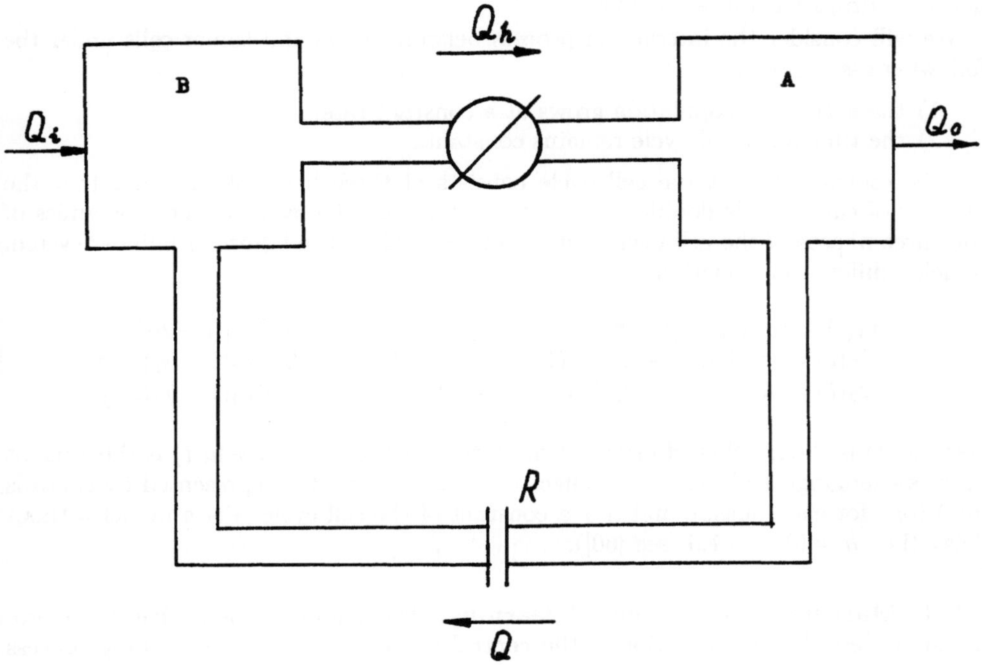

FIGURE 1.6.1. Scheme of the cardiac vasular system

More detailed models of arterial blood pressure regulation, allowing one to take into account features neglected in (6.1) and (6.2), have been considered in [202]. However, these models contain a larger number of parameters to be identified, and hence in actual applications their significance is less.

6.3. Cancer chemotherapy. There are few papers devoted to the effect of of antitumor drugs on the kinetics of the cell cycle. In them it is assumed that the drugs act instantaneously, and without aftereffect. However, in reality, to determine the time-varying effects of drugs on the cell cycle one has to take into account that the action of drugs has finite duration.

We will consider the interaction process between drugs and tumor cells under the following assumptions:

1) the entire cell population grows at a constant rate;
2) the time for a cell cycle remains constant.

It is known [60] that the cell cycle consists of three basic phases, and that the number of cancer cells doubles every two months, on the average. The dynamics of the three phases of the cell cycle can be represented by the following bilinear system of delay differential equations:

$$\left.\begin{aligned}\dot{x}_1(t) &= \alpha[a_3 + u_3(t-h_3)]x_3(t-h_3) - [a_1 + u_1(t-h_1)]x_1(t-h_1),\\ \dot{x}_2(t) &= [a_1 + u_1(t-h_1)]x_1(t-h_1) - [a_2 + u_2(t-h_2)]x_2(t-h_2),\\ \dot{x}_3(t) &= [a_2 + u_2(t-h_2)]x_2(t-h_2) - [a_3 + u_3(t-h_3)]x_3(t-h_3).\end{aligned}\right\}$$

Here, $x_i(t)$ is the number of cells at time t, the function $t \mapsto a_i + u_i(t)$ is the amount of cells changing at phase i. The effect of drugs at time t is represented by controls $u_i(t)$ (one for each phase), and α is a constant of the cell cycle. For a model without delay (i.e., $h = 0$) $\alpha = 1.6$ (see [60]).

6.4. Mathematical models of learning. The following model has been proposed to describe the behavior of the central nervous system in a learning process [449]

$$\dot{x}(t) = K[x(t) - x(t-1)][N - x(t)], \qquad t \geq 0, \tag{6.3}$$

$$x(t) = 0 \qquad (-1 \leq t < 0), \qquad x(0) = x_0. \tag{6.4}$$

Here, K and N are positive constants, $0 < x_0 < N$. Another learning model has been proposed in [215], using a system of nonlinear RDEs with discrete delays.

6.5. Mathematical models in immunology and epidemiology. FDEs can be efficiently used in the simulation of immune and epidemic processes. Here we will describe the model of virus disease [350, 351].

Let $V(t)$ be the number of viruses, $C(t)$ the number of plasmacytes producing antibodies, $m(t)$ the relative amount of damaged tissue, and $F(t)$ the amount of antibodies. Then the following system of FDEs holds:

$$\left.\begin{aligned} \dot{V}(t) &= [\beta - \gamma F(t)]V(t), \\ \dot{C}(t) &= \alpha\xi(m(t))F(t-h)V(t-h) - \mu_C[C(t) - C_0], \\ \dot{F}(t) &= \rho C(t) - \eta\gamma F(t)V(t) - \mu_f F(t), \\ \dot{m}(t) &= \delta V(t) - \mu_m m(t). \end{aligned}\right\} \tag{6.5}$$

Here, C_0 is a constant level of plasmacytes, the delay h is he time elapsing from the beginning of the stimulation of the lymphocytes to the beginning of the production of antibody mass, and $\xi(m)$ is a nonincreasing function. All other coefficients in (6.5) are constants.

The following RDE with delay depending on the unknown function was used in [134] to model epidemics:

$$\dot{x}(t) = ax(t - h(t, x(t))).$$

To model infection diseases, the following equation was used in [311]:

$$\dot{x}(t) = \left[\int_0^t K(t-\tau)\dot{x}(\tau)\,d\tau - f(t)\right].$$

6.6. Model of the human immunodeficiency virus (HIV) epidemic. To study the dynamics of HIV, which is the etiological agent for AIDS (acquired immunodeficiency syndrome), the following model has been proposed in [106]:

$$\left.\begin{aligned} &\dot{S}(t) = \Lambda - B(t) - S(t), \\ &\left(\tfrac{\partial}{\partial t} + \tfrac{\partial}{\partial \tau}\right) i(t,\tau) = -[1 + \alpha(\tau)]i(t,\tau), \\ &i(t,0) = B(t) = S(t)C(T(t))\tfrac{W(t)}{T(t)}, \\ &T = I + S, \qquad I(t) = \textstyle\int_0^\infty i(t,\tau)\,d\tau, \\ &W(t) = \textstyle\int_0^\infty \lambda(\tau)i(t,\tau)\,d\tau, \\ &\dot{A}(t) = \textstyle\int_0^\infty \alpha(\tau)i(t,\tau)\,d\tau - (1+\nu)A(t). \end{aligned}\right\} \tag{6.6}$$

This model was created to mimic the dynamics of HIV in a homogeneously mixed (male) homosexual population. In this population, three groups are distinguished: S (uninfected, but susceptible), I (HIV infected), and A (fully developed AIDS disease). It is supposed that members of I and S are sexually active and choose their partners at random, while members of A are sexually inactive. Also, the following notation is used: t is time, τ is time elapsed from the moment of infection, Λ is the (constant) rate of growth of the sexually active population, $B(t)$ is the number of new cases of infection per unit time, $i(t,\tau)$ is the infection-age density, so that the chance that a randomly selected partner was infected τ moments ago (has infection age τ) is $i(t,\tau)/[I(t)+S(t)]$, the function $\tau \mapsto \alpha(\tau)$ is the rate with which the infected persons with infection age τ stop being sexually active due to the disease, $C(T(t))$ is the average number of sexual contacts an average person has in unit time, ν is the average death rate of a person with AIDS, and $\tau \mapsto \lambda(\tau)$ is a given nonnegative function.

Another model for HIV in a homogeneously mixed single-gender group with distributed waiting times has been described in [107], using equations with distributed delay:

$$\left.\begin{aligned} \dot{S}(t) &= \Lambda - B(t) - \mu S(t), \\ I(t) &= I_0(t) + \textstyle\int_0^t B(s)e^{-\mu(t-s)}\rho(t-s)\,ds, \\ A(t) &= A_0(t) + A_1e^{-\epsilon t} + \textstyle\int_0^t d\tau \int_0^\tau B(s)e^{-\mu(t-s)}[-\dot{P}(\tau-s)e^{-\epsilon(t-\tau)}]\,ds. \end{aligned}\right\} \quad (6.7)$$

Here the functions $I_0(\cdot)$, $A_0(\cdot)$ and constants A_1, ϵ are given; μ is the rate with which sexual activity is stopped; $P(s)$ is the proportion of individuals infected at time t that are still alive at time $s+t$. The other notation in (6.7) is as in (6.6).

In [273] the authors suggest a model for describing the risk of infection by HIV as a function of time and an individual's history of sexually transmitted disease. The equation describing the epidemic progress is:

$$\frac{\partial H(t,k)}{\partial t} = \int_0^\infty \beta(C,k)P(C,k)\frac{N(k)-H(t,k)}{N(k)} \cdot \frac{H(t,C)}{N(C)}\,dC - \mu H(t,k).$$

Here $H(t,k)\,dk$ is the number of people with rate between k and $k+dk$ which are infected with HIV at time t; $\beta(C,k)$ is the probability that partnership between an uninfected person of rate k and an infected person of rate C results in transmission of the infection; $p(C,k)\,dC\,dk$ is the number of partnerships formed in unit time between people with rate between C and $C+dC$ and those with rate between k and $k+dk$. The number of people with rate between k and $k+dk$ is

$$N(k)\,dk = \frac{1}{k}\int_0^\infty \rho(C,k)\,dC\,dk.$$

Hence, $\int_0^\infty \rho(C,k)\,dC$ is the total number of partnerships formed by people with rate between k and $k+dk$. The constant μ is the rate with which sexual activity is stopped.

6.7. Model of survival of red blood cells. A model for the survival of red blood cells in an animal has been described in [500] by the equation

$$\dot{x}(t) = -ax(t) + be^{-\gamma x(t-h)}, \qquad t \geq t_0,$$

where $x(t)$ is the number of red blood cells at time t, a is the probability of death of a red blood cell, b, $\gamma > 0$ are constants related to the production of red blood cells per unit time, and the delay $h > 0$ is the time required to produce a red blood cells.

6.8. Vision process in the compound eye. This process has been modeled in [235] by

$$\dot{x}(t) + qx(t) + p\max\{x(\tau(t)), c\} = 0,$$

where $c < 0$, p, q are constants.

7. Aftereffect in economy and other sciences

Delay models are frequently encountered in economics and management science. To explain for price fluctuations, commodity supply fluctuations, trade cycles, etc., many authors use delay between beginning of a production decision and the delivery of goods to the market (see [44, 96, 104, 192, 197, 237, 308, 475]).

7.1. Optimal skill with retarded controls. In certain papers labor training, which is a necessary requirement for employment, is treated by the use of models with retarded controls. In such models, delays may occur because of differences between labor demanded and offered.

Let $x_1(t)$ be the fraction of labor force in training, $x_2(t)$ the fraction of unskilled labor force, and $x_3(t)$ the fraction of skilled labor force (all at time t). Then these variables can be described by the equations [439]

$$\left.\begin{array}{c} \dot{x}_1(t) = \beta\,[\,-x_1(t) - c + H(u(t)) + \\ + \int_{-\infty}^{t} e^{-\beta(t-\tau)} h(\max\{t-\tau, u(\tau)\})\psi_1(t-\tau, u(\tau))\,d\tau\Big], \\ \dot{x}_2(t) = \beta[1 - H(u(t)) - x_2(t)], \\ x_3(t) = 1 - x_1(t) - x_2(t). \end{array}\right\}$$

Here, β is a positive constant, and $h(x)$, $x \geq 0$, is a density function such that

$$H(u) = \int_0^u h(s)\,ds, \qquad H(\infty) = 1.$$

$H(u)$ is the proportion of new laborers that have been trained for at most u years (have untrainability index at most u), and $\psi_1(x, y) = 1$ if $x > y$ and $\psi_1(x, y) = 0$ if $x < 0$. The control action $u(t)$ is interpreted as the duration of time such that at time t any new laborer with untrainability index at most $u(t)$ will enter training. The restrictions on $x_i(t)$ and $u(t)$ are as follows:

$$x_1(t) \geq 0, \quad x_2(t) \geq 0, \quad x_1(t) + x_2(t) \leq 1, \quad u(t) \geq 0.$$

Finally, a cost functional, depending on $x_i(t)$, should represent the future output.

7.2. Optimal advertising policies. Several delay models have been proposed in which advertising expenditures are regarded as an investment in a goodwill stock $x(t)$.

Let $u(t)$ be the amount of advertising expenditures, and $t \mapsto f(t)$ a density function. Then [396]

$$\dot{x}(t) = g(u(t), x(t)) - \int_{-\infty}^{t} f(t-\tau) g(u(\tau), x(\tau))\,d\tau,$$

where g is the goodwill production function. The cost functional is the net profit stream, and the aim is to maximize it:

$$\int_0^T e^{-rt} R(u(t), x(t))\,dt + e^{-rT} S(x(T)) \to \max, \qquad u \geq 0,$$

where $r > 0$ is a (constant) amount of discount, R is the amount of profit, and S is a given function.

7.3. Commodity price fluctuations. Consider a single commodity market. Let $x(t)$ be the percentual deviation of the commodity price from equilibrium. The following equation has been proposed in [237]:

$$\dot{x}(t) = -ax(t) - b\int_0^\infty g(s)x(t-s)\,ds,$$

where a is proportional to the demand and b is proportional to the supply.

To explain the 4-year cycle in pork price, the following linear RDE with discrete constant delay h has been used in [308]:

$$\dot{x}(t) = k[x(t-h) - x_0].$$

A more complicated model of price fluctuation has been investigated in [44]. Let $x_1(t)$ be the current market price of the commodity, and $x_2(t)$ the current demanded price. Then [44]

$$\left.\begin{aligned} \dot{x}_1(t) &= [f_1(x_2(t)) - f_2(x_1(t-h(x_1(t))))]x_1(t), \\ x_2(t) &= \textstyle\int_{-\infty}^t K(t-s)x_1(s)\,ds. \end{aligned}\right\}$$

Here the kernel K is related to previous market prices, and the functions f_1, f_2 denote the demand and supply of the commodity.

7.4. Model of the fishing process. Let $P(t)$ be the average population of the resource, and $E(t)$ the fishing level. Then [289]

$$\left.\begin{aligned} \dot{P}(t) &= P(t)F(P(t)) - qH(E(t),P(t)), \\ \dot{E}(t) &= k[qH(E(t-rE(t)),P(t-rE(t))) - cE(t-rE(t))]. \end{aligned}\right\}$$

Here, $H(E,P)$ is the harvesting strategy, $F(\cdot)$ the relative growth rate, q the catchability coefficient, and c the cost per unit effort. All constants q, k, c, r are positive.

7.5. River pollution control. Let $z(t)$ and $q(t)$ be the concentrations per unit volume of biological oxygen demand (BOD) and dissolved oxygen (DO), respectively, at time t. It is assumed that the flow rate is constant, water is well mixed, and there exists $\tau > 0$ such that BOD and OD concentrations entering at time t are equal to the corresponding concentrations τ time units ago.

Using mass balance concentration, the following equations have been derived [313]:

$$\left.\begin{aligned} \dot{z}(t) &= -k_1(t)z(t) + \\ &+v^{-1}[Q_1(m+u_1(t)) + Qz(t-\tau) - (Q+Q_1)z(t)] + v_1(t), \\ \dot{q}(t) &= -k_3(t)z(t) + k_2(t)[q_0 - q(t)] + \\ &+v^{-1}[Qq(t-\tau) - (Q+Q_1)q(t)] + u_2(t) + v_2(t). \end{aligned}\right\}$$

Here, $k_i(\cdot)$ denote the BOD decay rate, the BO re-aereation rate, and the BOD deoxygenation rate; q_0 is the DO saturation concentration; Q and Q_1 are the stream flow rate and the effluent flow rate; v is the constant volume of water under consideration; $u_i(t)$ are controls; $v_i(\cdot)$ are random disturbances affecting the rates of change of BOD and DO; and m is a constant.

CHAPTER 2

General theory

1. Introduction. Method of steps

1.1. Notation. Let K be a subset of a set M, and f a function on M. Then $f|_K$ denotes the restriction of f to K.

If $x \in \mathbf{R}^n$, then $|x|$ is the euclidean norm of x. For $f\colon M \to \mathbf{R}^n$ we set, by definition, $\|f\| = \sup_{t\in M} |f(t)|$. We denote by $\mathbf{R}^{n\times n}$ the set of real square matrices of order n. For $A \in \mathbf{R}^{n\times n}$ we let $|A|$ be the euclidean norm of A:

$$|A| = \max_{\substack{x\in\mathbf{R}^n\\ |x|=1}} |Ax|.$$

A connected subset of $\mathbf{R}$ is called an *interval*. In the sequel we denote intervals by J, and set $\mathbf{R}_+ = [0,\infty)$.

For given J and n, the set of all continuous functions $f\colon J \to \mathbf{R}^n$ is denoted by $C(J,\mathbf{R}^n)$. It is a linear space with respect to the usual operations on functions. If J is closed, $C(J,\mathbf{R}^n)$ is a Banach space with respect to the norm $\|\cdot\|$. If n is clear from the context, we write $C(J)$ or CJ.

For a measurable function $f\colon J \to \mathbf{R}^n$ we define

$$\|f\|_p = \left[\int_J |f(t)|^p \, dt\right]^{1/p} \qquad (1 \le p < \infty),$$

$$\|f\|_\infty = \operatorname{vrai\,sup}_{t\in J} |f(t)| := \inf_M \sup_{t\in M} |f(t)|.$$

Here $M \subseteq J$ is an arbitrary set with $\operatorname{meas}(J\setminus M) = 0$. The symbol := means 'equality by definition'. For functions f defined almost everywhere on J one should take M such that f is defined on it.

For given J, n, p ($1 \le p \le \infty$), the set of all measurable functions $f\colon J \to \mathbf{R}^n$ such that $\|f|_{J_1}\|_p < \infty$ for every closed interval $J_1 \subseteq J$ is denoted $L_p(J,\mathbf{R}^n)$, $L_p(J)$, or L_pJ. It is a linear space, and if J is closed even a Banach space with respect to the norm $\|\cdot\|_p$. In addition, here we must assume that two functions differing only on a subset of J of measure zero represent the same element in L_pJ.

The linear space of all continuously differentiable functions $f\colon J \to \mathbf{R}^n$ is denoted

$C^1(J, \mathbf{R}^n)$ (or $C^1(J)$, $C^1 J$). If J is closed, it a Banach space with respect to the norm

$$\|f\|_{C^1} = \|f\| + \|\dot{f}\|.$$

The spaces $C^k(J, \mathbf{R}^n)$ $(k = 2, 3, \dots)$ are similarly defined.

The linear space of functions $f\colon J \to \mathbf{R}^n$ satisfying a Lipschitz condition on every closed interval $J_1 \subseteq J$ is denoted $\mathrm{Lip}(J, \mathbf{R}^n)$ (or $\mathrm{Lip}(J)$, $\mathrm{Lip}\, J$). If J is closed, it a Banach space with respect to the norm

$$\|f\|_{\mathrm{Lip}} = \|f\| + \|\dot{f}\|_\infty.$$

1.2. Cauchy problem for FDEs. The *Cauchy problem* (also called the *initial problem* or the *basic initial problem*) for a first order FDE is to find the solution of this FDE subject to a given initial function and initial value. For example, consider the RDE (1.1.7) with finite aftereffect, and let, for some t_0, the function $(t, \psi) \mapsto F(t, \psi)$ be defined for all $t \in [t_0, \infty)$, $\psi \in C(J_t)$, where $J_t = [-h(t), -g(t)] \subset (-\infty, 0]$. The point t_0 is called the *initial point* for the solution. For simplicity we assume that

$$\bar{t}_0 := \inf_{t \geq t_0} \{t - h(t)\} > -\infty.$$

In applications we usually have $\bar{t}_0 = t_0 - h(t_0)$. The *initial function* ϕ for (1.1.7) with prescribed initial point t_0 is given on the *initial interval* $[\bar{t}_0, t_0)$. Also, the *initial value* $x(t_0)$ of the solution must be given.

It is essential that the solution $x(t)$ is constructed in the direction of increasing t i.e. on an interval J_x with left end $t_0 \in J_x$. Here, x is regarded as a prolongation of the initial function, i.e. $x(t + \theta) := \phi(t + \theta)$ for $t + \theta < t_0$. Usually the initial value of the solution is included in the initial function, i.e. the latter is assumed to be given on the interval $[\bar{t}_0, t_0]$ and $\phi(t_0) = x(t_0)$. We adhere to this convention without further mentioning this.

In reality, t_0 does not always coincide with the moment of beginning of the process under investigation. Generally speaking, the hereditary differential law under consideration turns out to be applicable only from some stage onwards in the evolution of the process, and t_0 cannot be taken before this, but may be taken at a later stage.

The Cauchy problem for the NDE (1.1.8) or (1.1.9) is similarly posed. Further refinements are discussed in §3.

1.3. Step method for RDEs. For discrete delays, very often (almost always) the Cauchy problem on a finite interval (not too large, in practice) can be solved using a *step method*. We explain this method using the scalar equation

$$\dot{x}(t) = f(t, x(t), x(t-h)), \qquad t \geq t_0, \qquad h = \text{const} > 0. \tag{1.1}$$

Let $f\colon [t_0, \infty) \times \mathbf{R}^2 \to \mathbf{R}$ be a continuous function satisfying, for simplicity, a Lipschitz condition with respect to the second argument. We must prescribe the initial function ϕ for (1.1) on $[t_0 - h, t_0]$, and assume it is continuous.

If $t \in [t_0, t_0 + h]$ (this is the first step), then $t - h \in [t_0 - h, t_0]$. Therefore the RDE (1.1) becomes the ODE

$$\dot{x}(t) = f(t, x(t), \phi(t-h)), \qquad t_0 \leq t \leq t_0 + h.$$

Solving this equation under the initial condition $x(t_0) = \phi(t_0)$, we obtained the required solution on $[t_0, t_0 + h]$. In particular, the value $x(t_0 + h)$ is defined. If now, $t_0 + h \le t \le t_0 + 2h$ (this is the second step), then $t - h \in [t_0, t_0 + h]$, and so $x(t - h)$ is known from the first step. Hence (1.1) for $t_0 + h \le t \le t_0 + 2h$ once again becomes an ODE, which can be solved under the initial value $x(t_0 + h)$, defining the solution x on $[t_0 + h, t_0 + 2h]$. Now we consider the interval $[t_0 + 2h, t_0 + 3h]$, etc. In this way we can obtain the solution for arbitrarily large t (theoretically, for the semiaxis $[t_0, \infty)$).

For the simple case when the righthand side of (1.1) does not contain $x(t)$, at each step the solution leads to integration of a given function.

The degree of smoothness of the solution constructed presents an interesting problem. For simplicity, suppose f has continuous derivatives of all orders. Then (1.1) implies that $x \in C^1[t_0, \infty)$. However, even for an arbitrarily smooth function ϕ we obtain in general,

$$\dot{x}(t_0^+)(= f(t_0, \phi(t_0), \phi(t_0 + h))) \neq \dot{\phi}(t_0^-).$$

This means that if x is regarded as a prolongation of ϕ, then the composite function has a break for $t = t_0$ (a jump of the first derivative). From (1.1) we find that if $x \in C^1[t_0, \infty)$, then $\dot{x}|_{[t_0+h,\infty)} \in C^1[t_0 + h, \infty)$, i.e. $x|_{[t_0+h,\infty)} \in C^2[t_0 + h, \infty)$. But even for arbitrarily smooth ϕ, in general $\ddot{x}(t_0 + h^+) \neq \ddot{x}(t_0 + h^-)$. Using (1.1) again, we see that $x|_{[t_0+2h,\infty)} \in C^3[t_0 + 2h, \infty)$, etc.

This property of *solution smoothing* for increasing t is characteristic for RDEs. Together with some other properties it makes retarded functional differential equations resemble parabolic type partial differential equations. However, the reasons for this resemblance are not entirely clear.

Hardly ever does the step method lead to the exact solution of a problem. E.g., for (1.1.1) on an interval $[t_0, \infty)$ with $h > 0$ and initial function $t \mapsto \phi(t) = \phi_0 = \text{const}$ $(t_0 - h \le t \le t_0)$, this method gives

$$x(t) = \phi_0 \sum_{j=0}^{\infty} \frac{k^j}{j!}[t - t_0 - (j-1)h]^j H(t - t_0 - (j-1)h),$$
$$t_0 \le t < \infty,$$

where H is the Heaviside function $(H(t) := (1 + \operatorname{sgn} t)/2)$. Here the summation is in fact over finitely many terms, and the larger t, the larger the number of terms involved. For $t_0 - h < t \le t_0$, the formula gives the initial function.

We return to (1.1). Because at each step it is converted to an ODE, all methods for the numerical solution of ODEs (Runge–Kutta, etc.) can be applied to (1.1). In this case it is necessary to store the values of the solution at the present step, since they are used in the next step. If the number of steps is not too large, storage (filling the computer's memory) can be avoided by a simple method, which, apparently, was first proposed by R. Bellman.

For simplicity, assume that $t_0 = 0$, and put

$$x_j(t) = x(t + (j-1)h), \qquad 0 \le t \le h, \qquad j = 1, 2, \dots, \qquad x_0(t) = \phi(t - h).$$

Suppose we would like to perform N steps. Then the solution of (1.1) reduces to solving a sequence of N systems of ODEs, the jth of which has the form

$$\dot{x}_i(t) = f(t + (i-1)h, x_i(t), x_{i-1}(t)), \qquad 0 \le t \le h, \qquad i = 1, \dots, j,$$

with initial conditions

$$x_i(0) = x_{i-1}(h), \qquad i = 1, \dots, j,$$

taken from the solution of the $(j-1)$st system (for $j = 1$, $x_1(0) = \phi(0)$). So, only these conditions must be stored. Upon reaching $j = N$ we find the solution on the interval $[0, Nh]$.

A similar step method is applied to the more general equation (1.1.4) for positive continuous functions h_i (however, one of these is allowed to vanish identically). For simplicity we assume that all functions $t \mapsto t - h_i(t)$ tend to infinity as $t \to \infty$. Then we perform successive steps on intervals $[t_0, t_1], [t_1, t_2], \dots$, where

$$t_{j+1} = \inf\left\{t: \quad \max_i [t - h_i(t)] > t_j\right\}, \qquad j = 0, 1, \dots,$$

where in the maximum the index i for which $h_i(t) \equiv 0$ is not taken into account. At each step the RDE is converted to an ODE with regard to all previous steps. The totality of all steps leads, under natural assumptions on f and ϕ, to the construction of a solution on the semiaxis $[t_0, \infty)$, provided this solution does not tend to infinity for a finite t. The latter is possible if the Lipschitz condition, mentioned above for (1.1), is only locally fulfilled.

More general RDEs for which step methods can be used are:

$$\dot{x}(t) = F(t, x(t), x_t). \tag{1.2}$$

The meaning of x_t is as in (1.1.7), where the function g defining the right end of the interval J_t is positive and continuous.

1.4. Step methods for NDEs. Step methods for NDEs are used in much the same way as for RDEs. For example, let the defining scalar equation be

$$\dot{x}(t) = f(t, x(t), x(t-h), \dot{x}(t-h_1)), \qquad t_0 \le t < \infty,$$
$$h, h_1 = \text{const} > 0.$$

Sufficient conditions for applying a step method to this equation are: the continuous function $f\colon [t_0, \infty) \times \mathbf{R}^3 \to \mathbf{R}$ satisfies a Lipschitz condition with respect to the second argument, and the initial function $\phi \in C^1[t_0 - \max\{h, h_1\}, t_0]$. The scheme for applying the step method is as for (1.1), but with step $\Delta t = \min\{h, h_1\}$.

Usually, when applying step methods to (1.1.5) one assumes that delay zero is admitted in the unknown function, but not in its derivative, i.e. the equation is solved for $\dot{x}(t)$.

An essential feature of solutions of the Cauchy problem for NDEs is the absence of the smoothing property mentioned in subs. 1.3. The reason for this is obvious: At each step the solution of a 'proper' NDE (this is important, for every RDE is formally a NDE) must first be differentiated and is only then integrated.

To explain this, consider the simple scalar equation

$$\dot{x}(t) = \dot{x}(t-1), \qquad 0 \le t < \infty; \qquad x(t) = \phi(t), \qquad -1 \le t \le 0,$$

where $\phi \in C^1[-1,0]$ is a given initial function. The solution is

$$x(t) = \phi(t-j) + j[\phi(0) - \phi(-1)], \qquad j-1 \le t \le j, \qquad j = 1, 2, \ldots.$$

Hence, if $\dot{\phi}((-1)^+) \ne \dot{\phi}(0^-)$, then $\dot{x}$ has jumps at $t = 0, 1, \ldots$. If $\dot{\phi}((-1)^+) = \dot{\phi}(0^-)$ but $\phi \in C^2[-1,0]$ and $\ddot{\phi}((-1)^+) \ne \ddot{\phi}(0^-)$, then $\dot{x}$ is continuous and $\ddot{x}$ has jumps at $t = 0, 1, \ldots$, etc.

This example is generic: If the righthand side of an NDE (1.1.8) and its initial function ϕ are arbitrarily smooth, then the derivative of the (continuous) solution has jumps for arbitrarily large t. The jumps are absent if ϕ satisfies the necessary and sufficient *sewing condition*

$$\dot{\phi}(t_0^-) = F(t_0, \phi_{t_0}, \dot{\phi}_{t_0}).$$

Together with some other properties this property makes neutral functional differential equations resemble hyperbolic type partial differential equations.

1.5. Problems for a process with aftereffect renewal. We will illustrate by the step method some problems in solving differential equations with advanced argument (ADEs). We have already mentioned that such equations have only seldom appeared in applications. However, from the applied point of view, it is natural to regard the problem of renewal of a process with aftereffect given a final observation, which is equivalent to the problem of forward prolongation of solutions of ADEs.

For example, consider the simple equation

$$\dot{x}(t) = x(t-1), \qquad \infty < t \le 1.$$

The final values $x(t) = \psi(t)$, $0 \le t \le 1$, are given, where $\psi \in C^\infty[0,1]$. The problem is to renew $x|_{t<0}$. In the first step we set $t \in [0,1)$, and obtain

$$x(t) = \dot{\psi}(t+1), \qquad -1 \le t < 0.$$

Hence, if $\psi(0) \ne \dot{\psi}(1^-)$, then x has a jump at $t = 0$:

$$x(0^+) - x(0^-) = \psi(0) - \dot{\psi}(1^-).$$

Using distributions, subsequent steps give a formal solution of the problem:

$$x(t) = \sum_{j=1}^{\infty} \sum_{i=1}^{j} [\psi^{(i-1)}(0^+) - \psi^{(i)}(1^-)]\delta^{(j-i)}(t+j) + y(t).$$

Here

$$y(t) = \psi^{(j)}(t+j), \qquad -j \le k < -j+1, \qquad j = 1, 2, \ldots,$$

and δ is the delta-function.

Consequently, to avoid delta-summands one must impose the conditions

$$\psi^{(j)}(0^+) = \psi^{(j)}(1^-), \qquad j = 1, 2, \ldots.$$

The number of such conditions increases with the growth of the time-interval on which the solution is to be renewed. But even if these conditions hold, the renewal operator of a hereditary process with final observation is unbounded already at the first step since it is the differentiation operator (and a function can be small while having arbitrarily large derivative). So the renewal problem described above is ill-posed, like many other inverse problems, and corresponding methods must be used to solve it.

In relation to the similar problem for NDEs, the result depends on whether the equation is 'rear defended'. Thus, the inverse problem for the equation

$$\dot{x}(t) = \dot{x}(t-1) + x(t-2)$$

is ill-posed, but for the equation

$$\dot{x}(t) = x(t-1) + \dot{x}(t-2)$$

it is well-posed, as can be readily proved by the step method. The last equation is an example of a bilateral NDE (in Kamenskii's classification), which preserve their type under time inversion. Note that a sufficiently well developed general theory for this equation has not yet been developed.

2. Cauchy problem for RDEs

2.1. Basic solvability theorem. Here we give a widely-used version of a solvability theorem for the initial value problem for RDEs, assuming that the solution belongs to C^1 (i.e. is a continuously differentiable function of t) while the time of possible aftereffect is finite and constant. I.e. we consider the problem

$$\dot{x}(t) = F(t, x_t); \qquad x_t(\theta) := x(t+\theta), \qquad -h \le \theta \le 0, \tag{2.1}$$

$$x_{t_0} = \phi, \tag{2.2}$$

where $h = \text{const} \in [0, \infty)$, $x(t) \in \mathbf{R}^n$, $t_0 \in \mathbf{R}$, $\phi\colon [-h, 0] \to \mathbf{R}^n$, $n \ge 1$. We emphasize that, in contrast to §1, the initial function ϕ is now assumed to be defined on the 'canonical' interval $[-h, 0]$ instead of on $[t_0 - h, t_0]$, i.e. the solution $x(\cdot)$ is a prolongation of $\phi(t_0 + \cdot)$.

A function $x \in C^1(J_x)$ is said to be a solution of the problem (2.1)–(2.2) on an interval J_x with left end $t_0 \in J_x$ if it satisfies (2.1) and (2.2) at all points of J_x (i.e. $x_t(t+\theta) := \phi(t+\theta-t_0)$ for $t+\theta < t_0$, and $x(t_0) = \phi(0)$). Put $D_h := [t_0, \infty) \times C[-h, 0]$.

THEOREM 2.1. *Suppose $\phi \in C[-h, 0]$, and let $F\colon D_h \to \mathbf{R}^n$ be a continuous functional satisfying in some neighborhood of any point $(t, \psi) \in D_h$ a Lipschitz condition in the second argument (with Lipschitz constant depending, in general, on this point). Then there is $t_\phi (= t_{\phi, t_0, F}) \in (t_0, \infty]$ such that:*

a) *there is a solution x of (2.1)–(2.2) on the interval $[t_0, t_\phi)$;*
b) *on any subinterval $[t_0, t_1] \subset [t_0, t_\phi)$ the solution is unique;*
c) *if $t_\phi < \infty$, then $x(t)$ does not have a finite limit as $t \to t_\phi^-$*, i.e. $[t_0, t_\phi)$ is the maximal interval of existence of the solution of (2.1)–(2.2);
d) *the solution x depends continuously on F, ϕ.*

Assertion d) means that for any $t_1 \in (t_0, t_\phi)$, $\epsilon > 0$ there is a $\delta > 0$ such that if F, ϕ in (2.1)–(2.2) are replaced by $\overline{F}, \overline{\phi}$ with

$$|\overline{F}(t,\psi) - F(t,\psi)| < \delta \qquad \text{for } t_0 \le t \le t_1,\ \|\psi - x_t\| < \epsilon,$$
$$\|\overline{\phi} - \phi\| < \delta$$

and $\overline{F}, \overline{\phi}$ satisfy the same assumptions as do F, ϕ, then for the solution $\overline{x}$ of the transformed problem with maximal interval of existence of the solution $[t_0, \overline{t}_{\overline{\phi}})$ we have

$$\overline{t}_{\overline{\phi}} > t_1, \qquad |\overline{x}(t) - x(t)| < \epsilon, \qquad t_0 \le t \le t_1.$$

Taking $t - t_0$ as new independent variable, we can investigate the dependence of the solution on t_0, just as we have done for its dependence on F. This remark applies equally well to subsequent solvability theorems.

2.2. Variants. *A*. Applying thm. 2.1 at each step in the step method (subs. 1.3) proves that the first three assertions in this theorem are valid if the Lipschitz condition holds in the following weaker form: in some neighborhood of any point of D_h the inequality $|F(t,\psi_1) - F(t,\psi_2)| \le L\|\psi_1 - \psi_2\|$ holds for all $\psi_1, \psi_2 \in C[-h, 0]$ for which $\psi_1(\theta) \equiv \psi_2(\theta)$, $-h \le \theta \le -\epsilon$. Here, L and $\epsilon \in (0, h]$ may depend on the point chosen in D_h.

If $F(t, x_t) = f(t, x(t), x(t - h_1(t)), \dots, x(t - h_k(t)))$, then this remark applies to all arguments $f(t - h_i(t))$ for which h_i is continuous and (strictly) positive: for these arguments the Lipschitz condition is not required.

B. Let F be defined and satisfy the above stated condition only on a set $D \subset D_h$ open in D_h. Of course, we must require $(t_0, \phi) \in D$. In this case the first, second, and fourth assertions in thm. 2.1 hold. The third assertion should also include the possibility: $t_\phi < \infty$, $x(t)$ has a finite limit as $t \to t_\phi^-$, and

$$\lim_{t \to t_\phi^-} d((t, x_t), \partial D) = 0,$$

where d denotes the distance function and ∂D is the boundary of D. Hence, for this case the maximal interval of existence of (2.1)–(2.2) is also $[t_0, t_\phi)$.

C. Under additional, not very restrictive, conditions, the third assertion in thm. 2.1 can be sharpened. E.g. if F is bounded on every bounded subset in D_h (in practice this always holds, if the other conditions of thm. 2.1 hold), then $x(t)$ is unbounded as $t \to t_\phi^-$ and $t_\phi < \infty$. Moreover, there are important cases in which $|x(t)| \to \infty$ in this process. (It is known that for ODEs, i.e. when $h = 0$, all solutions that have a prolongation to only a finite t-interval have this property.) In particular, this is a case for which the step method is applicable in the form described in subs. 1.3, i.e. (2.1) takes the form (1.2) and the additional requirements for (1.2) are fulfilled. Examples in which we have

$$\limsup |x(t)| = \infty, \qquad \liminf |x(t)| < \infty \qquad \text{as } t \to t_\phi^- < \infty$$

while the requirements in thm. 2.1 hold, do not have real applications.

A solution x of (2.1)–(2.2) such that $|x(t)| \to \infty$ as $t \to t_\phi^- < \infty$ characterizes a *peaking regime*. In fact, such a solution either indicates a catastrophic evolution of the process, or the necessity to fundamentally change from some time onwards the mathematical model of the phenomenon.

D. What are the additional requirements to be imposed such that we would have, under the conditions of thm. 2.1, $t_\phi = \infty$? If, in particular, F is *sublinear*, i.e.

$$|F(t,\psi)| \leq A(t)\|\psi\| + B(t), \qquad \forall (t,\psi) \in D_h, \tag{2.3}$$

where $A, B \in C[t_0, \infty)$, then $t_\phi = \infty$. The inequality (2.3) holds if F satisfies on every bounded set $M \subset D_h$ a Lipschitz condition in the second argument, with Lipschitz constant depending on $\sup_{(t,u)\in M} t$.

A condition more general than (2.3) is the *Nagumo type condition*

$$|F(t,\psi)| \leq \Phi(t, \|\psi\|), \qquad \forall (t,\psi) \in D_h, \tag{2.4}$$

where $\Phi\colon [t_0, \infty) \times \mathbf{R}_+ \to (0, \infty)$ is a continuous function, nondecreasing in t, and such that

$$\int_0^\infty \frac{ds}{\Phi(t,s)} = \infty, \qquad t_0 \leq t < \infty.$$

Condition (2.4) is sufficient for having $t_\phi = \infty$.

More refined conditions for unrestricted prolongation of solutions of (2.1)–(2.2), taking into account the directions of the field F, are similar to such conditions in stability theory, using Liapunov functions or functionals. E.g., consider a simple example using a Razumkhin type stability theorem (see thm. 3.1.4). Suppose there is a function $V\colon \mathbf{R}^n \to \mathbf{R}$, $V \in C^1$, such that $V(y) \to \infty$ as $|y| \to \infty$, and for all sufficiently large $r > 0$,

$$\sup_{\substack{t_0 \leq t \leq r \\ \|\psi\| \leq r}} |F(t,u)| < \infty,$$

while if $\psi(0) = y$, $V(y) = r$, $V(\psi(\theta)) < r$, $-h \leq \theta < 0$, then

$$F'(t,\psi) \cdot \operatorname{grad} V(y) < 0$$

(here the prime denotes transposition). Then $t_\phi = \infty$ for any solution of (2.1)–(2.2).

E. Thm. 2.1 and its variants A–D can be generalized directly to RDEs with finite variable aftereffect which may even be unbounded as $t \to \infty$; more precisely, to equations like (1.1.7) with $g, h \in C[t_0, \infty)$. In fact, setting $h_1 = \max_{t_0 \leq t \leq t_1} h(t)$ for a fixed $t_1 \in (t_0, \infty)$, the functional $F(t, \cdot)$ can be defined on $C[-h_1, 0]$ for all $t \in [t_0, t_1]$ by the relation

$$F(t,\psi) = F(t, \psi|_{[-h(t),-g(t)]}), \qquad \forall \psi \in C[-h_1, 0].$$

In addition, the conditions of the theorem (and its variants) must be stated for this extended functional. Since t_1 is arbitrary, all assertions in the theorem hold. In addition, the initial function must be defined on the union of all intervals

$$[t - t_0 - h(t), t - t_0 - g(t)] \cap (-\infty, 0], \qquad t_0 \leq t < \infty.$$

2.3. Semigroup relation. Let $x(t;t_0,\phi)$ be the solution of (2.1)–(2.2) under the conditions of thm. 2.1. Assume that for any initial point $t = t_1 \geq t_0$ and any initial function $\phi \in C[-h,0]$ the solution of the corresponding Cauchy problem has a prolongation to the semiaxis $[t_1,\infty)$. Then thm. 2.1 implies that $x(\cdot;\cdot,\cdot)$ is continuous and satisfies the semigroup relation

$$x(t;t_2,x_{t_2}(\cdot;t_1,\phi)) = x(t;t_1,\phi), \qquad t \geq t_2 \geq t_1 \geq t_0, \qquad \phi \in C[-h,0],$$

which is obtained by *translation of the initial point* to t_2. This relation is very often met in the theory of equations with aftereffect in the context of the problem (2.1)–(2.2). It relies on the following simple argument. The solution $x\colon [t_0,t_\phi) \to \mathbf{R}^n$ restricted to an interval $[t_1,t_\phi)$ $(t_0 < t_1 < t_\phi)$ is also a solution of (2.1) with initial point t_1 and initial function x_{t_1}.

If (2.1) is an *autonomous equation*, i.e. has the form

$$\dot{x}(t) = F(x_t),$$

and the functional F satisfies the conditions of thm. 2.1, then according to this theorem, a solution of the Cauchy problem can be represented in the form $t \mapsto x(t-t_0;\phi) := x(t-t_0;0,\phi)$. Moreover, if every solution can be unrestrictedly extended, i.e. always $t_\phi = \infty$, then the semigroup relation takes the form

$$x(\tau_2;x_{\tau_1}(\cdot;\phi)) = x(\tau_1+\tau_2;\phi), \qquad \tau_1,\tau_2 \geq 0, \qquad \phi \in C[-h,0].$$

These relations may be interpreted in a way proposed by N. Krasovskii, and of fundamental significance for FDEs with aftereffect. Namely, assume that we associate to a solution of (2.1)–(2.2) not only its *graph* in $\mathbf{R}\times\mathbf{R}^n$, but also the *integral curve* in $\mathbf{R}\times C[-h,0]$ represented by the set of all pairs (t,x_t) with $t \in [t_0,t_\phi)$.

Consideration of the integral curve is convenient, because for (2.1) it is completely determined by the initial point (t_0,ϕ). Hence this curve has equation

$$x_t = T(t,t_0)\phi, \qquad t_0 \leq t < t_\phi,$$

where $T(t,t_0)\colon C[-h,0] \to C[-h,0]$ is the *shift operator* from t_0 to t corresponding to (2.1). We can similarly define the operator $T(t,t_1)$, where $t \geq t_1 \geq t_0$. It is obvious that $T(t,t) = I_{C[-h,0]}$ (the identity operator).

Under the assumption stated at the beginning of this subsection the operator $T(t,t_1)$ is defined and continuous for all $t_1 \geq t_0$, $t \geq t_1$. Moreover, if F is bounded on every bounded subset of D_h, then for $t \geq t_1 + h$, $T(t,t_1)$ is completely continuous, i.e. it is continuous and maps bounded sets of $C[-h,0]$ to compact sets.

In terms of the shift operator the semigroup relation takes the form

$$T(t,t_2)T(t_2,t_1) = T(t,t_1), \qquad t \geq t_2 \geq t_1 \geq t_0.$$

For autonomous equations we obtain $T(t,t_1) = T(t-t_1,0)$, and we may denote the righthand side by $T(t-t_0)$ (shift operator over time $t-t_0$). In this case the semigroup relation is

$$T(\tau_2)T(\tau_1) = T(\tau_1+\tau_2) \qquad (\tau_1,\tau_2 \geq 0).$$

2.4. Absolutely continuous solutions. By weakening the requirements for a solution of (2.1), one can enlarge the class of equations for which the Cauchy problem is solvable. One of the most widespread extension is as follows.

We say that x is an *absolutely continuous solution* (an AC solution) of (2.1)–(2.2) if x is defined on an interval J_x with left end $t_0 \in J_x$, is absolutely continuous in some neighborhood of any point of J_x, satisfies (2.1)–(2.2) for almost all $t \in J_x$, and equals $\phi(0)$ for $t = t_0$. Further, we say that a functional $(t, \psi) \mapsto F(t, \psi)$ satisfies the *Carathéodory condition* on D_h if it is continuous in ψ for almost all fixed $t \in [t_0, \infty)$, measurable in t for each fixed $\psi \in C[-h, 0]$, and if in a neighborhood of each point of D_h there is a Lebesgue integrable function $\lambda\colon [t_0, \infty) \to \mathbf{R}_+$ (depending on the neighborhood, in general) such that

$$|F(t,\psi)| \le \lambda(t), \qquad |F(t,\psi_1) - F(t,\psi_2)| \le \lambda(t)\|\psi_1 - \psi_2\|. \tag{2.5}$$

The inequalities in (2.5) must hold everywhere in this neighborhood.

The Carathéodory condition is especially useful because it covers linear RDEs with discontinuous but integrable coefficients.

THEOREM 2.2. *Let F be a functional satisfying on D_h the Carathéodory condition, and $\phi \in C[-h, 0]$. Then there is a $t_\phi \in (t_0, \infty]$ such that*

a) *there is an AC solution x of (2.1)–(2.2) on $[t_0, t_\phi)$;*
b) *on each interval $[t_0, t_1] \subset [t_0, t_\phi)$ this AC solution is unique;*
c) *if $t_\phi < \infty$, then $x(t)$ does not have a finite limit as $t \to t_\phi^-$;*
d) *the AC solution x depends continuously on F, ϕ.*

Assertion d) means that for any $t_1 \in (t_0, t_\phi)$, $\epsilon > 0$ there is a $\delta > 0$ such that if F, ϕ in (2.1)–(2.2) are replaced by $\overline{F}, \overline{\phi}$, where $\overline{F}, \overline{\phi}$ satisfy the same conditions as do F, ϕ, and

$$\int_{t_0}^{t_1} \sup_{\|\psi - x_t\| < \epsilon} |\overline{F}(t,\psi) - F(t,\psi)|\, dt < \delta, \qquad \|\overline{\phi} - \phi\| < \delta,$$

then for the AC solution $\overline{x}$ with maximal interval of existence $[t_0, t_{\overline{\phi}})$ we have

$$t_{\overline{\phi}} > t_1, \qquad |\overline{x}(t) - x(t)| < \epsilon, \qquad t_0 \le t \le t_1.$$

The variants of these assertions given in subs. 2.2 remain valid for AC solutions, except for the last assertion in D. Also, all considerations with respect to the semigroup relation are valid.

Note that an ordinary solution of (2.1) is also an AC solution. Also, if F and ϕ are continuous, then the AC solution of (2.1)–(2.2) coincides with the ordinary solution. Hence in this case a relaxation of the requirements does not lead to new solutions.

A further generalization lies in considering discontinuous initial functions. E.g., let $\phi\colon [-h, 0] \to \mathbf{R}^n$ be a fixed function. Let M_ϕ be the set of functions $y\colon [t_0 - h, \infty) \to \mathbf{R}^n$ such that $y_{t_0} = \phi$, $y|_{[t_0,\infty)} \in C[t_0, \infty)$, and let D_ϕ be the set of pairs (t, y_t) with $t_0 \le t < \infty$, $y \in M_\phi$, equipped with the distance function

$$D((t_1,\psi_1),(t_2,\psi_2)) = |t_2 - t_1| + \|\overline{\psi}_2 - \overline{\psi}_1\|.$$

Here $\overline{\psi}_i = \psi_i$ for $t_i \geq t_0 + h$, and if $t_0 \leq t_i < t_0 + h$, then

$$\overline{u}_i(\theta) = \begin{cases} u_i(\theta), & t_0 - t_i \leq \theta \leq 0, \\ u_i(t_0 - t_i), & -h \leq \theta < t_0 - t_i. \end{cases}$$

Assume that for any $y \in M_\phi$ the function $t \mapsto F(t, y_t)$ is defined for almost all $t \in [t_0, \infty)$ and belongs to $L_1[t_0, \infty)$. Let, moreover, F be continuous in ϕ for almost all t, and suppose that the conditions (2.5) hold in a neighborhood of each point of D_ϕ. Under these assumptions all assertions of subs. 2.4 hold for AC solutions of (2.1)–(2.2), except the assertion that the solution depends continuously on the initial function (which we have taken fixed). (It is easy to overcome this obstacle, but we prefer not to in this book.)

In fact, the generalization just mentioned is useful (and can be used) if, in particular, ϕ is uniformly continuous on $[-h, 0)$ but $\phi(0^-) \neq \phi(0)$.

2.5. RDEs with infinite delay. Another generalization of thm. 2.1 is obtained by considering RDEs with infinite delay:

$$\dot{x}(t) = F(t, x_t), \qquad t \geq t_0; \qquad x_t(\theta) := x(t+\theta), \qquad -\infty < \theta \leq 0, \tag{2.6}$$

$$x_{t_0} = \phi\colon (-\infty, 0] \to \mathbf{R}^n. \tag{2.7}$$

By imposing restrictions we have to regulate the behavior of the functions under consideration as $t \to -\infty$, by indicating a complete metric space $\mathcal{K} \subseteq C(-\infty, 0]$ containing the parts x_t of the solution of the Cauchy problem. We also introduce the set $\tilde{\mathcal{K}}$ of all functions $y \in C(\mathbf{R})$ such that $y|_{(-\infty,0]} \in \mathcal{K}$. Suppose that $\tilde{\mathcal{K}}$ is invariant under time shifts (in t), and that the metric d ($= d_\mathcal{K}$) in $\mathcal{K}$ satisfies the conditions:

if $y \in \tilde{\mathcal{K}}$, then the map $(t \mapsto y_t)\colon \mathbf{R} \to \mathcal{K}$ is continuous;

if $y, y_i \in \tilde{\mathcal{K}}$ $(i = 1, 2, \dots)$, $-\infty < t_1 < t_2 < \infty$, $d((y_i)_{t_1}, y_{t_1}) \to 0$, $\|(y_i - y)|_{[t_1,t_2]}\| \to 0$, $i \to \infty$, then $\max_{t_1 \leq t \leq t_2} d((y_i)_t, y_t) \to 0$ as $i \to \infty$.

Some examples of a set $\mathcal{K}$ are:

1) The Banach space of functions $\psi \in C(-\infty, 0]$ such that $e^{pt}\psi(t)$ is bounded and uniformly continuous, with norm

$$\|\psi\| := \sup_{-\infty < t \leq 0} e^{pt} |\psi(t)|.$$

Here, $p \in \mathbf{R}$ is fixed.

2) The Banach space of those functions as in 1), with $p = 0$, for which

$$\|\psi\| := \sup_{-\infty < t \leq 0} |\psi(t)| + \int_{-\infty}^{0} |\psi(t)|\, dt < \infty.$$

3) The space $C(-\infty, 0]$ equipped with the metric

$$d(\psi_1, \psi_2) = \sum_{i=1}^{\infty} \frac{1}{2^i} \frac{\|(\psi_1 - \psi_2)|_{[-i,0]}\|}{1 + \|(\psi_1 - \psi_2)|_{[-i,0]}\|}.$$

Etc.

THEOREM 2.3. *Let $\phi \in \mathcal{K}$, and let $F\colon D_\infty$ $(:= [t_0, \infty) \times \mathcal{K}) \to \mathbf{R}^n$ be a continuou functional such that for any $(t, \psi) \in D_\infty$ there are $\epsilon, L > 0$ for which the inequalit $|F(t+\tau, \psi_1) - F(t+\tau, \psi_2)| \le L\|u_1 - u_2\|$ holds if $\tau \in (0, \epsilon)$, $\psi_i \in \mathcal{K}$,*

$$(\psi_i)_{-\tau} = u, \qquad \max_{-\tau \le \theta \le 0} |\psi_i(\theta) - \psi_i(0)| < \epsilon, \qquad i = 1, 2.$$

Then there is a $t_\phi \in (t_0, \infty]$ such that

a) *there is a solution x of (2.6)–(2.7) on $[t_0, t_\phi)$;*
b) *on each interval $[t_0, t_1] \subset [t_0, t_\phi)$ this solution is unique;*
c) *if $t_\phi < \infty$, then $x(t)$ does not have a finite limit as $t \to t_\phi^-$;*
d) *the solution x depends continuously on F, ϕ.*

Assertion d) means that for any $t_1 \in (t_0, t_\phi)$, $\epsilon > 0$ there is a $\delta > 0$ such that if F, ϕ in (2.6)–(2.7) are replaced by $\overline{F}, \overline{\phi}$, where $\overline{F}, \overline{\phi}$ satisfy the same conditions as do F, ϕ and

$$|\overline{F}(t, \psi) - F(t, \psi)| < \delta, \qquad t_0 \le t \le t_1, \qquad d(\psi, x_t) < \epsilon,$$
$$d(\overline{\phi}, \phi) < \delta,$$

then for the solution $\overline{x}$ with maximal interval of existence $[t_0, \overline{t}_{\overline{\phi}})$ we have

$$\overline{t}_{\overline{\phi}} > t_1, \qquad |\overline{x}(t) - x(t)| < \epsilon, \qquad t_0 \le t \le t_1.$$

It is the indefiniteness of $F(t, x_t)$ that forces us to consider (2.6)–(2.7) with infinit delay on any finite time interval $[t_0, t_1]$ for a fixed initial function. If the behavio of $\phi(t)$ as $t \to -\infty$ implies that F is definite, then we can immediately reduce th problem under consideration to a similar problem with finite delay. In fact, for an pair $(t, \tilde{\psi})$, where $t_0 \le t < \infty$, $\tilde{\psi} \in C[t_0 - t, 0]$, $\tilde{\psi}(t_0 - t) = \phi(0)$, we can define

$$\tilde{F}(t, \tilde{\psi}) := F(t, \psi), \qquad \psi(\theta) = \begin{cases} \tilde{\psi}(\theta), & t_0 - t \le \theta \le 0, \\ \phi(\theta - t_0 + t), & -\infty < \theta \le t_0 - t. \end{cases}$$

With these notations, (2.6)–(2.7) is equivalent to the Cauchy problem for the equatio $\dot{x}(t) = \tilde{F}(t, x_t)$ with finite delay (though unbounded as $t \to \infty$) and single-point initia condition $x(t_0) = \phi(0)$.

This remark makes it possible to generalize the assertions of subs. 2.2, 2.3 to equa tions with unbounded delay.

2.6. Properties of the Cauchy problem for RDEs. We consider some distin guishing properties of the theory of solving the Cauchy problem for RDEs in contras to that for ODEs. It is believed that there are two general distinctions, and that al other distinctions can either be reduced to these or have a more special nature. Th two main ones are:

1) the necessity to pose not only the initial value of the solution, but also th initial function (if this function is not fixed, we obtain an infinite dimensiona solution manifold);
2) the existence of a solution of the Cauchy problem which has the usual propertie only forward in time.

Note, however, that although in general the dimension of the solution manifold is infinite, there is an important case in which it is n-dimensional, with n the dimension of the solution. The ***Cauchy problem without previous history*** has this property. For (1.1.7) with initial point t_0 this means that $h(t) \leq t - t_0$ for all $t \geq t_0$, i.e. the initial interval is empty and instead of the initial function we only have to pose an initial value $x(t_0)$. (The initial interval is said to degenerate to a point, and therefore only $\phi(0)$ must be given.) Aftereffect, for example, may be introduced as an integral of the unknown function between the limits t_0 and t.

Note that if we consider the Cauchy problem for such an equation for $t = \bar{t}_0 > t_0$, then, in general, the solution manifold is infinite dimensional. This once more proves the impossibility of extending the solution for decreasing t.

Formally we may assume that the initial function for the Cauchy problem without previous history and set up at $t = t_0$ is defined by a function ϕ: $[-h, 0] \to \mathbf{R}^n$ for any $h > 0$, because such a function also defines $\phi(0) = x(t_0)$. This method is sometimes used to obtain a uniform treatment. Of course it does not make the solution manifold infinite dimensional, because $\phi(\theta)$ for $-h \leq \theta < 0$ is not used, and may be ignored. Such *ignoring of nonused* (e.g., *too early*) *values of an initial function* is extensively used in the theory of FDEs. E.g., for variable delays the initial function may be given for all $t \leq t_0$, but in fact only its values influencing the solution should be taken into consideration.

Because the initial function is given on an interval with length equal to the time aftereffect, it can happen that two solutions coincide on a interval of smaller length, and then become different. This effect is related to the difference between the notions of 'graph' and 'integral curve' for solutions of RDEs, introduced in subs. 2.3.

The one-sided forward extendibility of solutions of a Cauchy problem makes it possible that solutions *stick* together. In contrast to ODEs, this interesting phenomenon may occur for RDEs with arbitrarily smooth righthand sides, and consists of the fact that solutions which are originally different will coincide after some moment of time. Next to the ill-posedness of problems for processes with aftereffect renewal (subs. 1.5) this phenomenon is the reason that the shift operators for RDEs, considered in subs. 2.3, do not form a group, but only a semigroup.

EXAMPLE 2.1. The Cauchy problem for the equation

$$\dot{x}(t) = x\left(t - \frac{t^2}{1+t^2}\right), \qquad n = 1,$$

with initial point $t_0 = 0$ is without previous history. Hence, if an arbitrary $t_0 < 0$ is taken as initial point, then two arbitrary solutions x_1 and x_2 with $x_1(0) = x_2(0)$ coincide (stick together) for all $t \geq 0$. Thus, after 'transition' through the point $t = 0$ the infinite dimensional solution manifold becomes 1-dimensional! □

EXAMPLE 2.2. Let the defining system of scalar equations be

$$\dot{x}_1(t) = 0, \qquad \dot{x}_i(t) = x_{i-1}(t-1), \qquad i = 2, \ldots, n, \qquad 0 \leq t < \infty, \tag{2.8}$$

with given continuous initial functions

$$t \mapsto x_i(t) = \phi_i(t), \qquad -1 \le t \le 0, \qquad i = 1, \dots, n.$$

By integrating these equations we successively obtain

$$x_i(t) = a_{i0}t^{i-1} + a_{i1}t^{i-2} + \cdots + a_{i,i-1}, \qquad i-1 \le t < \infty, \qquad i = 1, \dots, n.$$

Recurrence relations for the coefficients a_{is} follow from the identities

$$a_{i0}(i-1)t^{i-2} + a_{i1}(i-2)t^{i-3} + \cdots + a_{i,i-2} =$$
$$= a_{i-1,0}(t-1)^{i-2} + a_{i-1,1}(t-1)^{i-3} + \cdots + a_{i-1,i-2}, \qquad i = 2, \dots, n.$$

Each of these relations makes it possible to find $a_{i0}, \dots, a_{i,i-2}$ if $a_{i-1,0}, \dots, a_{i-1,i-2}$ ar known. The value of $a_{i,i-1}$ can be expressed in terms of the functions $\phi_1, \dots, \phi_i$ b formulas omitted here. So the infinite dimensional manifold of solutions of (2.8) on th interval $[0, n-1]$ becomes n dimensional for $t \ge n-1$. We can take $a_{10}, a_{21}, \dots, a_{n,n-}$ as parameters of this restricted solution. □

EXAMPLE 2.3. The following interesting example of a nonlinear autonomous equa tion with solutions sticking together was considered by H. Hausrath [238, §3.2]:

$$\dot{x}(t) = \beta \left[\max_{-1 \le \theta \le 0} |x(t+\theta)| - x(t) \right], \qquad 0 \le t < \infty,$$

where β = const > 0. (This RDE is often referred to as a *differential equatio with maxima*, i.e. and FDE involving the maximum or minimum of the unknow function.) It is obvious that every solution is nondecreasing for $t \ge 0$, and that ever nonnegative constant is a solution. Hence, for $\phi(0) \ge 0$ each solution sticks togethe with a constant, from some $t \in [0, 1)$ onwards.

For $\phi(0) < 0$ the possible behavior of solutions depends on whether $\beta > \beta_0$ o $\beta \le \beta_0$, where $\beta_0 \approx 0.278464$ is the root of the equation $\beta e^{1+\beta} = 1$. If $\beta > \beta_0$, the each solution sticks together with a constant, from some moment onwards which doe not exceed some $t_\beta > 0$ with $t_\beta \to \infty$ as $\beta \to \beta_0^+$. However, if $\beta \le \beta_0$, then there ar both solutions sticking together with constants and solutions not having this property Moreover, for each such β there are solutions for which the time at which stickin starts is arbitrarily large. □

3. Cauchy problem for NDEs

3.1. Smooth solutions. Consider the following problem:

$$\dot{x}(t) = F(t, x_t, \dot{x}_t), \qquad t \ge t_0, \qquad x_t(\theta) = x(t+\theta), \qquad -h \le \theta \le 0, \tag{3.1}$$

$$x_{t_0} = \phi, \qquad \dot{x}_{t_0} = \dot{\phi}, \tag{3.2}$$

where, as in subs. 2.1, the time $h > 0$ of possible aftereffect is assumed constant an finite. As already noted in §1.4, for an arbitrarily smooth functional F and initia function ϕ the solution x may fail to be continuously differentiable if the sewin condition is not fulfilled. This leads to certain complications in the theory of NDEs since in general it is not sufficient to use the space $C[0, h]$.

If the sewing condition is assumed, then the assertion about the solution of (3.1)–(3.2) is similar to thm. 2.1. In this case the notion of solution is the same as in subs. 2.1.

THEOREM 3.1. *Let $F\colon E_h$ $(:= [t_0,\infty)\times(C[-h,0])^2) \to \mathbf{R}^n$ be a continuous functional satisfying in some neighborhood of each point of E_h the condition*

$$|F(t,\psi_1,\chi_1)-F(t,\psi_2,\chi_2)| \le L\|\psi_1-\psi_2\| + l\|\chi_1-\chi_2\|$$

with constants $L\in[0,\infty)$, $l\in[0,1)$ (which may depend on the chosen point). Suppose that $\phi\in C^1[0,h]$ and that the sewing condition holds:

$$\dot\phi(0^-) = F(t,\phi,\dot\phi). \tag{3.3}$$

Then there is a $t_\phi\in(t_0,\infty]$ such that

a) *there is a solution x of (3.1)–(3.2) on $[t_0,t_\phi)$;*
b) *on each interval $[t_0,t_1]\subset[t_0,t_\phi)$ this solution is unique;*
c) *if $t_\phi<\infty$, then $\dot x(t)$ does not have a finite limit as $t\to t_\phi^-$;*
d) *the solution x depends, together with $\dot x$, continuously on F,ϕ.*

Assertion d) means that for any $t_1\in(t_0,t_\phi)$, $\epsilon>0$ there is a $\delta>0$ such that if F,ϕ in (3.1)–(3.2) are replaced by $\overline{F},\overline{\phi}$, where $\overline{F},\overline{\phi}$ satisfy the same conditions as do F,ϕ (including the sewing condition), and

$$|\overline{F}(t,\psi,\chi)-F(t,\psi,\chi)|<\delta,\qquad t_0\le t\le t_1,\qquad \|\psi-x_t\|<\epsilon,\qquad \|\chi-\dot x_t\|<\epsilon,$$

$$\|\overline{\phi}-\phi\|<\delta,\qquad \|\dot{\overline{\phi}}-\dot\phi\|<\delta,$$

then for the solution $\overline{x}$ with maximal interval of existence $[t_0,\overline{t}_{\overline{\phi}})$ we have

$$\overline{t}_{\overline{\phi}}>t_1,\qquad |\overline{x}(t)-x(t)|<\epsilon,\qquad |\dot{\overline{x}}(t)-\dot x(t)|<\epsilon,\qquad t_0\le t\le t_1.$$

Variants of these assertions similar to those considered in subs. 2.2, the considerations in subs. 2.3, and an analog of thm. 2.3 remain valid.

3.2. NDEs with functional of integral type. If the sewing condition (3.3) does not hold, then, in general, a solution does not have a continuous derivative, and certainly does not have a two-sided derivative at $t=t_0$. Of course, this considerably influences the assertions related to the solution of (3.1)–(3.2).

The simpler case is NDEs (3.1) with a functional F of integral type in $\dot x$. This means that the third argument of F may be considered as an element of $L_\infty[-h,0]$. But in this case we must keep in mind the following feature of the norm $\|\cdot\|_\infty$: if a function $u\in L_\infty[t_1,t_2]$ $(t_2>t_1+h)$ has an unremovable discontinuity for some $t_3\in[t_1,t_2]$, then the function $(t\mapsto u_t)\colon[t_1+h,t_2]\to L_\infty[-h,0]$ will be discontinuous for every $t\in[\max\{t_3,t_1+h\},\min\{t_2,t_3+h\}]$.

Therefore, for functions belonging to $L_\infty[-h,0]$ along with the norm $\|\cdot\|_\infty$ the norm $\|\cdot\|_1$ will be used. The convenience of the latter is determined by the following simple lemma.

LEMMA 3.1. *If $u\in L_1[-h-\epsilon,\epsilon]$ (in particular, if $u\in L_\infty[-h-\epsilon,\epsilon]$), where $\epsilon>0$, then $\|u_s-u_0\|_1\to 0$ as $s\to 0$.*

(Use of the two norms may be avoided if, e.g., the functional $(t, \psi, \chi) \mapsto F(t, \psi, \chi)$ is defined for all $\chi \in L_1[-h, 0]$. But for a nonlinear functional F this restricts in an essential way its growth in χ, which is undesirable for general existence theorems for the Cauchy problem.) We now say that a function x is a solution of (3.1)–(3.2) on an interval J_x containing the left end t_0 if x is of Lipschitz class and satisfies (3.1) and (3.2) almost everywhere on J_x, and $x(t_0) = \phi(0)$. For an arbitrary $N > 0$, we let $L_1^N[-h, 0]$ be the metric space of functions $\psi \in L_\infty[-h, 0]$ such that $\|\psi\|_\infty \leq N$ equipped with the metric

$$d(\psi_1, \psi_2) = \|\psi_1 - \psi_2\|_1.$$

THEOREM 3.2. *Let the functional $(t, \psi, \chi) \mapsto F(t, \psi, \chi)$ in (3.1) be defined on $[t_0, \infty) \times C[-h, 0] \times L_\infty[-h, 0]$ and be continuous on $[t_0, \infty) \times C[-h, 0] \times L_1^N[-h, 0]$ for all $N > 0$. Assume that $\phi \in \mathrm{Lip}[-h, 0]$, and that for each $t_1 \in [t_0, \infty)$ there is an $l \in [0, 1)$ such that for any bounded set $K \subset C[-h, 0] \times L_\infty[-h, 0]$ there are $\epsilon \in (0, h)$ $L \geq 0$ for which*

$$|F(t, \psi_1, \chi_1) - F(t, \psi_2, \chi_2)| \leq L(\|\psi_1 - \psi_2\| + \|(\chi_1 - \chi_2)|_{[-h, -\epsilon]}\|_\infty) + \\ + l\|(\chi_1 - \chi_2)|_{[-\epsilon, 0]}\|_\infty,$$

$t \in [\max\{t_0, t_1 - \epsilon\}, t_1 + \epsilon]$, $(\psi_i, \chi_i) \in K$, $i = 1, 2$.

Then there is a $t_\phi \in (t_0, \infty]$ such that

a) *there is a solution x of (3.1)–(3.2) on $[t_0, t_\phi)$;*
b) *on each interval $[t_0, t_1] \subset [t_0, t_\phi)$ this solution is unique;*
c) *$x \in C^1[t_0, t_\phi)$ and (3.1) holds for all $t \in [t_0, t_\phi)$;*
d) *if $t_\phi < \infty$, then $\|\dot{x}_t\|_\infty$ is unbounded as $t \to t_\phi^-$;*
e) *the solution x depends, together with $\dot{x}$, continuously on F, ϕ.*

Assertion e) means that for any $t_1 \in (t_0, t_\phi)$, $\epsilon \in (0, t_1 - t_0)$ there is a $\delta > 0$ such that if F, ϕ in (3.1)–(3.2) are replaced by $\overline{F}, \overline{\phi}$, where $\overline{F}, \overline{\phi}$ satisfy the same conditions as do F, ϕ, and

$$|\overline{F}(t, \psi, \chi) - F(t, \psi, \chi)| < \delta, \qquad t_0 \leq t \leq t_1, \qquad \|\psi - x_t\| < \epsilon, \qquad \|\chi - \dot{x}_t\| < \epsilon, \\ \|\overline{\phi} - \phi\|_{\mathrm{Lip}} < \delta,$$

then for the solution $\overline{x}$ with maximal interval of existence $[t_0, \overline{t}_{\overline{\phi}})$ we have

$$\overline{t}_{\overline{\phi}} > t_1, \qquad |\overline{x}(t) - x(t)| < \epsilon, \qquad t_0 \leq t \leq t_1, \\ |\dot{\overline{x}}(t) - \dot{x}(t)| < \epsilon, \qquad t_0 + \epsilon \leq t \leq t_1.$$

Variants of these assertions, similar to those in subs. 2.2, 2.3 and thm. 2.3, are also valid.

So, for a functional of integral type in the derivative of the unknown function, in comparison with thm. 3.1 there is no extension of the class of solutions, despite the fact that the class of initial functions is considerably larger and the sewing condition does not hold: the solutions remain smooth. This implies that equations of the type under consideration and with sufficiently smooth righthand side have the property of solution smoothing, mentioned in subs. 1.3. This property relates these equations and RDEs. Moreover, in certain cases we can immediately reduce such equations to RDEs. For example, if aftereffect occurs in the form of an integral

$$\int_{t-h}^{t} K(t,s)\dot{x}(s)\,ds,$$

with $K \in C^1$, then this reduction can be done by integration by parts.

3.3. Application of the step method. For NDEs with discrete delay in the derivative of the unknown function there arises a problem, because a functional depending on discrete function values cannot be regarded as given on a space (set) like L_∞, the functions in which are defined almost everywhere. There are various ways to overcome this obstacle. E.g., the initial function ϕ may be approximated by a sequence of functions ϕ_i satisfying the conditions of thm. 3.1, so that there is a sequence of solutions x_i. If F is such that the limit of the sequence x_i exists, then this limit is called a *generalized solution* of the original problem. However, this route may lead to cumbersome statements, which we omit.

Another way that is likely to lead to results which may be sufficient for applications, is to restrict the class of equations under investigation, so as to have the possibility of applying the step method with discrete delay in the derivative of the unknown function. E.g., suppose the functional F in (3.1) takes the form (compare with (1.1.5)):

$$F(t,x_t,\dot{x}_t) = f(t, x(t-h_1(t)),\dots,x(t-h_k(t)),\dot{x}(t-g_1(t)),\dots,\dot{x}(t-g_l(t))) + \\ +F_1(t,x_t,\dot{x}_t). \qquad (3.4)$$

Using the step method we then obtain the following theorem.

THEOREM 3.3. *Assume that:*

1) *all functions* $h_i\colon [t_0,\infty) \to \mathbf{R}_+$, $g_j\colon [t_0,\infty) \to (0,\infty)$ *are continuous, while on any interval* $[t_0,t_1]$ *the function* $t \mapsto t-g_j(t)$ *takes an arbitrary value at only finitely many points;*
2) *the function* $f\colon [t_0,\infty)\times \mathbf{R}^{nk+nl} \to \mathbf{R}$ *is continuous and satisfies a local Lipschitz condition in all arguments* $x(t-h_i(t))$ *for which the function* h_i *can take the value zero;*
3) *the functional* F *satisfies the conditions of thm.* 3.2;
4) *the initial function* ϕ *is piecewise smooth* (i.e. ϕ is continuous and $\dot{\phi}$ is continuous on the original interval except at a set of discontinuity points of the first kind not having limit points).

Then the assertions of thm. 3.2 *are valid for the solution* x *of* (3.1)–(3.2), *with the following modifications: a solution* $x\colon [t_0,t_\phi) \to \mathbf{R}^n$ *is a piecewise smooth function satisfying* (3.1) *for all* $t \in [t_0,t_\phi)$ *except on a set without limit points; the solution* x *depends, together with* $\dot{x}$, *continuously on* F_1,ϕ *(as in thm.* 3.2*), and also on* f, h_i *in the sense of the uniform metric.*

A solution of (3.1)–(3.2) is defined in much the same way as in subs. 3.2. Also, results similar to those in thm. 2.3 and subs. 2.2, 2.3 will be valid.

Concerning the dependence of the solution on g_j we note that variation of g_j usually induces a shift of the discontinuity points of $\dot{x}$. Hence, although x depends (as in thm. 3.2) continuously on g_j (in the uniform metric, on any interval $[t_0, t_1] \subset [t_0, t_\phi)$) for $\dot{x}$ this is in general not true near its discontinuity points.

To investigate the dependence of $\dot{x}$ on g_j we use the notion of 'set of possible breakpoints of a solution' $M_\phi \subset [t_0, \infty)$. It is defined as the minimal set containing t_0, and such that if $t \in (t_0, \infty)$ and, for some j, $t - g_j(t)$ either belongs to M_ϕ or is a discontinuity point of ϕ, then $t \in M_\phi$. The set M_ϕ does not have limit points, and the derivative of a solution x of (3.1)–(3.2) is continuous at every point of $[t_0, t_\phi) \setminus M_\phi$. For any $\epsilon > 0$ the values $\dot{x}(t)$, $t \in [t_0, t_1] \setminus U(M_\phi, \epsilon)$ (where $U(M_\phi, \epsilon)$ is the ϵ-neighborhood of M_ϕ), vary uniformly by an arbitrarily small amount under a uniformly sufficiently small variation of all g_j. If the set $U(M_\phi, \epsilon)$ is not excluded, the smallness of variation of $\dot{x}$ under such variation of g_j can only be stated in the metric of $L_1^N[t_0, t_1]$ for N sufficiently large.

Similar results can be proved in certain cases in which the integral and the discrete terms in (3.1) are related more tightly than in the functional (3.4).

3.4. Transition to an operator equation. There is another way to overcoming the obstacle arising in formulating general solvability theorems for Cauchy problems (3.1)–(3.2), if from the problem statement it makes sense to assume $\dot{x}_t \in L_\infty$ but the functional $(t, \psi, \chi) \mapsto F(t, \psi, \chi)$ cannot be regarded as defined on functions $\chi \in L_\infty$.

Note that essentially the same problem appears for ODEs with integrable terms E.g., consider the simple equation

$$\dot{x}(t) = f(t)x(t) \tag{3.5}$$

with an integrable function $f\colon [0,1] \to \mathbf{R}$. If f is regarded as an element of $L_1[0,1]$ then we must keep in mind that it is not a concrete function, but a class of integrable functions differing from each other on an arbitrary set of measure zero. With this interpretation it does not make sense to say that (3.5) is satisfied for a given t. It is more appropriate to regard both sides of (3.5) 'in the large', as elements of $L_1[0,1]$ because with this interpretation the problem with concreteness and also the distinction between functions differing on a set of measure zero are not essential. In other words (3.5) is interpreted as an equality of operator values of x, mapping the set of absolutely continuous functions on $[0,1]$ into $L_1[0,1]$, and not as equality of functional values for every or almost every $t \in [0,1]$.

During the last years this approach has been widely used for general FDEs, especially by N. Azbelev and his collaborators. It consists of interpretating the equation as equality of values of the operators defined by both sides of the equation under consideration, of course in an appropriate space. In particular, this approach allows us to investigate NDEs like (3.4) with discrete delays in the derivative of the unknown function, because with respect to the initial function the righthand side defines an operator acting from the space of absolutely continuous functions $x\colon [t_0, \infty) \to \mathbf{R}^n$ into $L_1[t_0, \infty)$. However, in the general case one should be careful, because of the appearance of composite functions. E.g., for the arguments $t - g_j(t)$ of $\dot{x}$ one must assume that the function $t \mapsto t - g_j(t)$ cannot map a set of positive measure to a set of measure zero. (If the function g_j is analytic or piecewise analytic (as often in applications), then the last condition means that the function $t \mapsto t - g_j(t)$ does not have intervals of constancy.) Moreover, if $x \notin \mathrm{Lip}[t_0, \infty)$, then we must impose an essential restriction on f (we will not be concerned with this).

We will now consider a variant of the solvability theorem for the Cauchy problem for NDEs, using the operator approach. We represent the equation as

$$(Dx)(t) = (\mathcal{F}x^\phi)(t). \tag{3.6}$$

Here, D is the differentiation operator, which maps $\mathrm{Lip}[t_0, \infty)$ into $L_\infty[t_0, \infty)$. The operator $\mathcal{F}$ defined by the righthand side of the FDE should map the set $\mathrm{Lip}[t_0 - h, \infty)$ into $L_\infty[t_0, \infty)$ and should be a Volterra operator, i.e. for any $t_1 \in (t_0, \infty)$ the equality $u_1(t) \equiv u_2(t)$, $t_0 - h \le t \le t_1$ ($u_1, u_2 \in \mathrm{Lip}[t_0 - h, \infty)$) implies $(\mathcal{F}u_1)(t) = (\mathcal{F}u_2)(t)$ for almost all $t \in [t_0, t_1]$. The function x^ϕ is the prolongation of a function $x\colon [t_0, T) \to \mathbf{R}^n$ (where, a priori, T is free) by the values $\phi(t_0 + \theta)$ ($-h \le \theta \le 0$) to the interval $[t_0 - h, T)$. It is assumed that $x \in \mathrm{Lip}[t_0, T)$, $\phi \in \mathrm{Lip}[t_0, t_0]$, $x(t_0) = \phi(t_0)$.

In contrast to (3.1), the righthand side of 93.6) does not directly contain $\dot{x}$. However, (3.6) includes NDEs, using the indication of the spaces to and from which $\mathcal{F}$ acts. Note that because $\mathcal{F}$ is a Volterra operator, it also acts in the spaces $\mathrm{Lip}[t_0 - h, t_1]$ for any $t_1 \in (t_0, \infty)$, mapping it into $L_\infty[t_0, t_1]$. Hence we may consider (3.6) on any interval containing its left end t_0.

A generalization of thm. 3.2 to possible discrete delays in the argument of $\dot{x}$ is as follows.

THEOREM 3.4. *Assume that all the above mentioned conditions for* (3.6) *hold, and that for all $t_1 \in (t_0, \infty)$ the operator $\mathcal{F}\colon \mathrm{Lip}[t_0 - h, t_1] \to L_\infty[t_0, t_1]$ is continuous. Suppose also that for any $t_1 \in [t_0, \infty)$ there is an $l \in [0, 1)$ such that for any $M > 0$ there exist $\epsilon \in (0, h)$, $L > 0$ such that if*

$$t \in [\max\{t_0, t_1 - \epsilon\}, t_1 + \epsilon], \qquad \psi_i \in \mathrm{Lip}[t_0 - h, t], \qquad \|\psi_i\|_{\mathrm{Lip}} < M \qquad (i = 1, 2),$$

then

$$\|\mathcal{F}\psi_1 - \mathcal{F}\psi_2\| \le L(\|\psi_1 - \psi_2\| + \|(\dot{\psi}_1 - \dot{\psi}_2)|_{[t_0-h,t-\epsilon]}\|_\infty) + \\ + l\|(\dot{\psi}_1 - \dot{\psi}_2)|_{[t-\epsilon,t]}\|_\infty.$$

Then there is a $t_\phi \in [t_0, \infty]$ such that

a) *there is a solution x of* (3.6) *on* $[t_0, t_\phi)$ (the initial condition (3.2) has been incorporated in this equation)*;*
b) *on any interval $[t_0, t_1] \subset [t_0, t_\phi)$ this solution is unique;*
c) *if $t_\phi < \infty$, then* $\operatorname{vrai\,sup}_{t_0 \le t < t_\phi} |\dot{x}(t)| = \infty$*;*
d) *the solution x depends, together with $\dot{x}$, continuously on $\mathcal{F}, \phi$.*

In d) the continuous dependence is in the following sense: for any $t_1 \in (t_0, t_\phi)$, $\epsilon > 0$ there is a $\delta > 0$ such that if $\mathcal{F}, \phi$ in (3.6) are replaced by $\overline{\mathcal{F}}, \overline{\phi}$ satisfying the same conditions as do $\mathcal{F}, \phi$ and with

$$\|\overline{\mathcal{F}}\psi - \mathcal{F}\psi\| < \delta, \qquad \psi \in \mathrm{Lip}[t_0 - h, t_1], \qquad \|(\psi - x^\phi)|_{[t_0-h,t_1]}\|_{\mathrm{Lip}} < \epsilon, \\ \|\overline{\phi} - \phi\|_{\mathrm{Lip}} < \delta,$$

then for the solution $\bar{x}$ with maximal interval of existence $[t_0, \bar{t}_{\bar{\phi}})$ we have

$$\bar{t}_{\bar{\phi}} > t_1, \qquad \|(\bar{x} - x)|_{[t_0,t_1]}\|_{\text{Lip}} < \epsilon.$$

Variants of these assertions similar to those in subs. 2.3 and thm. 2.3 are also valid.

3.5. Hale's form of NDEs. The form of NDEs proposed by J. Hale is

$$[x(t) - G(t, x_t)]^{\bullet} = F(t, x_t), \tag{3.7}$$

(see (1.1.9)) with the same initial condition (3.2). If the functional G is differentiable, then (3.7) can be reduced to (3.1). However, in general such a reduction is not assumed possible, and solvability theorems for (3.7) have a simpler form than those for (3.1). Therefore the above mentioned reduction will not always be possible.

Note that the converse reduction, from (3.1) to (3.7), is not possible, in general. A necessary condition on (3.1) for this reduction is ***quasilinearity***, i.e. linearity with respect to $\dot{x}_t$. The following examples show, however, that quasilinearity is not sufficient:

$$\dot{x}(t) = x(t-1)\dot{x}(t-1), \qquad \dot{x}(t) = x(t-1)\dot{x}(t-2).$$

The former can be brought to the form (3.7), the latter not. Nevertheless, many NDEs (including those important in applications), and even whole classes of them, can be brought to the form (3.7). Also, this form of the equation should follow from the actual sense of the problem. This and the simpler formulation of the theorems are responsible for the fact that the form (3.7) has been widely used during recent years. In the theory of NDEs the form (3.7) takes about the same place as does the divergent form (or form with divergent principal part) in the theory of partial differential equations.

Let us assume that the functionals F, G in (3.7) are defined on $[t_0, \infty) \times C[-h, 0]$ $(0 < h < \infty)$, and that $\phi \in C[-h, 0]$. A solution of (3.7), (3.2) is a continuous function x on some interval J_x with left end $t_0 \in J_x$ such that $x(t_0) = \phi(t_0)$ and the function $t \mapsto x(t) - G(t, (x^{\phi})_t)$ belongs to $C^1(J_x)$ and satisfies (3.7) for all $t \in J_x$. The function x^{ϕ} is the prolongation of x to the interval $[t_0 - h, t_0)$ by the values $\phi(t_0 + \theta)$ $(-h \leq \theta \leq 0)$. Note that differentiability of x is not assumed beforehand. Hence this notion of solution is similar to that of quasisolution in the theory of ODEs.

THEOREM 3.5. *Suppose that the functionals F, G: $[t_0, \infty) \times C[-h, 0] \to \mathbf{R}^n$ are continuous, and that for any $t_1 \in [t_0, \infty)$ there is an $l \in [0, 1)$ such that for any bounded set $K \subset C[-h, 0]$ there exist $\epsilon \in (0, h)$, $L > 0$ such that if $t \in [\max\{t_0, t_1 - \epsilon\}, t_1 + \epsilon]$, $\psi_1, \psi_2 \in K$, then*

$$|G(t, \psi_1) - G(t, \psi_2)| \leq L\|(\psi_1 - \psi_2)|_{[-h,-\epsilon]}\| + l\|(\psi_1 - \psi_2)|_{[-\epsilon,0]}\|,$$
$$|F(t, \psi_1) - F(t, \psi_2)| \leq L\|\psi_1 - \psi_2\|.$$

Then there is a $t_{\phi} \in [t_0, \infty]$ such that

a) *there is a solution x of (3.7), (3.2) on $[t_0, t_{\phi})$;*
b) *on any interval $[t_0, t_1] \subset [t_0, t_{\phi})$ this solution is unique;*
c) *if $t_{\phi} < \infty$, then $x(t)$ does not have a finite limit as $t \to t_{\phi}^-$;*
d) *the solution x depends continuously on F, G, ϕ.*

In d) the continuous dependence means that for any $t_1 \in (t_0, t_\phi)$, $\epsilon > 0$ there is a $\delta > 0$ such that if F, G, ϕ in (3.7), (3.2) are replaced by $\overline{F}, \overline{G}, \overline{\phi}$ satisfying the same conditions as do F, G, ϕ and with

$$|\overline{F}(t,\psi) - F(t,\psi)| < \delta, \qquad |\overline{G}(t,\psi) - G(t,\psi)| < \delta$$
$$(t_0 \le t \le t_1, \qquad \|\psi - x_t\| < \epsilon),$$
$$\|\overline{\phi} - \phi\| < \delta,$$

then for the solution $\overline{x}$ with maximal interval of existence $[t_0, t_{\overline{\phi}})$ we have

$$t_{\overline{\phi}} > t_1, \qquad |\overline{x}(t) - x(t)| < \epsilon, \qquad t_0 \le t \le t_1.$$

Variants of these assertions similar to those in subs. 2.2, 2.3 and thms. 2.2, 2.3 are also valid. In particular, if it is known that F is bounded on every bounded subset of $[t_0, \infty) \times C[-h, 0]$, then in c) above, 'does not have a finite limit' can be replaced by 'is unbounded'.

4. Differential inclusions of retarded type (RDIs)

4.1. Introduction. The general form of a first order retarded type differential inclusion (RDI) is

$$\dot{x}(t) \in \Phi(t, x_t). \tag{4.1}$$

The meaning of x_t is as before. $\Phi(t, x_t)$ is a bounded subset of $\mathbf{R}^n$, depending on t and x_t. Detailed assumptions about Φ will be given later. To be specific we assume that $x_t\colon [-h, 0] \to \mathbf{R}^n$.

First we list some natural appearances of RDIs.

A. Suppose the initial mathematical model has the form of the RDE (2.1), but with a certain uncertainty as to the righthand side, i.e. the righthand side is incompletely known. Then because of this the RDE (2.1) becomes the RDI (4.1). The exact solution of the RDE (e.g., under a given initial condition) is principally impossible, but guaranteed estimates of it can be obtained by investigating the RDI.

B. Consider the following control problem with aftereffect in the phase coordinates:

$$\dot{x}(t) = F(t, x_t, u(t)), \qquad u(t) \in U(t), \tag{4.2}$$

where $U(t)$ is a given set, depending on t.

For every $\psi \in C[-h, 0]$ we put

$$\Phi(t, \psi) = F(t, \psi, U(t)) := \{F(t, \psi, u) : u \in U(t)\}.$$

Now (4.2) can be transformed into (4.1), and investigation of the latter informs us about the possibilities of control.

C. Suppose the initial mathematical model is described by a retarded type *functional differential inequality*:

$$\Psi(t, \dot{x}(t), x_t) \le 0. \tag{4.3}$$

Clearly, (4.3) can be written in the form (4.1), so (4.1) includes this class of inequalities. Note, however, that the conditions on Φ stated below essentially restrict the class of inequalities (4.3) to which the theory of RDIs can be applied. E.g., these restrictions eliminate the inequalities that are linear in $\dot{x}(t)$.

D. RDIs (4.1) do appear if one wishes to define in a more correct manner the notion of solution of an RDE (2.1) with righthand side which is discontinuous in the second (functional) argument. This will be discussed below, in subs. 4.4. (Note that integrable discontinuities in the first argument are, in cases, are covered by the Carathéodory conditions.)

There are also other cases in which RDIs (4.1) naturally appear, for example in applications to mechanical systems with forced oscillations and combined dry and viscous damping, plasticity theory, electrical circuits theory, etc.

4.2. Multimaps. A *multimap* Φ from a set X into a set Y is a correspondence between these sets associating with each element $x \in X$ a subset $\Phi(x) \subseteq Y$, called the image of x (under Φ). Here we recall some notions related to multimaps which we will need in the sequel.

Let $K(\mathbf{R}^n)$ be the set of all compact (i.e. bounded closed) nonempty subsets of $\mathbf{R}^n$. Here we always assume that $\Phi(x) \in K(\mathbf{R}^n)$ for all $x \in X$. For a metric space X such a map Φ is called *upper semicontinuous* if for any $x_0 \in X$, $\epsilon > 0$ there is a $\delta > 0$ such that $x \in U(x_0, \delta)$ implies $\Phi(x) \subset U(F(x_0), \epsilon)$. (Here $U(a, \eta)$ and $U(A, \eta)$ are the η-neighborhood of the point a and set A, respectively.) In the particular case of a single-valued map Φ this definition becomes the usual definition of continuity.

An equivalent definition is 'due to Heine': a map $\Phi\colon X \to K(\mathbf{R}^n)$ is upper semicontinuous if it is locally bounded and $x_i \to \overline{x}$ $(i \to \infty)$, $y_i \in \Phi(x_i)$, $y_i \to \overline{y}$ imply $\overline{y} \in \Phi(\overline{x})$. (Here $\Phi(\overline{x})$ can contain 'redundant' points, being nonlimit points for any sequence $y_i \in \Phi(x_i)$. E.g., the multivalued function $\Phi\colon \mathbf{R} \to K(\mathbf{R})$ defined by

$$\Phi(t) = \begin{cases} \{0\}, & t \neq 0, \\ [0,1], & t = 0, \end{cases}$$

is upper semicontinuous.)

A multimap $\Phi\colon X \to K(\mathbf{R}^n)$, $X \subseteq \mathbf{R}$, is called *measurable* if the set X is measurable, $\Phi(x)$ is defined for almost all $x \in X$, and for any $\epsilon > 0$ there is a measurable set $X_\epsilon \subseteq X$ such that $\operatorname{meas}(X \setminus X_\epsilon) < \epsilon$ and $\Phi|_{X_\epsilon}$ is upper semicontinuous. If F is a single-valued map, then this definition becomes the well-known measurability definition of Lusin.

We say that a multimap $(t, \psi) \mapsto \Phi(t, \psi)$, $\Phi : [t_0, \infty) \times C[-h, 0] \to K(\mathbf{R}^n)$ satisfies the *Carathéodory condition* if for almost all $t \in [t_0, \infty)$ the map $\Phi(t, \cdot)\colon C[-h, 0] \to K(\mathbf{R}^n)$ is upper semicontinuous, for every $\psi \in C[-h, 0]$ the map $\Phi(\cdot, \psi)\colon [t_0, \infty) \to K(\mathbf{R}^n)$ is measurable, and on each bounded set $Q \subset [t_0, \infty) \times C[-h, 0]$ all values $|\Phi(t, \psi)|$ are not greaterthan $f_Q(t)$, where f_Q is an integrable function depending on Q.

(In subs. 2.2 the Carathéodory condition also includes a Lipschitz type inequality ensuring the uniqueness of the solution. In this case this is not necessary, because for RDIs the uniqueness problem does not arise.)

4.3. Solvability of the Cauchy problem for RDIs. Let $Kv(\mathbf{R}^n)$ be the family of all convex sets in $K(\mathbf{R}^n)$. In this subsection we assume that $\Phi\colon [t_0,\infty)\times C[-h,0] \to Kv(\mathbf{R}^n)$, while the initial condition is

$$x_{t_0} = \phi. \tag{4.4}$$

A solution of (4.1), (4.4) is a function x, defined on an interval J_x with left end $t_0 \in J_x$, which is absolutely continuous on every closed subinterval of J_x and satisfies (4.4) for $t = t_0$ and (4.1) for almost all $t \in J_x$ (where x is to be extended by the function $t \mapsto \phi(t-t_0)$ to the interval$[t_0-h, t_0)$).

THEOREM 4.1. *Suppose that in* (4.1) *the map* $\Phi\colon [t_0,\infty)\times C[-h,0] \to Kv(\mathbf{R}^n)$ *satisfies the Carathéodory condition, and* $\phi \in C[-h,0]$. *Then there is a* $t_\phi \in [t_0,\infty]$ *such that*

a) *there is at least one solution* x *of* (4.1), (4.4) *on* $[t_0,t_\phi)$;
b) *for any* $t_1 \in (t_0,t_\phi)$ *every solution of this problem defined on* $[t_0,t_1]$ *or* $[t_0,t_1)$ *can be extended to a solution defined on* $[t_0,t_\phi)$;
c) *if* $t_\phi < \infty$, *then there is at least one solution on* $[t_0,t_\phi)$ *that is unbounded as* $t \to t_\phi^-$; (so t_ϕ is the maximal value for which a) and b) are true);
d) *for any* $t_1 \in (t_0,t_\phi)$ *the set of all solutions of* (4.1), (4.4) *on* $[t_0,t_1]$, *regarded as a subspace of* $C[t_0,t_1]$, *is compact and connected;* (a closed set is said to be connected if it cannot be represented as the union of two nonempty disjoint closed sets);
e) *for any* $t_1 \in (t_0,t_\phi)$ *the solution set of* (4.1), (4.4) *on* $[t_0,t_1]$ *depends upper semicontinuously on the righthand side* Φ *and initial function* ϕ.

In e) the upper semicontinuous dependence means that for any $\epsilon > 0$ there is a $\delta > 0$ with the following property. Suppose that Φ,ϕ in (4.1), (4.4) are replaced by $\overline{\Phi},\overline{\phi}$ satisfying the same conditions as do Φ,ϕ, with

$$\|\overline{\phi}-\phi\| < \delta,$$

and for any solution x of the original problem on $[t_0,t_1]$ and any $t \in [t_0,t_1]$, $\psi \in C[-h,0]$ for which $\|\psi - x_t\| < \epsilon$, the inclusion

$$\overline{\Phi}(t,\psi) \subset U(\Phi(t,\psi),\delta)$$

holds. Then for the modified problem with maximal interval of existence $[t_0,t_{\overline{\phi}})$ we have $t_{\overline{\phi}} > t_1$, and for any solution $\overline{x}$ of the modified problem on $[t_0,t_1]$ there is a solution x of the original problem on $[t_0,t_1]$ for which

$$\|\overline{x}(t) - x(t)\| < \epsilon, \qquad t_0 \le t \le t_1.$$

Semicontinuity of the dependence of the solution set of the Cauchy problem (4.1), (4.4) on Φ,ϕ means, roughly speaking, that this set cannot be made essentially larger, but can be made essentially smaller, by slight variation of Φ,ϕ.

Thm. 4.1 can also be applied to (2.1)–(2.2) if in thms. 2.1, 2.2 the Lipschitz condition on F (which guarantees uniqueness of the solution) fails to hold.

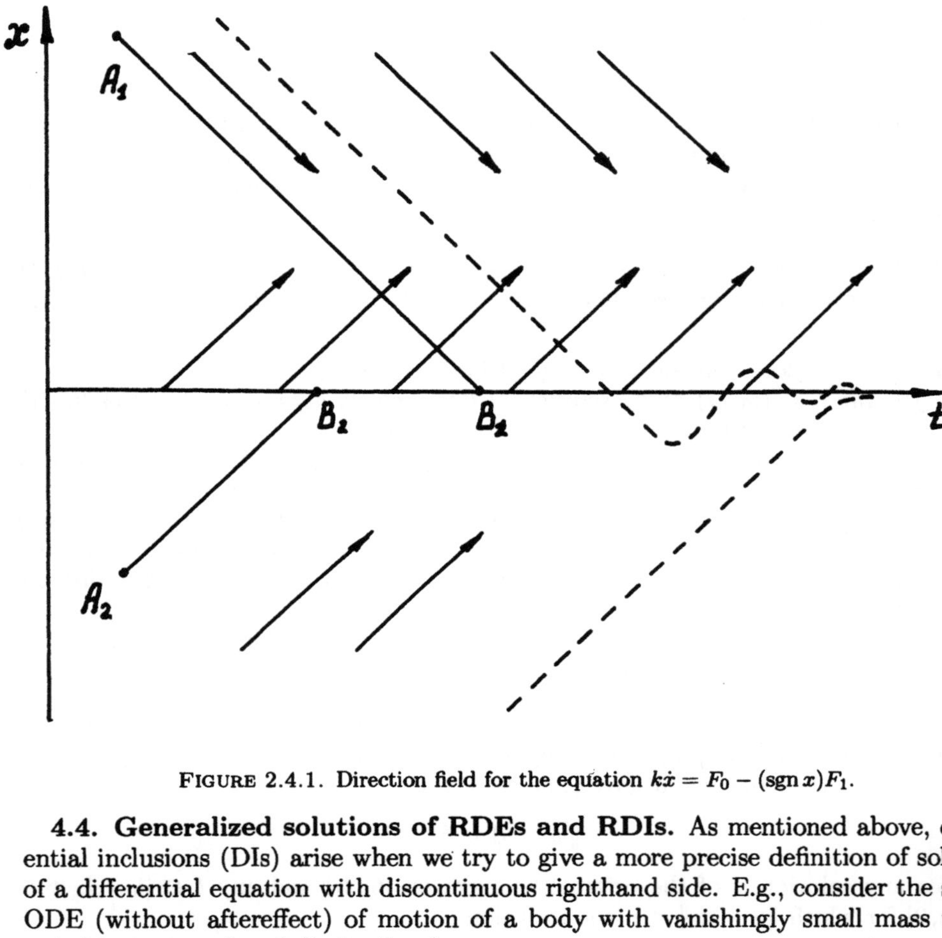

FIGURE 2.4.1. Direction field for the equation $k\dot{x} = F_0 - (\operatorname{sgn} x)F_1$.

4.4. Generalized solutions of RDEs and RDIs. As mentioned above, differential inclusions (DIs) arise when we try to give a more precise definition of solution of a differential equation with discontinuous righthand side. E.g., consider the scalar ODE (without aftereffect) of motion of a body with vanishingly small mass under switching external and dissipative forces:

$$k\dot{x} = F_0 - (\operatorname{sgn} x)F_1, \tag{4.5}$$

where $F_1 > F_0$, $k > 0$.

In Fig. 2.4.1 the corresponding direction field in the (x, t)-plane is shown for $x > 0$, $x < 0$, and $x = 0$. It is obvious that an integral curve starting above (in A_1) or below (in A_2) the t-axis, ends on this axis upon reaching it, since any extension of it would contradict the given direction field.

However, the actual process does continue in some way. So, for an adequate description of this process, the notion of solution of (4.5) should be revised. Note that an 'instantaneous' switching of the force from $F_0 - F_1$ (for $x > 0$) to $F_0 + F_1$ (for $x < 0$) means that in reality the force changes from one value to another in a negligibly small time interval, on which it must take all intermediate values. Hence the process can be described by the DI

$$k\dot{x}(t) \in \begin{cases} \{F_0 - (\operatorname{sgn} x)F_1\} & (x \neq 0), \\ [F_0 - F_1, F_0 + F_1] & (x = 0). \end{cases} \tag{4.6}$$

The solutions of (4.6) can be regarded as generalized solutions of (4.5), and have arbitrary long time prolongations. In particular, in Fig. 2.4.1, upon reaching B_i the integral curve of the DI starting at A_i ($i = 1, 2$) continues along the t-axis.

So the evolution of the process under consideration consists of two stages. Initially the motion is described by (4.5), and after reaching the level $x = 0$ (i.e. the discontinuity point of the righthand side of the equation) it is described by the DI (4.6). Of course, (4.6) can also be used for the first stage. For more complicated processes, described by ODEs with discontinuities on surfaces in phase space, the situation is very similar. Namely, such a process will consist of a number of stages, and is on certain stages represented by DEs, and on other stages (during which the representing point moves along the above mentioned discontinuity surfaces) by corresponding DIs.

Stages of the second kind are called *sliding modes*, because it is as if the point slides along the discontinuity surface, squeezed from both sides of this surface by the direction field.

Sliding modes are usually obtained as limits of a more detailed description of the process not using DIs. E.g., the solutions of (4.5) can be obtained as the limits as $m \to 0^+$ of the solutions of the equation

$$m\ddot{x} + k\dot{x} = F_0 - (\operatorname{sgn} x)F_1, \tag{4.7}$$

taking into account a small inertia, or as the limits as $a \to 0^+$ of the solutions of the equation

$$k\dot{x} = F_0 - \frac{2}{\pi}\left(\arctan\frac{x}{a}\right)F_1, \tag{4.8}$$

which models a fuzzy set of external force switching. Approximate graphs of solutions of these more detailed equations are indicated in Fig. 2.4.1 by dashed lines.

For RDIs we have a problem of precisely the type discussed above. Assume that in the problem (2.1)–(2.2) the functional $(t, \psi) \mapsto F(t, \psi)$ is measurable in t for any fixed $\psi \in C[-h, 0]$, is bounded in absolute value by an integrable function of t on every bounded subset of $[t_0, \infty) \times C[-h, 0]$, while at the same time there is no continuous dependence on ψ for almost all $t \in [t_0, \infty)$, as stipulated in thm. 2.2. Then, even for $\phi \in C[-h, 0]$ there cannot exist a solution to the problem. (See (4.5) with $x(t_0) = 0$.)

To overcome this obstacle we use models of actual phenomena in the same way as for (4.5). Let, for $(t, \psi) \in [t_0, \infty) \times C[-h, 0]$, $\Phi_F(t, \psi)$ be the convex hull of the limit sets of all convergent sequences $\{F(t, \psi_i)\}$, where $\psi_i \to \psi$ as $i \to \infty$. (Recall that the *convex hull* of a subset M of a linear space is the smallest convex set containing M.) Then the conditions of thm. 4.1 hold for the problem

$$\dot{x}(t) \in \Phi_F(t, x_t), \qquad x_{t_0} = \phi. \tag{4.9}$$

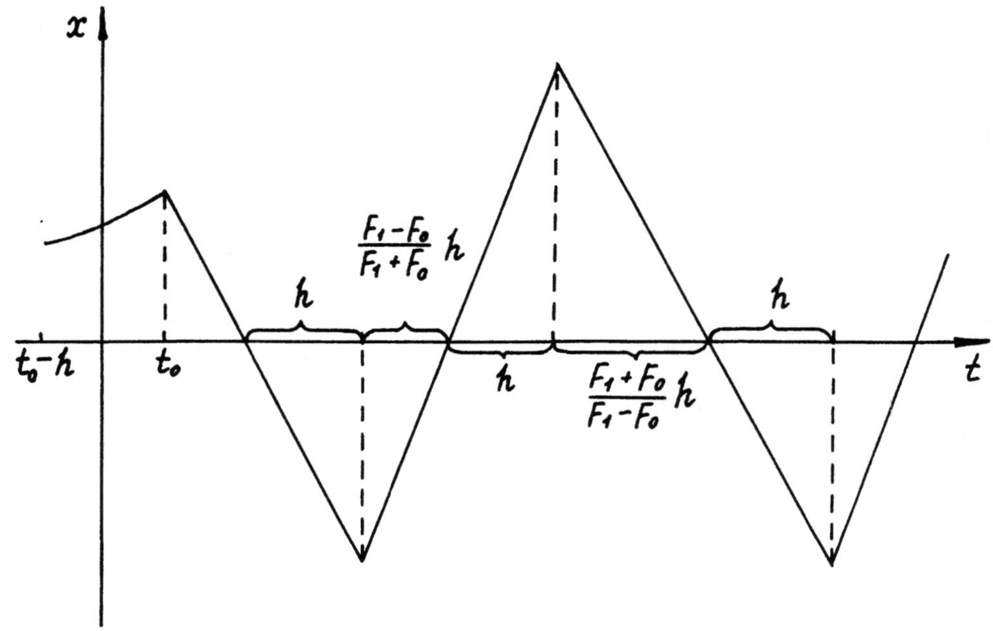

FIGURE 2.4.2. Trajectory of the equation $k\dot{x}(t) = F_0 - [\operatorname{sgn} x(t-h)]F_1$

It is natural to call solutions of (4.9) *generalized solutions* of (2.1)–(2.2). So, these generalized solutions have all the properties stated in thm. 4.1.

In certain cases the introduction of delays into an equation eliminates the sliding modes. E.g., in (4.5) let us introduce a delay $h > 0$, and obtain

$$k\dot{x}(t) = F_0 - [\operatorname{sgn} x(t-h)]F_1. \tag{4.10}$$

It is readily verified that if the initial function $\phi \in C[-h, 0]$ and $\phi(\theta) \neq 0$, then the generalized solution can only have isolated zeros (and must be periodic from some t onwards, see Fig. 2.4.2). So the generalized solution is a solution in the sense of subs. 2.2, although the conditions of thm. 2.2 do not hold. However, there are 'proper' generalized solutions of (4.10), e.g., $x(t) \equiv 0$. These solutions are in a certain sense exceptional, because they may be transformed into ordinary solutions by an arbitrarily small variation of the initial function.

This example is typical. Thus, if the righthand side of the RDE (1.1.4) depends discontinuously on $x(t - h_i(t))$ for $h_i(t) > 0$ only, then generalized solutions do not have sliding modes and are ordinary solutions in the sense of subs. 2.2. If, on the other hand, this dependence is discontinuous for $h_i(t) \equiv 0$, then sliding modes are an typical property of generalized solutions, and along them the corresponding RDI is satisfied but not (1.1.4).

We can similarly investigate RDIs (4.1) under the conditions of thm. 4.1 except the requirement of convexity of the set $\Phi(t,\psi)$. This case is a practical possibility. First of all we encounter the fact that, as for (4.5), the solution of (4.1), (4.4) need not exist. The simplest example of this kind is the ODI (without delay)

$$\dot{x}(t) \in \begin{cases} \{-\operatorname{sgn} x(t)\} & (x(t) \neq 0), \\ \{-1,1\} & (x(t) = 0), \end{cases}$$

with $x(t_0) = 0$. Even the additional requirements that the set $\Phi(t,\psi)$ be connected and depend continuously on (t,ψ) hardly help, although the counterexamples seem rather artificial. (A map Φ from a metric space X into $K(\mathbf{R}^n)$ is said to be continuous if for all $x_0 \in X$,

$$\inf\{\epsilon : \Phi(x) \subset U(\Phi(x_0),\epsilon),\ \Phi(x_0) \subset U(\Phi(x),\epsilon)\} \to 0 \qquad \text{as } x \to x_0.)$$

Moreover, if (4.1), (4.4) has solutions, then the solution set is, in general, nonclosed with respect to uniform convergence. This can be demonstrated using the following scalar DI (without delay) with constant righthand side:

$$\dot{x}(t) \in \{-1,1\}.$$

A solution of this DI is an absolutely continuous function with $|\dot{x}(t)| = 1$ almost everywhere. However, an arbitrary function satisfying the Lipschitz condition with constant 1 can be uniformly approximated to arbitrary accuracy using such functions. Thus the solution set is nonclosed.

These obstacles can be overcome as before. Namely, suppose that all conditions of thm. 4.1 except convexity of the sets $\Phi(t,\psi)$ hold for (4.1), (4.4). Let, for any $(t,\psi) \in [t_0,\infty) \times C[-h,0]$, $\operatorname{co}\Phi(t,\psi)$ be the convex hull of the set $\Phi(t,\psi)$. Then all conditions of thm. 4.1 hold for the problem

$$\dot{x}(t) \in \operatorname{co}\Phi(t,x_t), \qquad x_{t_0} = \phi. \tag{4.11}$$

In the case under consideration, the solutions of (4.11) are naturally called *generalized solutions* of (4.1), (4.4). So these generalized solutions have all the properties described in thm. 4.1.

Note that a generalized solution of (4.1), (4.4) can be approximated by 'proper' solutions of this problem to arbitrary accuracy by an arbitrarily small variation of the righthand side, e.g. by introducing an arbitrarily small additional delay. This fact, together with structural robustness of mathematical models of actual phenomena (i.e. stability of the description), give supplementary justification for using generalized solutions.

5. General linear equations with aftereffect

5.1. Cauchy problem for linear RDEs. In almost all applications linear RDEs can be represented in the following general form:

$$\dot{x}(t) = \sum_{j=1}^{m} A_j(t)x(t - h_j(t)) + \int_{-h(t)}^{0} A(t,\theta)x(t+\theta)\,d\theta + f(t), \tag{5.1}$$
$$t_0 \le t < \infty.$$

Here $x(\cdot)\colon \mathbf{R} \to \mathbf{R}^n$, $n \ge 1$, A_j, A are square n-order matrix functions, $h_j(t) \ge 0$, $j = 1,\dots,m$, $h(t) \ge 0$. Th Cauchy problem for (5.1) is to construct a solution satisfying

$$x(t) = \phi(t) \qquad (-\infty < t \le t_0), \tag{5.2}$$

where the interval $(-\infty, t_0]$ is chosen for simplicity reasons only. In fact the function ϕ may be given only on an interval J with righthand end $t_0 \in J$ and containing all values $t - h(t)$, $t - h_j(t) \le t_0$ ($j = 1,\dots,m$; $t \ge t_0$). Usually J is finite. There are two theorems about the solvability of (5.1)–(5.2) that are used most often. One of them is concerned with classical solutions, i.e. functions of class C^1 satisfying (5.1)–(5.2) for all t for which they are defined and equal to $\phi(t_0)$ at $t = t_0$. The other theorem, which deals with absolutely continuous solutions satisfying (5.1) only almost everywhere, covers equations with discontinuous locally summable coefficients.

THEOREM 5.1. *Assume that all* $A_j, h_j, h, f \in C[t_0,\infty)$, $\phi \in C(-\infty, t_0]$, *while* A *is continuous on the set* $\{(t,\theta)\colon t_0 \le t < \infty,\ -h(t) \le \theta \le 0\}$. *Then there is a classical solution of* (5.1)–(5.2) *on the interval* $[t_0,\infty)$. *On any interval* $[t_0,t_1] \subset [t_0,\infty)$ *this solution is unique and depends continuously on all prescribed functions.*

The last means that for any $t_1 \in (t_0,\infty)$, $\epsilon > 0$ there is a $\delta > 0$ such that when the functions A_j, h_j, A, h, f, ϕ in (5.1)–(5.2) are replaced by continuous functions $\overline{A}_j, \overline{h}_j, \overline{A}, \overline{h}, \overline{f}, \overline{\phi}$ satisfying

$$\overline{h}_j(t) \ge 0, \qquad \overline{h}(t) \ge 0,$$

$$\left.\begin{array}{c} \int_{t_0}^{t_1} |\overline{A}_j(t) - A_j(t)|\,dt < \delta, \qquad |\overline{h}_j(t) - h_j(t)| < \delta, \qquad |\overline{h}(t) - h(t)| < \delta \\ (t_0 \le t \le t_1, \quad j = 1,\dots,n), \\ \int_{t_0}^{t_1} |\overline{f}(t) - f(t)|\,dt < \delta, \qquad \int_{t_0}^{t_1} dt \int_{-\infty}^{0} |\tilde{\overline{A}}(t,\theta) - \tilde{A}(t,\theta)|\,d\theta < \delta, \\ |\overline{\phi}(t) - \phi(t)| < \delta \qquad (-\infty < t \le t_0), \end{array}\right\} \tag{5.3}$$

where

$$\tilde{A}(t,\theta) = \begin{cases} A(t,\theta), & -h(t) \le \theta \le 0, \\ 0, & -\infty < \theta < -h(t), \end{cases}$$

$$\tilde{\overline{A}}(t,\theta) = \begin{cases} \overline{A}(t,\theta), & -\overline{h}(t) \le \theta \le 0, \\ 0, & -\infty < \theta < -h(\theta), \end{cases}$$

we obtain $|\overline{x}(t) - x(t)| < \epsilon$ ($t_0 \le t \le t_1$).

THEOREM 5.2. *Assume that* $A_j, f \in L_1[t_0,\infty)$, $h_j, h \in C[t_0,\infty)$, *the function* A *is defined and measurable on the set* $\{(t,\theta)\colon t_0 \le t < \infty,\ -h(t) \le \theta \le 0\}$ *with* $(t \mapsto \int_{-h(t)}^{o} |A(t,\theta)|\,d\theta) \in L_1[t_0,\infty)$, *the function* ϕ *has only first kind discontinuity points, whose set is denoted by* $S_\phi \subset (-\infty, 0]$, *and the sets* S_ϕ *and all sets* $\{t\colon t - h_j(t) \in S_\phi\}$ *do not have limit points. Then there is an absolutely continuous solution of* (5.1)–(5.2) *on the interval* $[t_0,\infty)$. *On any interval* $[t_0, t_1] \subset [t_0,\infty)$ *this solution is unique and depends continuously on all prescribed functions.*

The last means that for any $t_1 \in (t_0,\infty)$, $\epsilon > 0$, $M > 0$ there is a $\delta > 0$ such that when the functions A_j, h_j, A, h, f, ϕ in (5.1)–(5.2) are replaced by functions $\overline{A}_j, \dots, \overline{\phi}$ satisfying the same requirements as do $A_j, \dots, \phi$, for which the inequality (5.3) holds, and also

$$\left.\begin{aligned} |\overline{\phi}(t) - \phi(t)| &< \delta \qquad (t \in (-\infty, t_0] \setminus U(S_\phi, \delta)),\\ |\overline{\phi}(t) - \phi(t)| &< M \qquad (t \in U(S_\phi, \delta) \cap (-\infty, t_0]), \end{aligned}\right\} \tag{5.4}$$

we obtain $|\overline{x}(t) - x(t)| < \epsilon$ $(t_0 \le t \le t_1)$.

5.2. Generalization. The linear RDE (5.1) admits a generalization which is widespread in mathematical papers and is founded on the use of the Stieltjes integral. We first recall some definitions.

The *variation* $V_J r$ *of a function* $r\colon J \to \mathbf{R}^n$, where J is an interval, is equal to

$$V_J r := \sup \sum_{i=1}^{k} |r(t_i) - r(t_{i-1})|.$$

Here the supremum is taken over all $k \ge 1$ and all $t_0 < t_1 < \cdots < t_k$ in J. If $V_J r < \infty$, then we say that r is of *bounded variation* (sometimes *finite variation*) on J. The set of functions $r\colon J \to \mathbf{R}^n$ with $V_{J_1} r < \infty$ for all closed intervals $J_1 \subseteq J$ is denoted by $\mathrm{BV}(J, \mathbf{R}^n)$ (or $\mathrm{BV}(J)$, $\mathrm{BV}\,J$). If J is a closed interval, $\mathrm{BV}\,J$ is a Banach space, with norm

$$\|r\|_{\mathrm{BV}(J)} := V_J r + |r(t_0)|,$$

where $t_0 \in J$.

A function $r \in \mathrm{BV}(J, \mathbf{R}^n)$ has a unique (up to constant terms) *Lebesgue decomposition*

$$r = r_a + r_d + r_s. \tag{5.5}$$

Here, r_a is absolutely continuous on every closed interval $J_1 \subseteq J$ and

$$V_J r_a = \int_J |\dot{r}_a(t)|\,dt.$$

The *jump function* r_d has the following structure:

$$r_d(t) = \begin{cases} -\sum_{j: t_j \in (t,\overline{t}]} \alpha_j - \sum_{j: t_j \in [t,\overline{t})} \beta_j + c & (t < \overline{t}),\\ c & (t = \overline{t}),\\ \sum_{j: t_j \in (\overline{t},t]} \alpha_j + \sum_{j: t_j \in [\overline{t},t)} \beta_j + c & (t > \overline{t}), \end{cases} \tag{5.6}$$

where $\{t_1, t_2, \dots\}$ is an at most countable set of points $t_j \in J$, $\bar{t} \in J$, $\alpha_j, \beta_j, c \in \mathbf{R}^n$, and the series are absolutely convergent. Moreover, $\alpha_j = 0$ ($\beta_j = 0$) if $t_j = \inf J$ ($t_j = \sup J$), and

$$V_J r_d = \sum_j (|\alpha_j| + |\beta_j|).$$

In particular, if the set $\{t_j\}$ does not have limit points, then r_d is piecewise constant. Finally, r_s is a *singular function*, i.e. $r_s \in C(J)$ and $\dot{r}_s(t) = 0$ for almost all $t \in J$. With the representation (5.5) we have the formula

$$V_J r = V_J r_a + V_J r_d + V_J r_s.$$

This implies that if the lefthand side is finite, then so are all terms at the righthand side.

In applications the decomposition (5.5) of the function $r \in \mathrm{BV}(J)$ usually contains only the first two terms.

The space $\mathrm{BV}(J, \mathbf{R}^{n\times m})$, where $\mathbf{R}^{n\times m}$ is the space of real $(n \times m)$-matrices, can be defined similarly to $\mathrm{BV}(J, \mathbf{R}^n)$.

If $f \in C(J, \mathbf{R}^n)$, $\sup_{t\in J} |f(t)| < \infty$, $r\colon J \to \mathbf{R}^{n\times n}$, $V_J r < \infty$, then the limit

$$\lim \sum_{i=1}^{k} [r(t_i) - r(t_{i-1})] f(\tau_i)$$

exists, where $t_0 < \tau_1 < t_1 < \tau_2 < \dots < \tau_k < t_k$ are points in J and the limit is taken for $k \to \infty$, $t_0 \to \inf J$, $t_k \to \sup J$, $\max_i (t_i - t_{i-1}) \to 0$. This limit is called the *Stieltjes integral of f with respect to r*, and is denoted by

$$\int_J [dr(t)] f(t).$$

The usual properties of the (Riemann) integral can be generalized to Stieltjes integrals in a natural way. We mention the estimate

$$\left| \int_J [dr(t)] f(t) \right| \le \int_J |f(t)| \cdot |dr(t)|,$$

where, by definition,

$$\int_J |f(t)| \cdot |dr(t)| := \liminf \sum_{i=1}^{k} |f(\tau_i)| \cdot |r(t_i) - r(t_{i-1})|.$$

If in (5.5) the function r_s is absent and r_d has the form (5.6), then

$$\int_J [dr(t)] f(t) = \int_J \dot{r}_a(t) f(t)\, dt + \sum_j (\alpha_j + \beta_j) f(t_j). \tag{5.7}$$

The generalization of (5.1) mentioned at the beginning of subs. 5.2 takes the form

$$\dot{x}(t) = \int_{-\infty}^{0} [d_\theta R(t, \theta)] x(t + \theta) + f(t), \qquad t_0 \le t < \infty, \tag{5.8}$$

under the initial condition (5.2). Here d_θ means that integration is with respect to θ; $f \in L_1[t_0, \infty)$, and $\phi \in C(-\infty, t_0]$ with $\sup |\phi(t)| < \infty$. It is also assumed that the following *condition A* be fulfilled:

Condition A. The *kernel* $R\colon [t_0,\infty)\times(-\infty,0]\to \mathbf{R}^{n\times n}$ is measurable and

$$V_{(-\infty,0]}R(t,\cdot)\le \lambda(t),\qquad t_0\le t<\infty, \tag{5.9}$$

where $\lambda\in L_1[t_0,\infty)$.
(Note that the requirements on ϕ may be relaxed when those on $R(t,\cdot)$ are made stronger, and vice versa, so as to ensure convergence of the integral in (5.8). E.g., the condition $\sup|\phi(t)|<\infty$ may be dropped if for any $t\in[t_0,\infty)$ the function $\theta\mapsto R(t,\theta)$ is constant for all $\theta\in(-\infty,-h(t))$, $h(t)\in[0,\infty)$. In this case the integral in (5.8) has the form $\int_{-h(t)}^{0}\cdot$.)

Under the above assumptions the problem (5.8), (5.2) has an absolutely continuous solution on the interval $[t_0,\infty)$, and on every interval $[t_0,t_1]\subset[t_0,\infty)$ this solution is unique and depends continuously on the kernel r and the functions f,ϕ (in a natural sense, not elaborated here). If, in addition, *condition B* below holds, then the solution is a classical solution:

Condition B. The functions $R(\cdot,0)$, λ (see (5.9)) and f belong to $C[t_0,\infty)$, the function ϕ is uniformly continuous, and

$$\forall t\in[t_0,\infty)\Rightarrow\int_{-\infty}^{0}|R(t,\theta)-R(t_1,\theta)|\,d\theta\to 0\quad\text{as } t_1\to t.$$

Using (5.7) the equation (5.1) follows from (5.8) if

$$R(t,\theta)=\int_{-\infty}^{\theta}\tilde A(t,\tau)\,d\tau+\sum_{j=1}^{m}H(\theta+h_j(t))A_j(t),$$
$$t_0\le t<\infty,\qquad -\infty<\theta\le 0,$$

where H is the Heaviside function and

$$\tilde A(t,\tau)=\begin{cases}A(t,\tau), & -h(t)\le\tau\le 0,\\ 0, & -\infty<\tau<-h(t).\end{cases}$$

5.3. Integral representation for the solution of the Cauchy problem (variation of constants formula). Here we assume that all functions given in (5.1)–(5.2) satisfy the conditions of thm. 5.2. Then the solution of this problem can be represented as $x=x_f+x_\phi$, where x_f (x_ϕ) is the solution of the problem (5.1)–(5.2) with $\phi=0$ (respectively, $f=0$). We will consider integral representations for x_f and x_ϕ.

Suppose that for any $s\in[t_0,\infty)$ the sets $\{t\colon t-h_i(t)=s\}$ do not have limit points. For such an s we denote by $(t\mapsto C(t;s))\colon[s,\infty)\to\mathbf{R}^{n\times n}$ the solution of the matrix equation

$$\frac{\partial C}{\partial t}=\sum_{j=1}^{m}A_j(t)C(t-h_j(t);s)+\int_{-h(t)}^{0}A(t,\theta)C(t+\theta;s)\,d\theta, \tag{5.10}$$
$$s\le t<\infty,$$

under the initial condition

$$C(t;s) = \begin{cases} 0, & -\infty < t < s, \\ I, & t = s, \end{cases}$$

where I is the identity matrix. Writing (5.10) out in columns, we obtain homogeneous equations of the type (5.1). Hence, by thm. 5.2 there is a unique solution of the problem stated, called the *Cauchy matrix* (or *fundamental matrix*) of the problem (5.1)–(5.2). It serves in the integral representation for x_f:

$$x_f(t) = \int_{t_0}^{t} C(t;s) f(s)\, ds, \qquad t_0 \le t < \infty. \tag{5.11}$$

Moreover, for any $t \in [t_0, \infty)$ the function $C(t; \cdot)\colon [t_0, t] \to \mathbf{R}^{n\times n}$ is bounded and measurable.

To obtain an integral representation for x_ϕ we assume in addition to the conditions of thm. 5.2 that for any $s \in (-\infty, t_0]$ the sets $\{t\colon t - h_j(t) = s\}$ do not have limit points and that the function ϕ has *compact support* (i.e. vanishes outside a finite interval), $\phi \in \mathrm{BV}(-\infty, t_0]$, and its Lebesgue decomposition (5.5) does not include a singular term. For a fixed $s \in (-\infty, t_0]$ we let $(t \mapsto G(t;s))\colon [t_0, \infty) \to \mathbf{R}^{n\times n}$ be the solution of (5.10) on the interval $t_0 \le t < \infty$ satisfying the initial condition

$$G(t;s) = \begin{cases} 0, & -\infty < t < s, \\ J, & s \le t \le t_0. \end{cases}$$

Then

$$\begin{aligned} x_\phi(t) = G(t;t_0)[\phi(t_0) - \phi(t_0^-)] + \sum_i G(t;s_i)[\phi(s_i^+) - \phi(s_i^-)] + \\ + \int_{-\infty}^{t_0} G(t;s)\dot\phi_a(s)\, ds. \end{aligned} \tag{5.12}$$

Here, summation is over all discontinuity points $s_i < t_0$ of the function ϕ, and ϕ_a is the absolutely continuous term in the Lebesgue decomposition of ϕ.

Under certain additional restrictions (e.g., all $h_j \in C^1[t_0, \infty)$ and $\dot h_j(t) < 1$) formula (5.12) can be brought to the form

$$x_\phi(t) = G(t;t_0)\phi(t_0) - \int_{-\infty}^{t_0} \frac{\partial G(t;s)}{\partial s} \phi(s)\, ds, \qquad t_0 \le t < \infty, \tag{5.13}$$

with an integrable integrand. Formula (5.13) can be used without the assumption that ϕ has compact support if all $\inf_{t \ge t_0}\{t - h_j(t)\}$, $\inf_{t\ge t_0}\{t - h(t)\}$ are finite (because then $\partial G(t;s)/\partial s = 0$ for all $s < 0$ with sufficiently large modulus). Note that $G(t;t_0) = C(t;t_0)$.

For the particular case when all h_j, h in (5.1) are independent of t, the righthand side of (5.13) can be written in terms of the function C:

$$x_\phi(t) = C(t;t_0)\phi(t_0) + \sum_{j=1}^{m} \int_{t_0-h_j}^{t_0} C(t;\alpha+h_j)A_j(\alpha+h_j)\phi(\alpha-t_0)\,d\alpha + $$
$$+ \int_{-t_0-h}^{t_0} \left[\int_{t_0}^{\alpha+h} C(t;\tau)A(\tau,\alpha-\tau)\,d\tau \right] \phi(\alpha-t_0)\,d\alpha, \qquad t_0 \le t < \infty.$$

5.4. Adjoint equation. Periodic solutions. Here we consider equation (5.1) with $f(t) \equiv 0$, where, for simplicity, all $h_j(\cdot)$ and $h(\cdot)$ are assumed to be constants:

$$\dot{x}(t) = \sum_{j=1}^{m} A_j(t)x(t-h_j) + \int_{-h}^{0} A(t,\theta)x(t+\theta)\,d\theta, \qquad t_0 \le t < \infty. \tag{5.14}$$

Suppose A_j and A satisfy the conditions of thm. 5.2.

Following A. Halanay, we call the equation

$$\dot{y}(t) = -\sum_{j=1}^{m} A_j'(t+h_j)y(t+h_j) - \int_{-h}^{0} A'(t-\theta,\theta)y(t-\theta)\,d\theta \tag{5.15}$$

the (Lagrange) *adjoint* of (5.14). (Here, as before, prime denotes transposition.) Hence (5.15) is an ADE (see (1.1.3)), which becomes an RDE when replacing t by $-t$. Consequently the Cauchy problem for (5.15) is well-posed for decreasing t direction.

A basic relation between solutions of both equations is as follows. Let x be a solution of (5.14) and y a solution of (5.15) on an interval J. Then

$$y'(t)x(t) + \sum_{j=1}^{m} \int_{t}^{t+h_j} y'(\theta)A_j(\theta)x(\theta-h_j)\,d\theta +$$
$$+ \int_{-h}^{0} d\tau \int_{t}^{t-\tau} y'(\theta)A(\theta,\tau)x(\theta+\tau)\,d\theta \equiv \text{const}, \qquad t \in J.$$

Adjoint equations are used for investigating boundary value problems for FDEs, but we do not discuss this here. We will consider only the important problem of the existence of a periodic solution of (5.1) under the assumption that all h_j, h are independent of t, and that

$$A_j(t+T) \equiv A_j(t) \qquad (j=1,\dots,m),$$
$$A(t+T,\theta) \equiv A(t,\theta), \qquad f(t+T) \equiv f(t)$$

for some $T > 0$ and all $t \in \mathbf{R}$. In this case the following modification of the Fredholm alternative holds.

THEOREM 5.3. *The linear spaces of T-periodic solutions of equations* (5.14) *and* (5.15) *have the same finite dimension. If this dimension is equal to zero* (i.e. these equations do not have nontrivial T-periodic solutions), *then for any T-periodic function f the equation* (5.1) *has only one T-periodic solution. If, on the other hand, this* *dimension is equal to* $d > 0$, *then* (5.1) *has* T*-periodic solutions if and only if*

$$\int_0^T y_i'(t) f(t)\, dt = 0, \qquad i = 1, \ldots, d.$$

Here $y_1, \ldots, y_d$ *is a basis in the space of* T*-periodic solutions of* (5.15).

Thm. 5.3 holds also for the general equation (5.8) with $r(t+T, \theta) \equiv r(t, \theta)$ and correspondingly defined adjoint equation.

For the approximate construction of a periodic solution one can use a finite-dimensional approximation. E.g., all functions under consideration may be represented as finite segments of T-periodic Fourier series. If (5.1) is an asymptotically stable equation, then it suffices to numerically construct a solution for any initial function because for such equations any solution tends to a T-periodic solution with exponential rate as $t \to \infty$.

5.5. Neutral type equations (NDEs). As already mentioned, nowadays there are two forms of NDEs in use, namely (3.1) and (3.7). Also, the latter can be reduced to the former only for 'sufficiently smooth' equations, while the former cannot, in general, be reduced to the latter, even in the presence of such smoothness. However the resulting linear forms, similar to (5.1), namely

$$\dot{x}(t) = \sum_{j=1}^{m} C_j(t) x(t - h_j(t)) + \sum_{j=1}^{m_1} D_j(t) \dot{x}(t - g_j(t)) + \\ + \int_{-h(t)}^{0} A(t, \theta) x(t + \theta)\, d\theta + f(t), \qquad t_0 \le t < \infty \tag{5.16}$$

(in which the integral term involving $\dot{x}$ is absent because it can be eliminated by integration by parts), and

$$\left[x(t) - \sum_{j=1}^{m_1} B_j(t) x(t - g_j(t)) \right]^{\bullet} = \sum_{j=1}^{m} A_j(t) x(t - h_j(t)) + \\ + \int_{-h(t)}^{0} A(t, \theta) x(t + \theta)\, d\theta + f(t), \qquad t_0 \le t < \infty, \tag{5.17}$$

can be reduced to one another under assumptions which are quite natural in applications. For a discontinuous initial function ϕ the solution of the Cauchy problem (5.17) (5.2) may depend on the interpretation of it, as discussed in subs. 1.1.2. Therefore (5.17) will be accepted as the main form of a linear NDE, and the solutions of (5.16) are defined to be the solutions of the equation (5.17) equivalent to (5.16).

A function $r \colon J \to \mathbf{R}^n$ (where $J \subseteq \mathbf{R}$ is an interval) is said to be *piecewise absolutely continuous* (PAC) if $r \in \mathrm{BV}(J)$, there is no singular term in the decomposition (5.5) and the set of discontinuity points of r does not have limit points in J. A function x defined on an interval J with left end $t_0 \in J$ is said to be a PAC solution of (5.17) (5.2) if x is a PAC function and the function

$$\left(t \mapsto x(t) - \sum_{j=1}^{m_1} B_j(t) x(t - g_j(t)) \right) : J \to \mathbf{R}^n \tag{5.18}$$

(where $x(t-g_j(t)) := \phi(t-g_j(t))$ if $t-g_j(t) < t_0$) is absolutely continuous, satisfies (5.17) for almost all $t \in J$, and also $x(t_0^+) = \phi(t_0)$.

THEOREM 5.4. *Assume that all B_j, ϕ are PAC functions, $g_j \in C^1[t_0,\infty)$, and $g_j(t) > 0$, $\dot{g}_j(t) < 1$, the functions A_j, h_j, A, h, f satisfy the conditions of thm.* 5.2, *and for any $s \in S_\phi \cup [t_0,\infty)$ the sets $\{t\colon\ t-h_j(t) = s\}$ do not have limit points. Then there is a PAC solution of* (5.17)–(5.2) *on the interval* $[t_0,\infty)$. *On any interval $[t_0,t_1] \subset [t_0,\infty)$ this solution is unique and depends continuously on all prescribed functions* (in a sense indicated below).

For more detailed information we denote by $S_B \subset [t_0,\infty)$ the set of discontinuity points of the functions B_j, and by $S \subset \mathbf{R}$ the smallest set containing S_ϕ, S_B and such that if $t-g_j(t) \in S$ for some $t \in (t_0,\infty)$ $j=1,\dots,k$, then $t \in S$. The continuity of the function(5.18) implies that $S \cap (t_0,\infty)$ is the set of all possible discontinuity points of the solution x.

Continuous dependence of the solution on all prescribed functions means that for any $t_1 \in (t_0,\infty)$, $\epsilon > 0$, $M > 0$ there are $\delta > 0$, $N > 0$ (with N independent of ϵ) such that when the functions $B_j, g_j, A_j, h_j, A, h, f, \phi$ in (5.17)–(5.2) are replaced by functions $\overline{B}_j,\dots,\overline{\phi}$ satisfying the same requirements as do $B_j,\dots,\phi$, the inequalities (5.3) and (5.4), and also

$$\left.\begin{array}{ll} |\overline{B}_j(t)-B_j(t)| < \delta & (t \in [t_0,\infty)\setminus U(S_B,\delta)), \\ |\overline{B}_j(t)-B_j(t)| < M & (t \in [t_0,\infty)\cap U(S_B,\delta)), \end{array}\right\} j=1,\dots,k,$$

then

$$\left[\overline{x}(t)-\sum_{j=1}^{m_1}\overline{B}_j(t)\overline{x}(t-\overline{g}_j(t))\right]-\left[x(t)-\sum_{j=1}^{m_1}B_j(t)x(t-g_j(t))\right] < \epsilon$$
$$(t_0 \le t \le t_1),$$

$$\left.\begin{array}{ll} |\overline{x}(t)-x(t)| < \epsilon & (t \in [t_0,t_1]\setminus U(S,\epsilon)), \\ |\overline{x}(t)-x(t)| < N & (t \in [t_0,t_1]\cap U(S,\epsilon)). \end{array}\right\} \tag{5.19}$$

If, in addition, all functions B_j, ϕ are assumed to be continuous, then S is empty, and instead of (5.19) we obtain

$$|\overline{x}(t)-x(t)| < \epsilon \qquad (t_0 \le t \le t_1).$$

In this case a PAC solution becomes an AC solution, and by writing out the expression between square brackets on the lefthand side of (5.17) one obtains (5.16). If, then, also A_j, A, f are continuous functions and $B_j, \phi \in C^1$, then the solution of (5.17), (5.2) is a classical solution.

The Cauchy function C and the function G can be introduced for the problem (5.17), (5.2) similarly as in subs. 5.3 (with the natural replacement of (5.10)). For this problem the integral representations (5.11) and (5.12) remain valid.

6. Linear autonomous equations

6.1. Exponential solutions of linear autonomous RDEs. According to subs 5.2, the general linear homogeneous autonomous RDE with finite delay has the form

$$\dot{x}(t) = \int_{-h}^{0} [dR(\theta)]x(t+\theta). \tag{6.1}$$

The corresponding inhomogeneous equation has the form

$$\dot{x}(t) = \int_{-h}^{0} [dR(\theta)]x(t+\theta) + f(t). \tag{6.2}$$

Here, $R\colon [-h, 0] \to \mathbf{R}^{n\times n}$ is a matrix-function of bounded variation, and $f \in L_1[t_0, \infty)$ For the Cauchy problem we must prescribe the initial condition:

$$x(t) = \phi(t) \qquad (t_0 - h \le t \le t_0), \tag{6.3}$$

where it is assumed that $\phi \in C[t_0 - h, t_0]$. By subs. 5.2 there is a unique solution o the problem (6.2)–(6.3) on $[t_0, \infty)$.

The main property of (6.1) is the presence of a sufficiently representative set o exponential solutions

$$x_k(t) = e^{z_k t} c_k, \qquad k = 1, 2, \ldots, \tag{6.4}$$

giving rise to a *normal motion* in real systems modeled by (6.1). We now assume tha $z_k \in \mathbf{C}$, $c_k \in \mathbf{C}^n$, $c_k \neq 0$. If $z_k \notin \mathbf{R}$, then, as for ODEs, the real and imaginary part of the solutions (6.4) do play a role.

Substitution of (6.4) into (6.1) shows that (6.1) has a solution of the form (6.4) fo a given z_k if and only if $z = z_k$ is an eigenvalue of the matrix $\int_{-h}^{0} e^{z\theta}\, dR(\theta)$, i.e. z_k must be a root of the *characteristic equation*

$$\det\left(zI - \int_{-h}^{0} e^{z\theta}\, dR(\theta)\right) = 0. \tag{6.5}$$

In this case c_k is the corresponding eigenvector. If the eigenvalue z_k has algebrai multiplicity $m_k > 1$, then next to the solutions (6.4) equation (6.1) can have solution $e^{z_k t} c_k(t)$, where $c_k(t)$ is a vector polynomial in t of degree lower than m_k. Thi polynomial can be found by the method of indefinite coefficients. Moreover, the tota number of linearly independent solutions of such a kind is, for a given z_k, equal t m_k.

We denote the lefthand side of (6.5) by $g(z)$. The function $z \mapsto g(z)$ is an entir function (i.e. it is analytic and single-valued for all $z \in \mathbf{C}$). In case (6.1) involves onl finitely many discrete delays, the expression $g(z)$ is a ***quasipolynomial***, i.e. a linea combination of products $z^k e^{\alpha_k z}$ (with integer $k \ge 0$). Therefore, in general $g(z)$ ma also be called a (generalized) ***characteristic quasipolynomial***. It can be shown that with the exception of the degenerate case (see below), for an RDE this quasipolynomia has infinitely many zeros $z_1, z_2, \ldots$, some of which may be multiple. In addition,

$$\operatorname{re} z_k \to -\infty \qquad \text{as } k \to \infty.$$

In the degenerate case $g(z)$ is a polynomial of degree n, and for arbitrary ϕ and sufficiently large t, the solution manifold of (6.1), (6.3) becomes n-dimensional; for such t (6.1) reduces to an ODE. E.g., in example 2.2 we have $g(z) = z^n$ for (2.8). For $t \geq n-1$, (2.8) implies the ODE $x_n^{(n)}(t) = 0$. Having solved this equation, we obtain the remaining components from the formulas $x_i(t) = x_n^{(n-i)}(t+n-i)$ $(i = 1, \dots, n-1;$ $i - 1 \leq t < \infty)$.

Suppose that the kernel, and hence the quasipolynomial g, depends on parameters $p_1, \dots, p_d$, where the function $(z; p_1, \dots, p_d) \mapsto g(z; p_1, \dots, p_d)$ is continuous. General theorems in analytic function theory imply that the roots z_k of (6.5) depend continuously on $p_1, \dots, p_d$. If the parameters tend to some set $p_1^0, \dots, p_d^0$, then some of these roots z_k may tend to infinity, and then $\operatorname{re} z_k \to -\infty$. This implies that if $\inf\{\operatorname{re} z\colon z \in G\} > -\infty$ for some open domain $G \subset \mathbf{C}$, then the number of roots of (6.5) belonging to G may vary under parameter variation only when one of these roots hits the boundary ∂G.

This property is the basis of the so-called *D-subdivision method*, in which the space of parameters on which the kernel R depends is partitioned into subsets on which the *instability degree* of (6.1) is constant. This degree is defined as the number of roots z_k of (6.5) with $\operatorname{re} z_k > 0$. Now we briefly indicate how this is done, assuming (to be specific) that $d = 2$ and p_1, p_2 are arbitrary real numbers. We represent z and g as

$$z = \alpha + i\beta, \qquad g(z; p_1, p_2) = g_1(\alpha, \beta; p_1, p_2) + i g_2(\alpha, \beta; p_1, p_2).$$

By virtue of the above mentioned property, a change in the instability degree $\kappa(p_1, p_2)$ of (6.1) in every neighborhood of certain values of p_1, p_2 is possible only if the equations

$$g_j(0, \beta; p_1, p_2) = 0, \qquad j = 1, 2, \tag{6.6}$$

have common solution β.

This condition defines in the parameter (p_1, p_2)-plane a collection of curves, dividing this plane into subdomains. This subdivision is called a D-subdivision. We will clarify how $\kappa(p_1, p_2)$ changes when (p_1, p_2) crosses a curve L dividing domains G_1 and G_2 in the D-subdivision. For this we must find the sign of the derivative $q^0 := (\partial g_1/\partial l)|_{(0,\beta^0;p_1^0,p_2^0)}$ with respect to some nontangential direction l to L (pointing from G_1 to G_2), with $(p_1^0, p_2^0) \in L$ chosen and β^0 as defined by (6.6) with $p_j = p_j^0$. If there is only one β^0 (which must equal zero in this case), then $\kappa(p_1, p_2)$ increases by 1 for $q^0 > 0$ and decreases by 1 for $q^0 < 0$ under transition through L from G_1 to G_2. If $q^0 = 0$, then higher-order derivatives must be used. If there are several β^0, then each of these should be verified. In particular, if there are two β^0's, then they differ only in sign and have the same q^0. Hence, for $q^0 \neq 0$ the values of $\kappa(p_1, p_2)$ on G_1 and G_2 differ by 2.

Now we consider (6.2) under the assumption that $f(t)$ is a quasipolynomial. Then a solution of quasipolynomial type can be found by the method of indefinite coefficients, similarly as for ODEs. First of all, since the equation is linear, it suffices to study the case

$$f(t) = e^{zt} P(t),$$

where $z \in \mathbf{C}$ and P is a vector polynomial of degree $m \geq 0$. If there is no *resonanc* (i.e. $g(z) \neq 0$), then (6.2) has exactly one particular solution of the form

$$x(t) = e^{zt}Q(t). \tag{6.7}$$

Here, the vector-polynomial Q also has degree m. In the presence of resonance th degree of Q exceeds m, but is not larger than $m+k$ (for a scalar equation (6.2) it i equal to $m+k$), where k is the algebraic multiplicity of z as a root of (6.5).

In particular, if $m = 0$ (i.e. $f(t) = e^{zt}c$) and $g(z) \neq 0$, then the solution of (6.7) ha the form

$$x(t) = e^{zt}\left(zI - \int_{-h}^{0} e^{z\theta}\, dR(\theta)\right)^{-1}.$$

From this we can derive the following assertion about periodic solutions of (6.2) Suppose f is periodic with period $T > 0$, and has Fourier expansion

$$f(t) = \sum_{k=-\infty}^{\infty} e^{(2k\pi i/T)t} c_k,$$

with

$$|c_k| = \mathcal{O}(k^{-\epsilon}) \qquad (k \to \infty)$$

for some $\epsilon > 0$ (this condition can be weakened). Then in the nonresonance case (i.e when $g(2k\pi i/T) \neq 0$ for all k) equation (6.2) has

$$x(t) = \sum_{k=-\infty}^{\infty} e^{(2k\pi i/T)t}\left(\frac{2k\pi i}{T}I - \int_{-h}^{0} e^{(2k\pi i/T)\theta}\, dR(\theta)\right)^{-1} c_k \tag{6.8}$$

as unique T-periodic solution.

However, if $g(2k\pi i/T) = 0$ for some k, then for the existence of a T-periodic solutior it is necessary and sufficient that the corresponding $c_k = 0$. If this condition holds then in (6.8) for the resonance values of k we must take an arbitrary eigenvector o the matrix $\int_{-h}^{0} e^{(2k\pi i/T)\theta}\, dR(\theta)$ (or the zero-vector) instead of the factors

$$\left(\frac{2k\pi i}{T}I - \int_{-h}^{0} e^{(2k\pi i/T)\theta}\, dR(\theta)\right)^{-1} c_k.$$

6.2. Solution of the Cauchy problem. Using the Laplace integral transform i is easy to obtain an explicit formula for the solution of the problem (6.2)–(6.3). Recal that for an arbitrary function $h \in \mathrm{AC}[0,\infty)$ satisfying for some $\alpha_0 \in \mathbf{R}$ the estimate

$$\int_0^{\infty} |h(t)|e^{-\alpha t}\, dt < \infty \qquad (\forall \alpha > \alpha_0),$$

its *Laplace transform* is the following function of the complex variable z:

$$z \to \tilde{h}(z) = (Lh)(z) := \int_0^{\infty} e^{-zt}h(t)\, dt.$$

This function is analytic in the halfplane $\alpha(:= \operatorname{re} z) > \alpha_0$. Conversely, a function h is determined by its Laplace transform $\tilde{h}$ by the formula

$$h(t) = (L^{-1}\tilde{h})(t) := \int_{\alpha-i\infty}^{\alpha+i\infty} e^{zt}\tilde{h}(z)\, dz,$$

where the integration is over a straight line parallel to the β-axis, and $\alpha > \alpha_0$ is arbitrary.

Suppose that the function f in (6.2) satisfies the estimate $|f(t)| = \mathcal{O}(e^{\gamma t})$ ($t \to \infty$). Then the solution of (6.2)–(6.3) satisfies the exponential estimate $|x(t)| = \mathcal{O}(e^{\max\{\gamma,v\}t})$, where $v = V_{[-h,0]}R$ and for $\gamma = v$ the righthand side of this estimate must be multiplied by the factor t. Replacing in (6.2) t by $t+t_0$ and applying Laplace transformation to both sides of the equation, we can find the Laplace transform of $x(\cdot+t_0)$. Applying the inversion formula, passing from $t+t_0$ to t, and using the initial condition (6.3), we arrive at a formula for the solution of (6.2)–(6.3):

$$\begin{aligned} x(t) = \frac{-1}{2\pi i}\int_{\alpha-i\infty}^{\alpha+i\infty} e^{zt}\left[\int_{-h}^{0} e^{z\theta}\,dR(\theta) - zI\right]^{-1} \times \\ \times\left\{e^{-zt_0}\phi(t_0) + \int_{-h}^{0}[dR(\theta)]\left[e^{z\theta}\int_{t_0+\theta}^{t_0} e^{-z\tau}\phi(\tau)\,d\tau\right] + \right. \\ \left. + \int_{t_0}^{\infty} e^{-z\tau}f(\tau)\,d\tau\right\}dz, \\ t_0 \le t < \infty, \end{aligned} \tag{6.9}$$

where $\alpha > \max\{\gamma, v\}$.

Comparing (6.9) and (5.11), we obtain the Cauchy matrix for (6.2)–(6.3):

$$C(t;s) = \frac{-1}{2\pi i}\int_{\alpha-i\infty}^{\alpha+i\infty} e^{z(t-s)}\left[\int_{-h}^{0} e^{z\theta}\,dR(\theta) - zI\right]^{-1} dz, \qquad \alpha > v.$$

This identity holds for all $s,t \in \mathbf{R}$, $t \neq s$ (if $t = s$, then the righthand side equals $I/2$). An explicit expression for the matrix $G(t;s)$ in subs. 5.3 can be given in much the same way.

Formula (6.9) can be transformed by distinguishing at the righthand side the principal terms as $t \to \infty$. For simplicity we assume that $f(t) \equiv 0$, and put

$$\begin{aligned} \Phi(z) = \left[\int_{-h}^{0} e^{z\theta}\,dR(\theta) - zI\right]^{-1} \times \\ \times\left\{e^{-zt_0}\phi(t_0) + \int_{-h}^{0}[dR(\theta)]\left[e^{z\theta}\int_{t_0+\theta}^{t_0} e^{-z\tau}\phi(\tau)\,d\tau\right]\right\}. \end{aligned} \tag{6.10}$$

This is an analytic function of z in the whole complex plane with the exception of the zeros $z_1, z_2, \ldots$, of the quasipolynomial $g(z)$ (see subs. 6.1), at which it has poles; also $\alpha_k (= \operatorname{re} z_k) \le v$.

Having chosen an arbitrary $\tilde\alpha < \alpha$, $\tilde\alpha \notin \{\alpha_k\}$, and using standard methods of function theory, we obtain

$$\begin{aligned} x(t) = -\frac{1}{2\pi i}\int_{\alpha-i\infty}^{\alpha+i\infty} e^{zt}\Phi(z)\,dz = \\ = -\sum_{\substack{k:\\ \tilde\alpha<\alpha_k<\alpha}} \operatorname{Res}_{z=z_k}[e^{zt}\Phi(z)] - \frac{1}{2\pi i}\int_{\tilde\alpha-i\infty}^{\tilde\alpha+i\infty} e^{zt}\Phi(z)\,dz. \end{aligned} \tag{6.11}$$

Here, $\operatorname{Res}_{z=z_k}[e^{zt}\Phi(z)]$ is the *residue* of the analytic function $z \to e^{zt}\Phi(z)$ at its singularity z_k. If z_k is a pole of g of order $l \geq 1$, then this residue can be computed by the formula

$$\operatorname{Res}_{z=z_k}[e^{zt}\Phi(z)] = \frac{1}{(l-1)!}\lim_{z\to z_k}\frac{d^{l-1}}{dz^{l-1}}[(z-z_k)^l e^{zt}\Phi(z)]. \tag{6.12}$$

For the most important case $l = 1$ this formula becomes

$$\operatorname{Res}_{z=z_k}[e^{zt}\Phi(z)] = \lim_{z\to z_k}[(z-z_k)\Phi(z)]\cdot e^{z_k t}. \tag{6.13}$$

The last term in (6.11) satisfies the estimate

$$\left|\frac{1}{2\pi i}\int_{\tilde{\alpha}-i\infty}^{\tilde{\alpha}+i\infty} e^{zt}\Phi(z)\,dz\right| = \mathcal{O}(e^{\tilde{\alpha}t}) \qquad (t\to\infty). \tag{6.14}$$

The formulas (6.10)–(6.14) show that for any of the above $\tilde{\alpha}$ the solution of the problem (6.1), (6.3) can be represented as

$$x(t) = \sum_{\substack{k:\\ \alpha_k>\tilde{\alpha}}} Q_k(t)e^{z_k t} + \mathcal{O}(e^{\tilde{\alpha}t}) \qquad (t\to\infty). \tag{6.15}$$

Here, the vector-polynomial Q_k has degree less than l_k and can be given in explicit form, using (6.12), and l_k is the multiplicity of z_k as a root of (6.5). Under certain additional assumptions one can pass to the limit in (6.15) as $\tilde{\alpha} \to -\infty$, and obtain a representation of the solution as the sum of a series of exponential solutions of (6.1) and the quasipolynomial solutions adjoint to them. However, in applications this series is not commonly used, as the most intrinsic properties of the solution (estimates as $t \to \infty$, oscillation, etc.) are in fact determined by the (really existing) summands on the righthand side of (6.15) that have largest α_k. (The remark about the existence is made because the polynomials Q_k are determined by the initial function ϕ, and so some of them may prove to vanish identically.)

We list all roots z_k of g according to nonincrease of real parts, taking into account algebraic multiplicity, i.e. $\alpha_1 \geq \alpha_2 \geq \alpha_3 \geq \ldots$. Then there are two 'rough' (structurally stable) cases relative to arbitrary (sufficiently small) variations of the kernel R. Namely, $\alpha_1 > \alpha_2$ (then $\beta_1 := \operatorname{im} z_1 = 0$) and $\alpha_1 = \alpha_2 > \alpha_3$ (then $\beta_1 = -\beta_2 \neq 0$). In the first case all 'typical' real solutions have the form $ce^{\alpha_1 t} + o(e^{\alpha_1 t})$ ($c \neq 0$), and in the second $ce^{\alpha_1 t}\sin(\beta_1 t + \psi) + o(e^{\alpha_1 t})$ ($c \neq 0$) as $t \to \infty$. This implies oscillation properties of the solutions. In particular, it makes it possible to determine z_1 approximately by numerically solving the Cauchy problem (6.1), (6.3) for various initial functions ϕ on a time interval of sufficient length.

6.3. Example of a showering person. Consider again equation (1.1.10), modeling the behavior of a showering person. We set $x(t) = T_m(t) - T_d$, $\gamma = k\kappa > 0$. Then (1.1.10) takes the form

$$\dot{x}(t) = -\gamma x(t-h). \tag{6.16}$$

Suppose the water temperature before regulation is constant, i.e. the initial condition is given by the relation

$$x(t) = x_0 = \text{const} \neq 0 \qquad (t_0 - h \leq t \leq t_0). \tag{6.17}$$

For small $t - t_0$ the solution can be obtained by the step method (see subs. 1.3). We will consider the behavior for increasing t.

Equation (6.16) is obtained from (6.1) if

$$n = 1, \qquad r(\theta) = \begin{cases} 0, & t = -h, \\ -\gamma, & -h < t \leq 0. \end{cases}$$

hence in this case the characteristic equation (6.5) takes the form

$$-\gamma e^{-hz} - z = 0, \qquad \text{or} \quad z = -\gamma e^{-hz}. \tag{6.18}$$

It is easy to show that (6.18) has a multiple (double) root if and only if $\gamma h = 1/e$. Depicting the graphs of the lefthand and righthand sides of (6.18) for $z \in \mathbf{R}$, we see that for $\gamma h < 1/e$ (6.18) has two real roots, z_1, z_2, $z_2 < z_1 < 0$. For a fixed $h > 0$ ($\gamma > 0$), these roots depend monotonically on γ (h); moreover, $z_1 \to 0$ ($z_1 \to -\gamma$), $z_2 \to -\infty$ as $\gamma \to 0$ (as $h \to 0$), and $z_1 \to -h^{-1}$, $z_2 \to -h^{-1}$ as $\gamma \to [(eh)^{-1}]^-$ (as $h \to [(e\gamma)^{-1}]^-$). For $\gamma h = 1/e$ we obtain $z_1 = z_2 = -h^{-1}$; for $\alpha h > 1/e$ there are no real roots.

Put $z = \alpha + i\beta$. We verify that for $\gamma h \leq 1/e$ all roots except z_1, z_2 satisfy the inequality $\alpha < z_2$. In fact, (6.18) implies that for such roots

$$|\alpha| < |z| = \gamma e^{-h\alpha}, \qquad \text{i.e. } \alpha > -\gamma e^{-h\alpha}.$$

Examination of the above described graphs shows that if we would have $\alpha \geq z_2$ for some $\gamma \leq 1/(eh)$, then $\alpha > z_1$. Since the root under consideration depend continuously on γ, the last inequality remains valid as γ decreases to 0. But this is impossible, since for $\gamma = 0$ (6.18) has only the root $z_1 = 0$.

Hence, for $\gamma \leq 1/e$ the listing of the roots z_1, z_2 corresponds to the rule given at the end of subs. 6.2.

We show that in the representation

$$x(t) = ce^{z_1 t} + o(e^{z_1 t}), \qquad t \to \infty,$$

for the solution of (6.16)–(6.17) $c \neq 0$ for $\gamma < 1/e$. In fact, (6.10) implies

$$\Phi(z) = -\frac{x_0}{z}(\gamma e^{-hz} + z)^{-1}(z + \gamma e^{-zt} - \gamma)e^{-t_0 z}.$$

This, (6.11), (6.13), and (6.18) imply

$$c = -\frac{x_0 \gamma}{z_1(1 + hz_1)} e^{-z_1 t_0} \neq 0. \qquad \text{Q.E.D.}$$

Therefore, for $\gamma h < 1/e$, from some moment onwards the regulated temperature will monotonically tend to the desired value. It is interesting to note that as long as the last inequality is valid, z_1 is a decreasing function of both γ and h, i.e. with the increase in sensitivity or time delay, the stability of the system improves, contrary to the existing common opinion.

When γh increases and passes through the value $1/e$, the roots z_1, z_2 become pure imaginary conjugates. For a further investigation of the change of the roots we separate the real and imaginary parts in (6.18). We obtain

$$\alpha = -\gamma e^{-\alpha h} \cos \beta h, \qquad \beta = \gamma e^{-\alpha h} \sin \beta h. \tag{6.19}$$

This implies that $\alpha_k = \alpha_l$ if and only if $\beta_k = \pm\beta_l$. Hence under continuous variation of α and h the ordering of the α_k remains the same: $\alpha_1 = \alpha_2 > \alpha_3 = \alpha_4 > \dots$ and stability is determined by the roots $z_{1,2}$, which were initially real. Equation (6.19) leads to the conclusion that the stability of the system deteriorates when h passes through $1/(e\gamma)$. The main term of the solution of the problem (6.16)–(6.17) now takes the form

$$\begin{aligned} x(t) &= -2 \operatorname{re} \operatorname{Res}_{z=z_1} [e^{zt}\Phi(z)] + o(e^{\alpha_1 t}) = \\ &= -2\gamma x_0 [(\alpha_1 + h\alpha_1^2 - h\beta_1^2)^2 + (\beta_1 + 2h\alpha_1\beta_1)^2]^{-1} \times \\ &\times [(\alpha_1 + h\alpha_1^2 - h\beta_1^2) \cos \beta_1(t - t_0) + (\beta_1 + 2h\alpha_1\beta_1) \sin \beta_1(t - t_0)] \times \\ &\times e^{\alpha_1(t-t_0)} + o(e^{\alpha_1 t}) \qquad (t \to \infty). \end{aligned} \tag{6.20}$$

Thus, as long as $\alpha_1 < 0$, the regulated temperature, being a damped oscillation, asymptotically tends to the desired value.

It is well known that first-order autonomous ODEs cannot have oscillating solutions. Therefore the *appearance of oscillations for* $h > (e\gamma)^{-1}$ *is a pure delay effect.* Since a considerable delay of information of water temperature takes place at the mixer output, a person regulating this temperature will endlessly turn the mixer handle through the optimal position from one side to the other.

Equation (6.19) also implies that the D-subdivision (see subs. 6.1) of the parameter (γ, h)-plane (more exactly, of the first quadrant of this plane) is performed by the lines

$$\gamma h = \frac{\pi}{2}(4j - 3) \qquad (j = 1, 2, \dots).$$

Moreover, the readily verified equality

$$\gamma(1 + h^2\beta^2)\, d\alpha|_{\alpha=0} = \beta^2\, d(\gamma h)$$

implies that each root z_k lying on the imaginary axis for some γ, h moves to the right halfplane as γh increases, and this transition is always accomplished simultaneously for conjugate pairs of roots. hence, for subsequent domains in the D-division,

$$0 < \gamma h < \frac{\pi}{2}, \quad \frac{\pi}{2} < \gamma h < 5\frac{\pi}{2}, \quad 5\frac{\pi}{2} < \gamma h < 9\frac{\pi}{2}, \dots,$$

the instability degree of the system equals $0, 2, 4, \dots$, respectively.

In particular, by fixing γ and increasing h we can see that the critical value h after which the system is unstable equals $\pi/(2\gamma)$. The information delay then becomes so large that a person regulating the temperature will twist the mixer handle further and further from one side to the other, and then jumps out from the shower. The same result appears if the delay is 'moderate' but the person has a hot temperament.

It is interesting that if we keep γ constant and increase h further, the value of $\alpha_1(h)$ tends to zero, after achieving a possible maximum, i.e. instability relaxes in an exponential rate as $h \to \infty$.

This is understandable, because for sufficiently large delays the rate of change of a process is influenced by its previous states when the oscillations are relatively small. However, the situation does not really improve, because for the same reason the amplitude of oscillation becomes very large even in the first oscillation cycle. (Increase of the coefficient at the slowly growing exponential also follows from (6.20).)

Note that from the analysis given here some important conclusions follow about the possibility of a principal influence of even small time delays on the solutions of the equations of mathematical physics. Consider the simplest initial boundary value problem for the heat equation

$$\frac{\partial u}{\partial t} = \eta \frac{\partial^2 u}{\partial x^2} \qquad (0 < t < \infty, \quad 0 \le x \le l; \quad \eta = \text{const} > 0),$$
$$u|_{x=0,l} = 0 \quad (0 \le t < \infty); \qquad u|_{t=0} = u_0 \in C[0,l].$$

It is well known that the solution of this problem exists and is exponentially stable (with exponent $\pi^2\eta/l^2$) with respect to u_0. Assume now that we would like to take into consideration delays anyhow, and have changed the equation as follows:

$$\frac{\partial u(t,x)}{\partial t} = \eta \frac{\partial^2 u(t-h,x)}{\partial x^2} \qquad (h > 0).$$

As for ODEs, the continuous initial function u_0 will be prescribed on $[-h,0] \times [0,l]$. First of all we can conclude that the solution can be obtained by the method of steps,

$$u(t,x) = u_0(0,x) + \eta \int_0^t \frac{\partial^2 u_0(\tau - h, x)}{\partial x^2}\, d\tau \qquad (0 \le t \le h),$$

etc., where the boundary condition must be ignored. Being continued with respect to t, the smoothness of the solution with respect to x becomes worse (the faster the smaller h). Therefore, for the existence of continuous solutions on $[0,\infty) \times [0,l]$ we must assume that u_0 has continuous derivatives with respect to x of arbitrary orders. To satisfy the boundary condition we must have

$$\left.\frac{\partial^{2k} u_0}{\partial x^{2k}}\right|_{x=0,l} = 0 \qquad (-h \le t \le 0; \quad k = 0, 1, \dots).$$

Assume that these conditions hold. The particular solutions of the form $u(x,t) = cv(t)\sin(n\pi x/l)$ $(n = 1, 2, \dots)$ lead to the following equation for v:

$$\dot v(t) = -\eta \frac{n^2\pi^2}{l^2} v(t-h),$$

which is of the form (6.16). By investigating this equation we find that for $n > l/\sqrt{2\pi\eta h}$ it has an oscillating solution with exponentially increasing amplitude. But n is arbitrary. Hence the solution of the initial-boundary value problem is always unstable.

It is interesting that under the additional requirement $\sup_{x,t,k}|\partial^k u_0(x,t)/\partial x^k| < \infty$ on any fixed time interval, the solution of the problem with delay h tends to the solution of the original initial-boundary value problem for the heat equation as $h \to 0$. Hence, for small fixed $h > 0$ at the initial stage of the process the solution decays, but in its spectral expansion there is redistribution of amplitudes on behalf of high frequency summands. Moreover, under unrestricted development of the process these summands become dominant, generating instability.

All these complications could not have appeared if the delay is taken into account in another way, e.g. if the heat equation is replaced by

$$\frac{\partial u(x,t)}{\partial t} = \eta \frac{\partial^2 u(t,x)}{\partial x^2} + \alpha \frac{\partial^2 u(t-h,x)}{\partial x^2},$$

where $|\alpha| < \eta$ (see subs. 3.2.3.1).

6.4. Linear autonomous NDEs. We do not aim at studying linear autonomous NDEs in the most common form, but rather restrict ourselves to the autonomous homogeneous and inhomogeneous equations (5.17):

$$\left[x(t) - \sum_{j=1}^{m_1} B_j x(t-g_j)\right]^{\bullet} = \\ = \sum_{j=1}^{m} A_j x(t-h_j) + \int_{-h}^{0} A(\theta) x(t+\theta)\, d\theta, \tag{6.21}$$

$$\left[x(t) - \sum_{j=1}^{m_1} B_j x(t-g_j)\right]^{\bullet} = \\ = \sum_{j=1}^{m} A_j x(t-h_j) + \int_{-h}^{0} A(\theta) x(t+\theta)\, d\theta + f(t). \tag{6.22}$$

It is assumed that all $g_j > 0$, $h_j \geq 0$, $A(\cdot) \in L_1[-h,0]$, $f \in L_1[t_0,\infty)$; without loss of generality we can also assume that all $g_j \leq h$, $h_1 \leq h$ (if this does not hold, increase h). The initial condition has the form (6.3), where ϕ is supposed to be a PAC function.

By thm. 5.4 there is a unique PAC solution of the problem (6.22)–(6.3) on $[t_0,\infty)$.

To find the exponential solutions (6.4) of (6.21) we have to consider the characteristic equation

$$\det\left(z\sum_{j=1}^{m_1} e^{-g_j z} B_j + \sum_{j=1}^{m} e^{-h_j z} A_j + \int_{-h}^{0} e^{\theta z} A(\theta)\, d\theta - zI\right) = 0.$$

In the nondegenerate case, this equation has, similarly to (6.5), an infinite number of roots $\{z_k\}$ without finite limit points. However, in contrast to (6.5), the set $\{\operatorname{re} z_k\}$, which is bounded above, does have finite limit point. For example, such is the case when all g_j are different and $\det B_j \neq 0$ for at least one j.

Suppose we have an estimate $|f(t)| = \mathcal{O}(e^{\gamma t})$ $(t \to \infty)$. Then the solution of (6.22), (6.3) also satisfies an exponential estimate. Hence we can apply the Laplace transformation, and we obtain, similarly to subs. 6.2,

$$x(t) = -\frac{1}{2\pi i} \lim_{\beta\to\infty} \int_{\alpha-i\beta}^{\alpha+i\beta} e^{zt}\Phi(z)\,dz +$$
$$-\frac{1}{2\pi i} \int_{\alpha-i\infty}^{\alpha+i\infty} e^{zt}[\Psi(z)]^{-1} \left[\int_{t_0}^{\infty} e^{-z\tau} f(\tau)\,d\tau\right] dz \qquad (t_0 < t < \infty), \tag{6.23}$$

where

$$\Psi(z) := z\sum_{j=1}^{m_1} e^{-g_j z}B_j + \sum_{j=1}^{m} e^{-h_j z}A_j + \int_{-h}^{0} e^{z\theta}A(\theta)\,d\theta - zI,$$

$$\Phi(z) := [\Psi(z)]^{-1}\left\{e^{-t_0 z}\left[\phi(t_0) - \sum_{j=1}^{m_1} B_1\phi(t_0 - g_j)\right] + \right.$$
$$+z\sum_{j=1}^{m_1} e^{-g_j z}B_j \int_{t_0-g_j}^{t_0} e^{-z\tau}\phi(\tau)\,d\tau + \sum_{j=1}^{m} e^{-h_j z}A_j \int_{t_0-h_j}^{t_0} e^{-z\tau}\phi(\tau)\,d\tau +$$
$$\left. + \int_{-h}^{0} e^{z\theta}A(\theta)\left[\int_{t_0+\theta}^{t_0} e^{-z\tau}\phi(\tau)\,d\tau\right] d\theta\right\}$$

and $\alpha > \max\{\gamma, \sup_k \alpha_k\}$. The limit sign on the righthand side of (6.23) is necessary because for a discontinuous function ϕ the solution x will also have discontinuity points. At these points the corresponding integral with infinite integration limits diverges, but its so-called *principal value*, given in (6.23), exists. For any $t > t_0$ the value of the righthand side of (6.23) is equal to $[x(t^-) + x(t^+)]/2$. This means that at continuity points of the solution it is equal to $x(t)$, while at discontinuity points, which can be of the first kind only, it equals the arithmetic mean of the left and right limit values of x at t. For $t = t_0$ the righthand side equals $[\phi(t_0^-) + \phi(t_0)]/2$.

For $f(t) \equiv 0$ there is a transition similarly to (6.11), giving an estimate like (6.13), but without 'comments' if $\tilde\alpha \notin \{\alpha_k\}$ and the strip $\tilde\alpha - \epsilon < \operatorname{re} z < \alpha$ ($\epsilon > 0$) contains only finitely many roots z_k. In this case, if the set of such roots is nonempty, one can distinguish the principal exponential or quasipolynomial terms of the solution as $t \to \infty$ (except in the case when all coefficients turn out to be zero). If the given strip contains infinitely many roots z_k, then in the analog of (6.11) the sums become infinite series, and as a rule application of it is not very expedient. In particular, it may happen that only a subsequence of the partial sums of this series is convergent.

In contrast to RDEs, for NDEs the estimate $|x(t)| = \mathcal{O}(e^{\tilde\alpha t})$ $(t \to \infty)$ does not follow from the absence of roots z_k in the strip $\tilde\alpha \le \operatorname{re} z < \alpha$. E.g., W. Brumley [79] has constructed an NDE for which all $\operatorname{re} z_k < 0$ but for which not all solutions are bounded. P. Gromova and A. Zverkin [214] have constructed an NDE for which all roots z_k are simple and $\operatorname{re} z_k = 0$, but for which not all solutions are bounded. Consequently, if it is impossible to distinguish finitely many principal terms of the solution, then assertions about its asymptotic behavior must be given with great care.

7. Hopf bifurcation

7.1. Introduction. Here we consider the autonomous vector RDE

$$\dot{x}(t) = F(x_t; \alpha); \qquad x_t(\theta) := x(t+\theta), \qquad -h \le \theta \le 0, \tag{7.1}$$

with sufficiently smooth righthand side F depending on the parameter $\alpha \in (\underline{\alpha}, \overline{\alpha})$. W assume that $F(0; \alpha) \equiv 0$, i.e. for all α the zero function $t \mapsto 0$ is a solution of (7.1) In less formal language one could say that the system under consideration, whicl depends on a parameter, has a fixed equilibrium position for all parameter values Extracting the linear part from (7.1), we can rewrite it as

$$\dot{x}(t) = L(x_t; \alpha) + f(x_t; \alpha),$$

where $L(\cdot; \alpha)$ is a linear functional and $|f(u; \alpha)| = o(\|u\|)$ $(\|u\| \to 0)$ for any fixed α

Here we will use the basic notations connected with the stability of this equilibriun position. The respective definitions are quite the same as for ODEs, and will b discussed in detail in chapt. 3. In particular, an analog of Liapunov's stability theoren for first approximation is valid: if for some α all roots λ_k of the characteristic equatioı

$$\det[\lambda I - P(\lambda; \alpha)] = 0, \tag{7.2}$$

where

$$P(\lambda; \alpha) := e^{-\lambda t} L(t \mapsto e^{\lambda t}; \alpha),$$

are such that $\operatorname{re} \lambda_k < 0$, then for this α the rest point is asymptotically stable; however if $\operatorname{re} \lambda_k > 0$ for a single k, then it is unstable.

Now we assume that there is an $\alpha_0 \in (\underline{\alpha}, \overline{\alpha})$ such that for $\alpha < \alpha_0$ the first case holds and for $\alpha > \alpha_0$ the second. Then the parameter value α_0 is called a *critical value* if α increases and crosses the value α_0, then there is stability loss of the equilibriun position under consideration. This can happen in various ways.

E.g., it may happen that despite the stability loss for $\alpha > \alpha_0$ of the trivial solution the solution of the Cauchy problem for (7.1), posed for $t_0 - h \le t \le t_0$, will b arbitrarily small on the whole interval $[t_0, \infty)$ if the initial function and the value o $\alpha - \alpha_0$ are sufficiently small. Loosely speaking, small initial disturbances cannot lea to a principal change of the picture; in particular, this cannot lead to a catastroph under small transitions of the parameter through the critical value. This kind o stability loss is called *safe*. In contrast to this, in the case of a *dangerous stabilit loss* an arbitrary small transition through the critical value and arbitrarily smal disturbances do lead to essential consequences.

Another classification of the types of stability loss reflects the behavior of the root of (7.2) when α passes through the critical value. In practice, the two following case are most often met:

A. If α increases and passes through α_0, one root of (7.2) turns from the negativ to positive while all other roots belong to the open left halfplane (i.e. fo them $\operatorname{re} \lambda_k < 0$ for all sufficiently small $|\alpha - \alpha_0|$). In this case one says tha the *principle of stability change* holds for the system with parameter unde consideration. Arbitrarily small initial disturbances lead to the emergence o an accelerated aperiodic process, carrying the system away from its initia equilibrium position. (For a special set of initial disturbances, in theory, the process asymptotically tends to this position, but such processes are unstable and do not change the general picture.)

B. If α increases and passes through α_0, two conjugate nonreal roots of (7.2) turn from the open left halfplane into the open right halfplane, while the remaining roots remain in the open left halfplane. After such a transition, stability loss takes place in the following manner: arbitrarily small initial disturbances generate an oscillatory process with increasing amplitude and (circular) frequency near ω_0, where $\pm\omega_0 i$ are the values of the above-mentioned roots for $\alpha = \alpha_0$.

The latter type of stability loss occurs under Hopf bifurcation, and we discuss it below.

More complicated ways of stability loss are described in stability theory.

7.2. Example. We will illustrate the main idea in the study of *Hopf bifurcation* by means of the following scalar ODE:

$$\ddot{x} - \alpha\dot{x} + \dot{x}^3 + x + x^2 + x^3 = 0, \tag{7.3}$$

involving a parameter α. Equation (7.3) is equivalent to the following system of first-order ODEs:

$$\dot{x}_1 = x_2, \qquad \dot{x}_2 = -x_1 - x_1^2 - x_1^3 + \alpha x_2 - x_2^3. \tag{7.4}$$

One could say that this represents an oscillator with additional nonlinear dissipation and nonlinear restoring force which, as compared to the linear ones, are weaker for small velocities and deviations and stronger for high velocities and large deviations. There is only one rest point, $x = 0$, of the system, for any α.

It is clear that for small absolute values of $\alpha < 0$, i.e. when the linear dissipative term of the equation corresponds to small positive friction, the system has damped oscillation when taken off the rest point. This also follows from the expression for the roots $\lambda_{1,2}$ of the characteristic equation corresponding to (7.3), linearized about the origin (the nonlinear terms $\dot{x}^3$ and $x^2 + x^3$ are dropped in this linearization):

$$\lambda_{1,2} = \frac{1}{2}\alpha \pm i\sqrt{1 - \frac{1}{4}\alpha^2}. \tag{7.5}$$

It is clear that the decrease in damping of the damped oscillation tends to zero near the rest point, while the frequency tends to $\omega_0 = 1$ as $\alpha \to 0^-$.

When α passes through $\alpha_0 = 0$ and becomes positive, the friction becomes negative for small velocities, and the rest point $x = 0$ is unstable. This is also obvious from (7.5), because the roots $\lambda_{1,2}$ pass from the left to the right halfplane. Hence, for small initial disturbances there emerge oscillations with increasing amplitudes in the system. However, this increase cannot be unbounded, because for large velocities the nonlinear dissipative term annihilates the effect of the negative linear friction. Using results from the theory of oscillation of autonomous systems with one degree of freedom, this implies that the oscillations mentioned above tend to periodic oscillations as $t \to \infty$. An approximate phase portrait corresponding to (7.4) is depicted in Fig. 2.7.1. We see that a *limit cycle*, represented by the heavy line, appears in the phase plane. It contracts towards a point as $\alpha \to 0^+$. Thus, one can say that a limit cycle emerges

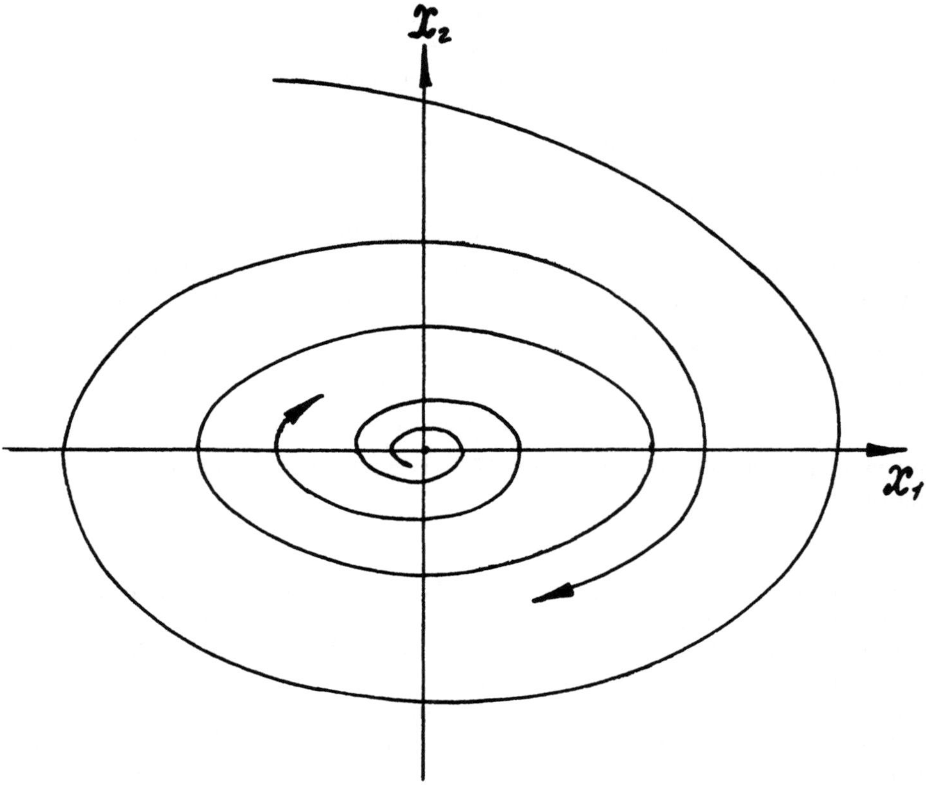

FIGURE 2.7.1. Approximate phase portrait of the system $\dot{x}_1 = x_2$, $\dot{x}_2 = -x_1 - x_1^2 - x_1^3 + \alpha x_2 - x_2^3$

from the rest point $(0,0)$ in the phase plane when α passes through the value $\alpha_0 = 0$ In other words, it appears as if there is dichotomy (*bifurcation*) from the stable rest point to an unstable rest point and a limit cycle. (In this example it is possible to verify that there is only one limit cycle.)

For not too large values of α an asymptotic representation of the emerging cycle can be obtained by the method of the small parameter. However, the following feature must be taken into consideration. Let us look for a solution of (7.3) as the sum of a series:

$$x(t) = \alpha x_1(t) + \alpha^2 x_2(t) + \alpha^3 x_3(t) + \dots \tag{7.6}$$

(the term $x_0(t)$ is absent here, since for $\alpha = 0$ we must have $x(t) \equiv 0$). Substituting this expansion into (7.6) and equating coefficients at equal powers of α, we arrive at the system of equations

$$\left.\begin{aligned} &\ddot{x}_1 + x_1 = 0, \qquad \ddot{x}_2 + x_2 = \dot{x}_1 - x_1^2, \\ &\ddot{x}_3 + x_3 = \dot{x}_2 - \dot{x}_1{}^3 - 2x_1x_2 - x_1^3, \dots . \end{aligned}\right\} \tag{7.7}$$

The first equation implies the quite natural formula $x_1 = C_1 \cos t + C_2 \sin t$. However, substitution of this formula in the next equation gives

$$x_2 = -\frac{1}{2}t(C_1 \cos t + C_2 \sin t) - \frac{1}{6}(C_1^2 + C_2^2) + \\ -\frac{1}{6}(C_1^2 - C_2^2)\cos 2t + \frac{1}{3}C_1C_2 \sin 2t + C_3 \cos t + C_4 \sin t.$$

This expression is not acceptable as a representation of the cycle, because its righthand side is unbounded as $t \to \infty$. Of course, we can set $C_1 = C_2 = 0$, but then the same complication would arise for x_3, etc. As a result we would obtain the solution $x(t) \equiv 0$.

A similar difficulty always arises when we would like to use the method of the small parameter to describe a periodic process whose frequency is only known in the zeroth approximation. The reason for this can be understood by the following simple example. Assume that purely harmonic oscillations change frequency by a small amount, from ω to $\omega + \alpha$ ($|\alpha| \ll \omega$), and that we would like to represent the new process using expansions in the small parameter α, taking the original process as zero approximation. Then we obtain the expansion

$$M \sin(\omega + \alpha)t = M(\sin \omega t \cdot \cos \alpha t + \cos \omega t \cdot \sin \alpha t) = \\ = M\left(\sin \omega t + \alpha t \cos \omega t - \frac{1}{2}\alpha^2 t^2 \sin \omega t - \dots\right).$$

The partial sums of this series expansion uniformly approximate the function $t \mapsto M \sin(\omega + \alpha)t$ on any given finite interval, but describe incorrectly the behavior of this function on the infinite t-interval. The reason for this is that the partial sums of the Taylor series of the function $t \mapsto \sin t$ describe incorrectly its behavior as $t \to \infty$.

To avoid this difficulty we make the linear change $t = c\tau$ of the variable t so that the unknown oscillation frequency at the 'new time' τ is independent of α. Then there appears an unknown coefficient c, but we have got rid of the expansions of periodic functions by Taylor series in their argument.

Because the oscillation frequency of the cycle tends to 1 as $\alpha \to 0^+$, the coefficient c is assumed to be such that this oscillation frequency is equal to 1 at time τ for all $\alpha > 0$. This means that $c|_{\alpha=0} = 1$. Further, the first equation in (7.7) implies that we can accept as first approximation, up to phase shift (i.e. time shift), the function $t \mapsto \xi \cos \tau$, where ξ is related to α and $\xi|_{\alpha=0} = 0$.

It turns out to be more convenient to consider expansion with respect to ξ rather than with respect to α. The reason for this is as follows. The change $\xi \mapsto -\xi$ is equivalent to phase shift of the oscillation by π, which preserves the invariance of the cycle and hence does not change the values of α and c. Hence $\alpha \sim \xi^2$, and $\xi \sim \sqrt{\alpha}$. Thus, in the expansion in powers of α, not only integer but also half-integer powers are needed; in the expansion in powers of ξ, however, only integer powers are needed. (A similar situation takes place in all examples in applications. However, it is theoretically possible that in the expansion in powers of α we must use *Puiseux series*, i.e. series of the form $\sum_{r=0}^{\infty} a_r \alpha^{r/m}$, where $m \in \mathbf{N}$ may be larger than 2.)

Upon changing $t = c\tau$, equation (7.3) becomes

$$c\frac{d^2x}{d\tau^2} - \alpha c^2\frac{dx}{d\tau} + \left(\frac{dx}{d\tau}\right)^3 + c^3(x + x^2 + x^3) = 0. \tag{7.8}$$

Keeping in mind the previous section we look for a solution and constants in the form

$$\left.\begin{aligned} x &= \xi\cos\tau + \xi^2 x_2(t) + \xi^3 x_3(t) + \dots, \\ c &= 1 + c_2\xi^2 + c_4\xi^4 + \dots, \\ \alpha &= \alpha_2\xi^2 + \alpha_4\xi^4 + \dots \end{aligned}\right\} \tag{7.9}$$

(of course, here $x_2, x_3, \dots$ are different from those in (7.6)). Substituting these expansions into (7.8) and equating coefficient in front of equal powers of ξ, we arrive at the sequence of equations

$$\left.\begin{aligned} \tfrac{d^2x_2}{d\tau^2} + x_2 &= -\cos^2\tau, \\ \tfrac{d^2x_3}{d\tau^2} + x_3 &= -2c_2\cos\tau - \alpha_2\sin\tau + \sin^3\tau - 2x_2\cos\tau - \cos^3\tau, \\ &\dots\dots\dots\dots \end{aligned}\right\} \tag{7.10}$$

All functions $x_2, x_3, \dots$ must be periodic with period 1. It is possible to assume that the Fourier expansions of these functions do not contain terms with $\sin\tau$ and $\cos\tau$ (since we can change the parameter ξ and make an additional phase shift). Consequently, the first of the equations (7.10) gives

$$x_2(\tau) = -\frac{1}{2} + \frac{1}{6}\cos 2\tau.$$

Substituting this into the second of equations (7.10) and writing the righthand side of this equation as a trigonometric polynomial, we obtain

$$\frac{d^2x_3}{d\tau^2} + x_3 =$$
$$= \left(-2c_2 + \frac{1}{12}\right)\cos\tau + \left(-\alpha_2 + \frac{3}{4}\right)\sin\tau - \frac{5}{12}\cos 3\tau - \frac{1}{4}\sin 3\tau.$$

The Fourier expansion of the righthand side should not contain terms with $\sin\tau$ and $\cos\tau$, because otherwise the solution would have nonperiodic resonance terms. This implies that

$$c_2 = \frac{1}{24}, \quad \alpha_2 = \frac{3}{4}, \quad x_3(\tau) = \frac{5}{96}\cos 3\tau + \frac{1}{32}\sin 3\tau.$$

This procedure allows us to continue the expansion (7.9) indefinitely, in principle. In it, $x_k(\tau)$ represents a linear combination of the functions $\cos j\tau$ and $\sin j\tau$ with $j \le k$ and j and k of the same parity. In constructing $x_k(\tau)$ with k odd, the values of c_{k-1} and α_{k-1} are determined from the condition that resonance is absent. Similar assertions can be proved to hold in the construction of a generating cycle for general RDEs (see subs. 7.3).

We restrict ourselves to the values obtained. Then the generating cycle can, for small $\alpha > 0$, be represented as

$$x = \xi \cos\tau + \xi^2 \left(-\frac{1}{2} + \frac{1}{6}\cos 2\tau\right) + \\ +\xi^3 \left(\frac{5}{96}\cos 3\tau + \frac{1}{32}\sin 3\tau\right) + \mathcal{O}(\xi^4),$$

where

$$\tau = \left[1 + \frac{1}{24}\xi^2 + \mathcal{O}(\xi^4)\right] t, \qquad \alpha = \frac{3}{4}\xi^2 + \mathcal{O}(\xi^4).$$

Elimination of ξ gives a representation of the cycle in terms of the basic parameter α:

$$x = \frac{2}{\sqrt{3}}\sqrt{\alpha}\cos\tau + \left(-\frac{2}{3} + \frac{2}{9}\cos 2\tau\right)\alpha + \mathcal{O}(\alpha^{3/2}),$$

$$\tau = \left[1 + \frac{1}{18}\alpha + \mathcal{O}(\alpha^2)\right] t.$$

In conclusion we note that there is a safe stability loss of the rest point in passage through the value $\alpha = 0$ (see subs. 7.1). This can be easily proved by contradiction.

7.3. General case. Here we describe a method for the construction of a limit cycle from a stable rest point under passage of the parameter of the system through a critical value, for the general scalar nth order autonomous RDE

$$\left.\begin{aligned} x^{(n)}(t) &= F(x_t, \dot{x}_t, \ldots, x_t^{(n-1)}; \alpha); \\ x_t^{(j)}(\theta) &:= x^{(j)}(t+\theta), \qquad -h \le \theta \le 0, \end{aligned}\right\} \tag{7.11}$$

with sufficiently smooth righthand side. (This last requirement can be made more precise, but we do not need this precization.) As above we assume that $F(0, \ldots, 0; \alpha) \equiv 0$, i.e. the zero function, corresponds to the equilibrium point of the system under consideration, and hence is a solution of (7.11) for all parameter values α. Then we can rewrite (7.11) as

$$x^{(n)}(t) = L(x_t, \dot{x}_t, \ldots, x_t^{(n-1)}; \alpha) + f(x_t, \dot{x}_t, \ldots, x_t^{(n-1)}; \alpha),$$

where L is a linear functional depending on the parameter α, and

$$f(u_1, \ldots, u_n; \alpha) = o(\|u_1\| + \cdots + \|u_n\|) \qquad (\|u_1\| + \cdots + \|u_n\| \to 0).$$

The characteristic equation of the linear approximation to (7.11) has the form

$$\lambda^n = e^{-\lambda t} L(t \mapsto e^{\lambda t}, t \mapsto \lambda e^{\lambda t}, \ldots, t \mapsto \lambda^{n-1} e^{\lambda t}; \alpha)$$

(it is obvious that t does not really participate at the righthand side). We assume that there is α_0 such that for $\alpha < \alpha_0$ all roots λ_k of the characteristic equation belong to the open left halfplane, whereas for $\alpha = \alpha_0$,

$$\left.\begin{aligned} &\lambda_{1,2}\big|_{\alpha=\alpha_0} = \pm i\omega_0; \qquad \omega_0 > 0; \\ &\frac{d \operatorname{re} \lambda_{1,2}(\alpha)}{d\alpha}\bigg|_{\alpha=\alpha_0} > 0, \\ &\operatorname{re} \lambda_j < 0 \qquad (j > 2). \end{aligned}\right\} \tag{7.12}$$

If, under these conditions, α increases and passes through the value α_0, then the zero solution, which was stable, becomes unstable, i.e $\alpha = \alpha_0$ is a critical value. In addition, as a rule (exact assumptions will be given later), when α passes through α_0 from either side, a periodic solution (the limit cycle in phase space) separates from zero. This solution is stable if it arises for $\alpha > \alpha_0$ and unstable in the opposite case. Theoretically it is possible that several cycles separate from zero.

To construct, as in subs. 7.2, a generating cycle we change the independent variable $t = c\tau$ in (7.11) in such a way that the oscillation period $2\pi/\omega_0$ corresponding to this cycle at the 'new time' τ is fixed. We also put $x(c\tau) = y(\tau)$. The a priori unknown functional relationship between c and α must be defined such that $c|_{\alpha=\alpha_0} = 1$. After this we obtain the equation

$$\left.\begin{aligned} y^{(n)}(\tau) &= c^n F_c(y_\tau, c^{-1}y_\tau^{(1)}, \ldots, c^{-n+1}y_\tau^{(n-1)}; \alpha); \\ y_\tau^{(j)}(\theta) &:= y^{(j)}(\tau+\theta) \qquad (c^{-1}h \le \theta \le 0) \end{aligned}\right\} \tag{7.13}$$

where

$$F_c(u_1, \ldots, u_n; \alpha) := F(u_1^c, \ldots, u_n^c; \alpha);$$
$$u_j^c(\tau) := u_j(c\tau) \qquad (c^{-1}h \le \tau \le 0).$$

As in subs. 7.2 we look for a periodic solution of (7.13) in the form (up to time shift)

$$y(\tau) = \xi \cos\omega_0\tau + \xi^2 y_2(\tau) + \xi^3 y_3(\tau) + \ldots. \tag{7.14}$$

The factor c of the oscillation period and the parameter α are looked for in the form

$$c = 1 + c_2\xi^2 + c_4\xi^4 + \ldots; \qquad \alpha = \alpha_0 + \alpha_2\xi^2 + \alpha_4\xi^4 + \ldots. \tag{7.15}$$

Substituting these expressions into (7.13), expanding the righthand sides in powers of ξ, and equation coefficients at equal powers of ξ gives the sequence of equations of the form

$$\begin{aligned} y_{2j}^{(n)}(\tau) = L((y_{2j})_\tau, (y_{2j}^{(1)})_\tau, \ldots, (y_{2j}^{(n-1)})_\tau; \alpha_0) + \\ +\phi_{2j}(y_1, y_2, \ldots, y_{2j-1}; c_0, c_2, \ldots, c_{2j-2}, \\ \alpha_0, \alpha_2, \ldots \alpha_{2j-2}), \end{aligned} \tag{7.16}$$

$$\begin{aligned} y_{2j+1}^{(n)}(\tau) = L((y_{2j+1})_\tau, (y_{2j+1}^{(1)})_\tau, \ldots, (y_{2j+1}^{(n-1)})_\tau; \alpha_0) + \\ +\phi_{2j+1}(y_1, y_2, \ldots, y_{2j}; c_0, c_2, \ldots, c_{2j}, \\ \alpha_0, \alpha_2, \ldots \alpha_{2j}), \end{aligned} \tag{7.17}$$

where $y_1(\tau) := \cos\omega_0\tau$, $c_0 := 1$, and ϕ_{2j}, ϕ_{2j+1} are nonlinear functional differential operators transforming the set of $(2\pi/\omega_0)$-periodic functions into itself and depending on the given parameters. (In concrete examples these operators can be given explicitly, as has been done in (7.10).)

The equations (7.16) and (7.17) allow us to successively determine the functions $y_2, y_3, \ldots$ similarly as in subs. 7.2; using the absence of terms $\cos\omega_0\tau$ and $\sin\omega_0\tau$ in their Fourier expansions, these functions are determined uniquely. In addition, it turns out that the Fourier expansion of each next function ϕ_k includes only the terms

$\cos m\omega_0\tau$ and $\sin m\omega_0\tau$ with $m \leq k$ and m and k of the same parity. Hence the Fourier expansion of y_k has this property too. However, to construct a periodic solution of (7.17) there should be no resonance at all (there can be no resonance in (7.16)), i.e. the expansion of ϕ_{2j+1} should not contain $\cos\omega_0\tau$ and $\sin\omega_0\tau$. This restriction can be written as

$$\left.\begin{aligned}
&\textstyle\int_0^{2\pi/\omega_0} \phi_{2j+1}(y_1, y_2, \dots, y_{2j}; c_0, c_2, \dots, c_{2j}, \alpha_0, \alpha_2, \dots, \alpha_{2j})(\tau)\times \\
&\qquad\qquad \times \cos\omega_0\tau\, d\tau = 0, \\
&\textstyle\int_0^{2\pi/\omega_0} \phi_{2j+1}(y_1, y_2, \dots, y_{2j}; c_0, c_2, \dots, c_{2j}, \alpha_0, \alpha_2, \dots, \alpha_{2j})(\tau)\times \\
&\qquad\qquad \times \sin\omega_0\tau\, d\tau = 0.
\end{aligned}\right\} \tag{7.18}$$

Because all functions and constants on the lefthand side are assumed to be known from the previous steps, except for c_{2j} and α_{2j}, (7.18) is a system of equations in two unknowns. One can show that both equations are linear, and that the system has nonzero determinant. This makes it possible to determine c_{2j}, α_{2j} in this step, and to determine y_{2j+1}. Thus, the expansions (7.14) and (7.15) can be indefinitely continued, according to the scheme

$$(y_1, c_0, \alpha_0) \to y_2 \to (c_2, \alpha_2) \to y_3 \to y_4 \to (c_4, \alpha_4) \to \dots.$$

THEOREM 7.1. *Assume that all conditions* (7.12) *are met, and that* $\alpha_2 \neq 0$. *If* $\alpha_2 > 0$ *(*$\alpha_2 < 0$*) and* α *increases (decreases) and passes through* $\alpha = \alpha_0$, *then exactly one periodic solution (up to time shift) of equation* (7.11) *arises near the solution* $x(t) \equiv 0$ *of the same equation. This solution is stable (unstable), and has asymptotic representation as* $\alpha \to \alpha_0^+$ *(*$\alpha \to \alpha_0^-$*)*

$$x(t) = \left(\frac{\alpha - \alpha_0}{\alpha_2}\right)^{1/2} \cos\frac{\omega_0}{c}t + \mathcal{O}(|\alpha - \alpha_0|),$$

$$c = 1 + \frac{c_2}{\alpha_2}(\alpha - \alpha_2) + \mathcal{O}((\alpha - \alpha_0)^2)$$

(the same is true for $x^{(1)}(t), \dots, x^{(n-1)}(t)$). *In addition, any solution of* (7.11) *on the whole t-axis and sufficiently close to zero tends either asymptotically to zero or asymptotically to the periodic solution as* $t \to \infty$ *and* $t \to -\infty$.

Here, closeness of solutions is understood in the sense of uniform closeness of the functions and their derivatives up to order $n - 1$, inclusively.

7.4. Variants. *A. Equations with several parameters.* Assume that the righthand side of (7.1) contains, besides α, parameters $\beta_1, \dots, \beta_k$ ($k \geq 1$), varying near values $\beta_{10}, \dots, \beta_{k0}$. Without loss of generality we may assume that all $\beta_{j0} = 0$, and that for any value $\beta = (\beta_1, \dots, \beta_k)$ the critical value $\alpha_0 = 0$.

If the functional F depends sufficiently smoothly on all its functions and parameters, and for $\beta = 0$ condition (7.12) holds, then for any β which is sufficiently small (in modulus) we can construct expansions (7.14) and (7.15), with coefficients depending sufficiently smoothly on β. In particular, the second expansion in (7.15) now takes the form

$$\alpha = \alpha_2(\beta)\xi^2 + \alpha_4(\beta)\xi^4 + \dots. \tag{7.19}$$

It turns out that for sufficiently small $|\alpha|, |\beta|$ and for every sufficiently small $\xi > 0$ for which (7.19) is valid, there is (up to time shift) a unique solution of (7.11) of the form (7.14) for given α, β. In addition, distinct ξ give distinct solutions. This makes it possible to infer assertions about the number of periodic solutions bifurcating from the zero solution, by investigating (7.19) as an equation in α, β, ξ in a neighborhood of the origin of the space $\mathbf{R}^{1+k+1}$ (more exactly, in the halfspace, since $\xi > 0$).

The most important case in applications is when $\alpha_2(0) \neq 0$. Direct use of the implicit function theorem shows that only one periodic solution emerges, and only if $\operatorname{sgn}\alpha = \operatorname{sgn}\alpha_2(0)$. However, if $\alpha_2(0) = 0$, then the picture is more complicated. For example, if the righthand side of (7.19) has the form $\beta_1\xi^2 - \xi^4$, then for $\alpha < 0$ only one periodic solution emerges from the zero solution; if $\alpha > 0$, the relation between α and β_1 plays a role: if $\alpha < \beta_1^2/4$ then there arise two such solutions, while for $\alpha > \beta_1^2/4$ none.

A detailed theoretical analysis of the general case, as well as some examples of applications, are given in [280].

B. Vector RDEs. The study of (7.1) under the condition (7.12) and when $\{\lambda_k\}$ is the set of roots of (7.2) is done similarly as in subs. 7.3. The expansion (7.14) must be replaced by

$$y(\tau) = \xi \operatorname{re}(a_0 e^{i\omega_0\tau}) + \xi^2 y_2(\tau) + \xi^3 y_3(\tau) + \dots,$$

where a_0 is an eigenvector of the matrix $P(i\omega_0, \alpha_0)$ (see (7.2)) corresponding to the eigenvalue $i\omega_0$, and $y_2, y_3, \dots$ are real-valued vector functions. The form of the expansions (7.15) does not change. Instead of (7.18), here we have the following condition for absence of resonance:

$$\int_0^{2\pi/\omega_0} \tilde{a}_0' e^{-i\omega_0\tau} \phi_{2j+1}(y_1, y_2, \dots, y_{2j}; c_0, c_2, \dots, c_{2j}, \\ \alpha_0, \alpha_2, \dots, \alpha_{2j})(\tau)\, d\tau = 0. \tag{7.20}$$

Here $\tilde{a}_0$ is an eigenvector of the matrix $[P(i\omega_0, \alpha_0)]'$ corresponding to the eigenvalue $i\omega_0$, and the prime denotes transposition, as before. Since the first two factors in the integrand are complex-valued functions, (7.20) is a system of two real equations for determining c_{2j} and a_{2j}.

C. Bifurcation for NDEs. In [280] it has been shown that all main assertions of subs. 7.3 and 7.4 hold for vector-valued autonomous NDEs in Hale's form:

$$\left[x(t) - \sum_{j=1}^{m} A_j(\alpha)x(t-h_j) - f(x(t-h_1), \dots, x(t-h_k); \alpha)\right]^{\bullet} = \\ = \sum_{j=1}^{m} B_j(\alpha)x(t-h_j) + g(x(t-h_1), \dots, x(t-h_k); \alpha). \tag{7.21}$$

It is assumed that all matrix functions A_j, B_j and vector functions f, g are analytic; $0 < h_1 < \dots < h_m$; $\alpha \in (\underline{\alpha}, \overline{\alpha})$;

$$|f(u_1, \dots, u_m; \alpha)| + |g(u_1, \dots, u_m; \alpha)| = o(|u_1| + \dots + |u_m|) \\ (|u_1| + \dots + |u_m| \to 0);$$

for some $\alpha_0 \in (\underline{\alpha}, \overline{\alpha})$ we have $\lambda_{1,2}|_{\alpha=\alpha_0} = \pm i\omega_0$ ($\omega_0 > 0$), $d\,\mathrm{re}\,\lambda_{1,2}(\alpha)/d\alpha|_{\alpha=\alpha_0} > 0$; and

$$\sup_{\substack{k>2\\ \alpha\in(\underline{\alpha},\overline{\alpha})}} \lambda_k(\alpha) < 0.$$

Here $\{\lambda_1, \lambda_2, \dots\}$ is the set of roots of the characteristic equation

$$\det\left(\lambda I - \sum_{j=1}^{m}[\lambda A_j(\alpha_0) + B_j(\alpha_0)]\right) = 0$$

of the linear approximation of (7.21).

7.5. Example of an RDE with constant delay: intraspecial struggle for a common food. In subs. 7.5–7.7 we will consider three substantive examples related to the emergence of periodic solutions, following [280].

The simplest model of intraspecial struggle for a common food when there are limited self-renewing food resources has the form (1.5.1):

$$\dot N(t) = \gamma[1 - K^{-1}N(t-h)]N(t).$$

Here $N(t)$ is the volume of the population at time t, γ is the Malthus coefficient, K is the equilibrium value of the population, h is the average time of development of food resources. Because the system under consideration is near its equilibrium position $N(t) \equiv K$, we introduce a new function x by $N = K(1+x)$, and to reduce the number of parameters we introduce dimensionless variables $t_l = t/h$, $\alpha = \gamma h$. We thus obtain

$$\dot x(t_l) = -\alpha x(t_l - 1) - \alpha x(t_l)x(t_l - 1). \tag{7.22}$$

By subs. 6.3, the critical value for the linearized equation is $\alpha = \alpha_0 := \pi/2$, and the function $x = C\cos(\frac{\pi}{2}t_l + \text{const})$ is a periodic solution. Using the algorithm in subs. 7.3 we obtain

$$\begin{aligned}
x(t_l) &= \xi\cos\frac{\pi}{2}\tau + \xi^2 x_2(\tau) + \xi^3 x_3(\tau) + \dots,\\
t_l &= (1 + c_2\xi^2 + c_4\xi^4 + \dots)\tau,\\
x(t_l - 1) &= \xi\cos\frac{\pi}{2}(\tau - (1 + c_2\xi^2 + \dots)^{-1}) +\\
&\quad + \xi^2 x_2(\tau - (1 + c_2\xi^2 + \dots)^{-1}) + \dots,\\
\alpha &= \frac{\pi}{2} + \alpha_2\xi^2 + \alpha_4\xi^4 + \dots.
\end{aligned}$$

(In the expression for $x(t_l - 1)$ there arises an additional expansion in powers of ξ, due to the transition to time τ.) Substituting these expressions into (7.22) and equating coefficients at equal powers of ξ gives the sequence of equations

$$\dot{x}_2(\tau) = -\frac{\pi}{2}x_2(\tau-1) - \frac{\pi}{2}\cos\frac{\pi}{2}\tau\cdot\sin\frac{\pi}{2}\tau,$$
$$\dot{x}_3(\tau) = -\frac{\pi}{2}x_3(\tau-1) +$$
$$-\frac{\pi}{2}[x_2(\tau-1)\cos\frac{\pi}{2}\tau + x_2(\tau)\sin\frac{\pi}{2}\tau + c_2\sin\frac{\pi}{2}\tau] +$$
$$-\frac{\pi^2}{4}c_2\cos\frac{\pi}{2}\tau - \alpha_2\sin\frac{\pi}{2}\tau,$$
$$\cdots\cdots\cdots\cdot$$

The first equation implies that

$$x_2(\tau) = 0.2\cos\pi\tau + 0.1\sin\pi\tau.$$

After substituting this expression into the second equation, the condition of absence of resonance (i.e. absence of terms $\cos\frac{\pi}{2}\tau$ and $\sin\frac{\pi}{2}\tau$ in the Fourier expansion of the righthand side) implies $c_2 = 0.1/\pi$, $\alpha_2 = 0.075\pi - 0.05$. So $\alpha_2 > 0$, and thm. 7.1 asserts that there emerges a cycle for $\alpha > \alpha_0$, i.e. $\gamma h > \pi/2$. This cycle is stable, and can be represented as (after elimination of ξ):

$$N = K\left[1 + (0.075\pi - 0.05)^{-1/2}\left(\gamma h - \frac{\pi}{2}\right)^{1/2}\cos\frac{\pi}{2}\tau + \right.$$
$$\left. +(0.075\pi - 0.05)^{-1}\left(\gamma h - \frac{\pi}{2}\right)(0.2\cos\pi\tau + 0.1\sin\pi\tau) + \mathcal{O}\left(\left(\gamma h - \frac{\pi}{2}\right)^{3/2}\right)\right], \tag{7.23}$$

where

$$\tau = \frac{1}{h}\left[1 - \frac{0.1}{\pi}(0.075\pi - 0.05)^{-1}\left(\gamma h - \frac{\pi}{2}\right) + \mathcal{O}\left(\left(\gamma h - \frac{\pi}{2}\right)^2\right)\right]t.$$

Therefore, if the coefficient γ of species reproduction passes through the critical value $\pi/2h$, then the equilibrium position fails to be stable, and periodic oscillations as described by (7.23) arise. Of course, this is a pure delay effect, because for $h = 0$ the solutions cannot oscillate. Such oscillations have been actually discovered in experiments, and the coincidence of theory and experiments proved to be sufficiently good even for large amplitude oscillations [280]. However, the parameter values of the problem had to be adjusted to the experimental results.

7.6. Example of an RDE with autoregulative delay: combustion in the chamber of a turbojet engine. Because we are only interested in the mathematical formalism, we do not give a derivation of the basic equation, nor the actual meaning of all the quantities involved (see, e.g., [280] for this information). The equation under consideration is:

$$\ddot{x}(t) + (\alpha + \beta p)\dot{x}(t) + \alpha\beta p x(t) + \gamma x(t - \Delta) =$$
$$= \delta_1 x(t)\dot{x}(t) + \delta_2 x^2(t). \tag{7.24}$$

Here, $x(t)$ is a small deviation in the velocity $v(t)$ of the liquid fuel in the line supply from its stationary value v_0, $p > 0$ is the stationary pressure in the combustion chamber (this is taken to be the basic parameter of the process), $\alpha, \beta, \gamma, \delta_1, \delta_2$ are fixed parameters, characterizing the fuel and the chamber, Δ is the time interval from the formation of a portion of the combustion product to the moment of outflow through the nozzle.

The essential distinguishing feature of this example is the dependence of Δ on the unknown solution x. It is possible to assume that

$$\begin{aligned}\Delta = \Delta(p) + h_1 p\Delta'(p)x(t) + h_2\Delta'(p)\dot{x}(t) + \\ +0.5[h_1^2p^2\Delta''(p) + h_3\Delta'(p)]x^2(t) + \\ +h_4p\Delta''(p)x(t)\dot{x}(t) + 0.5h_2^2\Delta''(p)\dot{x}^2(t),\end{aligned}$$

where h_j are given parameters, $p \mapsto \Delta(p)$ is an empirical function whose graph is similar to that of $\Delta = pe^{-p}$, and Δ' and Δ'' are its derivatives. The corresponding linearized equation is

$$\ddot{x}(t) + (\alpha + \beta p)\dot{x}(t) + \alpha\beta x(t) + \gamma x(t - \Delta(p)) = 0. \tag{7.25}$$

Investigation of the roots of the characteristic equation shows that (7.25) is stable for $0 \leq p \leq p_l$ and $p_h \leq p < \infty$, where $p_l > 0$ and $p_h > p_l$ are obtained by solving a transcendental equation in p involving the quantity $\Delta(p)$. For $p = p_l$ ($p = p_h$) the characteristic equation has a pair of conjugate purely imaginary roots $\pm i\omega_l$ (resp., $\pm i\omega_h$).

The study of the complete equation (7.24), which could in principle be done as in subs. 7.5, is more cumbersome, in particular because of the complicated structure of the delay. For p near p_l we set in (7.24):

$$\begin{aligned}x(t) = \xi\cos\omega_l\tau + \xi^2x_2(\tau) + \xi^3x_3(\tau) + \dots, \\ t = (1 + c_2\xi^2 + c_4\xi^4 + \dots)\tau, \\ p = p_l + p_2\xi^2 + p_4\xi^4 + \dots,\end{aligned}$$

and perform the transition from $\dot{x}, \ddot{x}$ to derivatives with respect to τ, and make a corresponding change in $x(t - \Delta)$. Then we expand all terms in powers of ξ, and construct the periodic solutions of the derived equations. The rather cumbersome expressions for c_2, p_2, and hence of the periodic solutions of the first approximation, have been written out in [280].

We stress that the coefficients which are derived do not only contain the empirical function $p \mapsto \Delta(p)$, which is known with little accuracy, but also its derivatives Δ' and Δ''. Therefore the final result is likely to be unreliable, and certain conclusions can only be given on the basis of empirical evidence. It seems that for increasing p there appear first low frequency oscillations, these becoming high frequency oscillations until for p sufficiently large the combustion becomes stationary. These experimental conclusions are in agreement with the mathematical theory.

7.7. Example NDE: auto-oscillation in a long line with tunnel diod. The model equations of this system have been described in subs. 1.3.4, and take the form (see (1.3.4)):

$$\dot{x}(t_l) - K\dot{x}(t_l - h) + (1-\alpha)x(t_l) + K(1+\alpha)x(t_l - h) = \\ = [x(t_l) - Kx(t_l - h)]^2. \tag{7.26}$$

In the sequel it is assumed that $K > 0$, $\alpha < 1$, since otherwise the stationary solution is certainly unstable. Study of the roots of the characteristic equation corresponding to the linearization of (7.26) shows that stability loss in (7.26) occurs when K passes through

$$K_0 = [\omega_0^2 + (1-\alpha)^2]^{1/2}[\omega_0^2 + (1+\alpha)^2]^{-1/2}.$$

Here ω_0 is the smallest positive root of the equation

$$\cos\omega_0 h = [\omega_0^2 - (1-\alpha)^2][\omega_0^2 + (1-\alpha)^2]^{-1/2}[\omega_0^2 + (1+\alpha)^2]^{-1/2}.$$

For $K = K_0$ the characteristic equation has two purely imaginary roots, $\pm i\omega_0$, and the set of real parts of the remaining roots has negative least upper bound.

The study of the complete equation (7.26) for K near K_0 can be done along the same lines as before, using the expansions

$$x(t_l) = \xi\cos\omega_0\tau + \xi^2 x_2(\tau) + \xi^3 x_3(\tau) + \dots,$$
$$t_l = (1 + c_2\xi^2 + c_4\xi^4 + \dots)\tau,$$
$$K = K_0 + K_2\xi^2 + K_4\xi^4 + \dots.$$

Substitution of these expansions into (7.26) allows us to obtain explicit expressions for $x_2(\tau)$, c_2, and K_2 (here we do not give these, see [280]). In particular, the expression for K_2 implies $K_2 < 0$ for sufficiently small fixed $\alpha > 0$, i.e. for such α there emerges a cycle for $K < K_0$, and this cycle is unstable.

The total bifurcation picture for $\alpha \in (0,1)$ has not yet been obtained. It has only been shown that $K_2|_{\alpha=0} = K_4|_{\alpha=0} = 0$. It is conjectured in [280] that K_2 becomes positive when α passes through some value $\alpha_0 \in (0,1)$, i.e. that auto-oscillations arise in the system.

8. Stochastic retarded differential equations (SRDEs)

The functioning of many actual delay systems are subject to unknown disturbances, which may be regarded as stochastic. For the description of such systems we can use SRDEs.

An example of this kind is the model of a motion with viscous aftereffect given in subs. 1.2.2. Another example is the average of a stock market, which depends on many random factors, as well as on the previous history. Furthermore, in many actual situations the value of the delay is imperfectly known, and may be regarded as being random.

In addition, the solution of a stochastic ODE can be approximated by those of SRDEs with small delays. E.g., in [46] the solution of the equation $dx(t) = g(x(t))\,d\xi(t)$

is approximated by $\{x^n(t)\}$, where $dx^n(t) = g(x^n(t-1/n))\,d\xi(t)$. Of course, a similar procedure can be used for deterministic equations.

8.1. Initial value problem. We will consider SRDEs of the form

$$\left.\begin{aligned} dx(t) &= a(t,x_t)\,dt + b(t,x_t)\,d\xi(t), \qquad t \geq 0, \\ x_t(\theta) &:= x(t+\theta), \qquad -\infty < \theta \leq 0. \end{aligned}\right\} \tag{8.1}$$

(Here we assume, without loss of generality, that $t_0 = 0$.) In this equation a, b: $[0,\infty)\times C(-\infty,0] \to \mathbf{R}$ are continuous functionals (see subs. 2.5) and $t \mapsto \xi(t) \in \mathbf{R}^l$ is the standard Wiener process, defined on a probability space $(\Omega, \mathcal{A}, P)$. We recall some basic definitions (see, e.g., [256]). Points $\omega \in \Omega$ are called *elementary events*. $\mathcal{A}$ is a sigma-algebra of subsets of Ω; these subsets are called *random events*. The measure P is defined on (the elements of) $\mathcal{A}$, such that $P(\Omega) = 1$; it is called a *probability measure*. A real-valued function $\omega \to \zeta(\omega)$ from Ω to $\mathbf{R}$ which is measurable with respect to $\mathcal{A}$ (i.e. $\{\omega \in \Omega\colon\ \zeta(\omega) \leq x\} \in \Omega$ for all $x \in \mathbf{R}$) is called a *random variable*. The *distribution function* F of ζ is defined by

$$F(x) = P\{\omega \in \Omega : \ \zeta(\omega) \leq x\}, \qquad x \in \mathbf{R}.$$

A function $\omega \mapsto \zeta(\omega) \in \mathbf{R}^n$ is called a *random vector* if every component ζ_i, $i = 1,\dots,n$, is a random variable. The distribution function F of a random vector $\omega \mapsto \zeta(\omega) \in \mathbf{R}^n$ is the *joint distribution function* of the random variables $\zeta_1,\dots\zeta_n$, defined as follows:

$$F(x_1,\dots,x_n) = P\{\omega \in \Omega : \ \zeta_1(\omega) \leq x_1,\dots,\zeta_n(\omega) \leq x_n\},$$
$$x_i \in \mathbf{R}, \qquad i = 1,\dots,n.$$

A family of random variables $\omega \mapsto \zeta(t,\omega) \in \mathbf{R}^n$ defined on some t-interval J and, in general, correlated with each other for distinct t, is called a *stochastic process* or a *random process*. This correlation gives rise to the probability space $(\overline{\Omega}, \overline{\mathcal{A}}, \overline{P})$, where $\overline{\Omega}$ is the space of functions $J \to \Omega$ and $\overline{\mathcal{A}}, \overline{P}$ are the σ-algebra and measure on $\overline{\Omega}$. We do not consider the details of this construction. Thus, a random process is formally defined as a function $(\overline{\omega} \mapsto \zeta(\cdot,\overline{\omega}))\colon \overline{\Omega} \to \overline{\Omega}$, i.e. $((t,\overline{\omega}) \mapsto \zeta(t,\overline{\omega}))\colon J \times \overline{\Omega} \to \mathbf{R}^n$. The argument $\overline{\omega}$ is usually omitted, and one simply writes $t \mapsto \zeta(t)$. For a fixed $t \in J$ the quantity $\zeta(t)$ can be identified with the stochastic quantity $\omega \mapsto \zeta(t,\omega)$, which is known as a *section of the random process* under consideration. As a function of t for fixed ω, the process $t \mapsto \zeta(t,\omega)$ is called a *trajectory of the random process*.

A process $t \mapsto \zeta(t)$ is said to be a *random process with independent increments* if for any finite set $(t_1,\dots,t_k)$, $k \geq 3$, such that $t_1 < t_2 < \dots < t_k$ the random variables $\zeta(t_2) - \zeta(t_1),\dots,\zeta(t_k) - \zeta(t_{k-1})$ are independent. The *standard Wiener process* $t \mapsto \xi(t)$, $t \geq 0$, is the random process with independent stationary gaussian increments such that

$$\xi(0) = 0, \qquad \mathcal{E}\xi(t) = 0,$$
$$\mathcal{E}[(\xi(t) - \xi(s))(\xi(t) - \xi(s))'] = (t-s)I, \qquad t \geq s \geq 0.$$

Here $\mathcal{E}$ denotes mathematical expectation, I is the identity matrix, and the prime denotes transposition. The Wiener process $t \mapsto \xi(t)$ is completely determined by these conditions as a concrete random process. The trajectories of the Wiener process $t \mapsto \xi(t)$ are continuous functions (of t) that are, with probability 1, nowhere differentiable, satisfy the Hölder condition with index $< 1/2$, and have finite variation on any bounded time interval. This process was introduced by N. Wiener to model Brownian motion of physical particles suspended in a liquid. The properties of the trajectories of the process $t \mapsto \xi(t)$ mean that in this model the particles do not have velocities. A natural way to overcome this deficiency is to take into consideration viscous aftereffect, which is always present in actual motions, i.e. to use FDEs (see subs. 1.2.2).

Let $B_{t_1t_2}(d\xi) \subset \mathcal{A}$ be the minimal σ-algebra with respect to which all random variables $\xi(s) - \xi(t)$ are measurable, for all $s, t \in [t_1, t_2]$.

We return to equation (8.1). As initial condition we take

$$x(\theta) = \phi(\theta) \qquad (-\infty < \theta \le 0), \tag{8.2}$$

where ϕ is a process with continuous trajectories. It is assumed that the process $\theta \mapsto \phi(\theta)$, $-\infty < \theta \le 0$, is independent of the increments of the Wiener process $t \mapsto \xi(t)$, i.e. the σ-algebra $B_{-\infty 0}(\phi)$ is independent of $B_{0\infty}(d\xi)$. We say that the process $t \mapsto x(t)$ is a solution of (8.1)–(8.2) if $B_{-\infty t}(x) \cup B_{0t}(d\xi)$ is independent of $B_{t\infty}(d\xi)$ and if with probability 1 the following equality holds for all $t \ge 0$:

$$x(t) = x(0) + \int_0^t a(s, x_s)\, ds + \int_0^t b(s, x_s)\, d\xi(s). \tag{8.3}$$

The second integral in (8.3) is called *Itô's stochastic integral*, and is defined by a special limiting procedure. The following two properties of this integral are used quite often ($T > 0$):

$$\mathcal{E} \int_0^T f_1(t)\, d\xi(t) = 0,$$

$$\mathcal{E}\left\{\int_0^T f_1(t)\, d\xi(t) \left[\int_0^T f_2(t)\, d\xi(t)\right]'\right\} = \int_0^T \mathcal{E}[f_1(t) f_2'(t)]\, dt.$$

Here, the continuous processes $f_i \colon [0, T] \to \mathbf{R}^{n\times l}$ should be measurable with respect to $B_{0t}(d\xi)$ and

$$\int_0^T \mathcal{E}|f_i(t)|^2\, dt < \infty.$$

8.2. Existence and uniqueness of solutions. Suppose the continuous functionals a and b in (8.1) satisfy for arbitrary continuous functions $\psi, \chi \in C(-\infty, 0]$ and all $t \in [0, \infty)$ the conditions

$$|a(t,\psi) - a(t,\chi)|^2 \le \int_{-\infty}^0 |\psi(\theta) - \chi(\theta)|^2\, dK_1(\theta),$$

$$|b(t,\psi) - b(t,\chi)|^2 \le \int_{-\infty}^0 |\psi(\theta) - \chi(\theta)|^2\, dK_2(\theta).$$

Here $K_{1,2}$ are nondecreasing bounded functions. Then [256] there is for any initial process ϕ with

$$\sup_{\theta \le 0} \mathcal{E}|\phi(\theta)|^4 < \infty$$

a unique solution of the problem (8.1)–(8.2) on $[0, \infty)$. This solution has bounded fourth moment on any finite time interval $[0, t]$ and is measurable with respect to $\phi(\theta)$, $\theta \le 0$, and $\xi(s)$, $0 \le s \le t$, i.e. $B_{-\infty t}(x) \subset B_{-\infty 0}(\phi) \cup B_{0t}(d\xi)$.

Note that the functionals a and b defined on $[0, \infty) \times C(-\infty, 0]$ may be regarded as Volterra type operators mapping the random function x to the random function $t \mapsto a_i(t, x_t)$ ($a_1 = a$, $a_2 = b$), and the probability distributions of the latter are induced by the probability distribution of x.

8.3. Some characteristics of solutions of linear equations. Here we consider the linear SRDE with discrete delay of the form (8.1):

$$dx(t) = A(t)x(t-h)\,dt + \sigma(t)\,d\xi(t), \qquad 0 \le t < \infty,$$

where $h > 0$, A and σ are given continuous matrix functions. It is obvious that the function $t \mapsto m(t) = \mathcal{E}x(t)$ satisfies the relation

$$\dot{m}(t) = A(t)m(t-h), \quad 0 \le t < \infty; \qquad m(\theta) = \mathcal{E}\phi(\theta), \quad \theta \le 0.$$

Given a time moment $T > 0$ and a matrix $H \in \mathbf{R}^{n\times n}$, the quantity $\delta = \mathcal{E}x'(T)Hx(T)$ is defined by the formula

$$\delta = \mathcal{E}\left[\phi'(0)P\phi(0) + \phi'(0)\int_{-h}^{0} Q(\theta)\phi(\theta)\,d\theta + \right.$$
$$\left. + \int_{-h}^{0} \phi'(\theta)\theta'(\theta)\phi(0)\,d\theta + \int_{-h}^{0} d\theta \int_{-h}^{0} \phi'(\theta)R(\theta,\tau)\phi(\tau)\,d\tau\right] +$$
$$+ \int_{0}^{T} \mathrm{Tr}[B'(t)HB(t)\sigma(t)\sigma'(t)]\,dt.$$

Here the matrix $B(t)$ is a solution of the problem

$$\dot{B}(t) = -B(t+h)A(t+h), \qquad B(T) = I, \qquad B(t) = 0, \quad t > T.$$

The matrices P, Q, R are given by:

$$P = B'(0)HB(0), \qquad Q(t) = B'(0)H\dot{B}(t), \qquad R(t,\tau) = \dot{B}'(t)H\dot{B}'(\tau).$$

CHAPTER 3

Stability of retarded differential equations

The following three chapters deal with stability problems for FDEs.

Stability of a process (in particular, of a stationary state) is the ability of the process to resist a priori unknown, small influences (disturbances). A process is said to be stable if such disturbances do not essentially change it. This property turns out to be of utmost importance. We emphasize that an individual predictable process can be physically realized only if it is stable or quasistable in the corresponding natural sense.

The great variety of types of processes and disturbances has led to numerous mathematical notions of stability, among them stability for hereditary processes [2, 8, 25, 38, 68, 114, 122, 142, 151, 152, 211, 213, 257, 275–277, 286, 291, 299, 451, 534, 535].

In the sequel we will give our main attention to notions and assertions that are likely to be useful from the point of view of applications. In particular, we give attention to linear autonomous equations and the corresponding quasipolynomials, Liapunov's direct method, and concrete hereditary systems, as well as to the notions of absolute, stochastic, etc. stability.

1. Liapunov's direct method

1.1. Stability definitions. We begin with the basic stability definitions for RDEs with bounded delay. Consider the following Cauchy problem for $x\colon J_x \to \mathbf{R}^n$:

$$\left.\begin{aligned} \dot{x}(t) = f(t, x_t) \quad (t \geq t_0), \qquad x_t(\theta) &= x(t+\theta), \quad -h \leq T \leq 0, \\ x_{t_0} &= \phi. \end{aligned}\right\} \tag{1.1}$$

Here, $f\colon \mathbf{R} \times C[-h,0] \to \mathbf{R}^n$ is a continuous functional which is locally Lipschitz in the second argument; $\phi \in C[-h,0]$.

As usual, the study of the stability of any solution $x^0\colon [t_0,\infty) \to \mathbf{R}^n$ of (1.1), corresponding to ϕ^0, can be reduced to the study of the stability of the trivial (zero) solution, by changing the unknown and initial functions: $x(t) = x^0(t) + y(t)$, $\phi(\theta) = \phi^0(\theta) + \psi(\theta)$. For y we obtain

$$\dot{y}(t) = f(t, x_t^0 + y_t) - f(t, x_t^0) =: \tilde{f}(t, y_t) \quad (\Rightarrow \tilde{f}(t,0) \equiv 0),$$
$$y_{t_0} = \psi.$$

Hence in the sequel we may take $f(t,0) \equiv 0$ in (1.1). Thus, we have to study only th stability of the trivial solution of (1.1) corresponding to the zero initial function. Le $t \to x(t;t_0,\phi)$ be the solution of (1.1), and let

$$Q_\delta := \{u \in C[-h,0] : \|u\| < \delta\}.$$

The trivial solution of (1.1) is called

A) *stable* for a given $t_0 \in \mathbf{R}$ if for any $\epsilon > 0$ there is a $\delta = \delta(\epsilon, t_0)$ such tha $|x(t;t_0,\phi)| \le \epsilon$ for any initial function $\phi \in Q_\delta$ and $t \in [t_0,\infty)$; in the opposit case the solution is called *unstable*;
B) *uniformly stable* if for any $\epsilon > 0$ there is a $\delta(\epsilon) > 0$ independent of t_0 such tha $|x(t;t_0,\phi)| \le \epsilon$ for any initial function $\phi \in Q_\delta$, $t_0 \in \mathbf{R}$, $t \in [t_0,\infty)$;
C) *asymptotically stable* for a given $t_0 \in \mathbf{R}$ if it is stable and there is a $\Delta =$ $\Delta(t_0) > 0$ such that $\lim_{t\to\infty} x(t;t_0,\phi) = 0$ for any initial function $\phi \in Q_\Delta$ the set of initial functions ϕ for which $\lim_{t\to\infty} x(t;t_0,\phi) = 0$ is called th *attraction domain* of the trivial solution for initial time t_0, and is denoted b $\Omega(t_0) \subseteq C[-h,0]$;
D) *uniformly asymptotically stable* if it is uniformly stable and there is a $H >$ (such that for any $\gamma > 0$ there is a $T(\gamma) > 0$ such that $|x(t;t_0,\phi)| \le \gamma$ for an $t_0 \in \mathbf{R}$, $t \ge t_0 + T(\gamma)$ and $\phi \in Q_H$.

Note that $Q_H \subseteq \Omega(t_0)$ for all $t_0 \in \mathbf{R}$.

Other, infrequently used, notions of stability, such as stability under persistentl acting disturbances, global stability, exponential stability, stability under delay dis turbances, stability for part of the variables, etc., can be similarly given. E.g., th trivial solution of (1.1) is *exponentially stable* if there are constants $H > 0$, $C_1 > 0$ $C_2 > 0$ such that for any $\phi \in Q_H$ the solution $x(t;t_0,\phi)$ of this equation satisfies th inequality

$$|x(t;t_0,\phi)| \le C_1\|\phi\|\exp[-C_2(t-t_0)], \qquad t_0 \le t < \infty.$$

REMARK 1.1. The requirement of uniform asymptotic stability is stronger tha that of asymptotic stability. But for autonomous or periodic equations (1.1), asymp totic stability of the trivial solution implies uniform asymptotic stability of this solu tion. A similar relation holds for stability and uniform stability. □

EXAMPLE 1.1. The solution of the Cauchy problem

$$\dot{x}(t) = e^{-t}x(t), \qquad x(t_0) = x_0,$$

is

$$x(t) = x_0 \exp(e^{-t_0} - e^{-t}).$$

Therefore, for any $x_0 \ne 0$ we find

$$|x(t)| < |x_0| \exp e^{-t_0}, \qquad t_0 \le t < \infty,$$

and for any $\epsilon > 0$ the optimal value of $\delta(\epsilon, t_0)$ is $\epsilon \exp(-e^{-t_0})$. We find that the trivia solution is stable, but not uniformly stable because $\delta(\epsilon,t_0) \to 0$ as $t_0 \to -\infty$. □

REMARK 1.2. This remark is concerned with the dependence of stability on the initial moment t_0. By thm. 2.2.1 the trivial solution, which can be indefinitely extended forward in time, depends continuously on the initial function, on every bounded interval. hence, if the trivial solution is stable with respect to disturbances of the initial function for time t_0, then it must be stable with respect to disturbances for any previous initial moment $t_1 < t_0$. However, since not all solutions of RDEs can be indefinitely extended backwards, the trivial solution may become unstable at a future moment $t_1 > t_0$. Luckily, such asymmetry in the stability property is artificial and is unlikely to be met in practical applications. □

EXAMPLE 1.2. Consider the scalar equation

$$\dot{x}(t) = a(t)x(t - h(t))$$

with $a, h \in C^1(\mathbf{R})$, $h(t) \geq 0$, $\dot{h}(t) < 1$, whose zero solution is asymptotically stable for arbitrary $t_0 \leq 1$ and unstable for any $t_0 \geq 0$. Take for h a $C^1(\mathbf{R})$-function such that

$$h(t) = 0, \qquad -\infty < t \leq -1, \quad 0 \leq t < \infty,$$
$$h(t) = (t^2 \ln t^{-2})(1 + t \ln t^{-2})^{-1}, \qquad -e^{-2} \leq t < 0.$$

Let the function $x_0 \in C^2(-\infty, 0]$ be defined by

$$x_0(t) > 0, \quad -\infty < t < 0; \qquad x_0(t) \equiv 1, \quad -\infty < t \leq -1;$$
$$x_0(t) = \exp t^{-1}, \qquad -0.5 \leq t < 0.$$

Finally we choose $a \in C^1(\mathbf{R})$ such that

$$a(t) = \dot{x}_0(t)[x_0(t - h(t))]^{-1}, \qquad -\infty < t < 0,$$
$$a(t) = 1, \qquad 1 \leq t < \infty.$$

It is possible to choose such an $a(t)$, since

$$a(t) \equiv -1, \qquad -e^{-2} \leq t < 0.$$

If $t_0 \leq -1$, then the solution x of the Cauchy problem for $t_0 \leq t < 0$ is proportional to $x_0(t)$. Consequently, $x(0) = 0$, and also $x|_{[0,\infty)} \equiv 0$. But if $t_0 \geq 0$, then it is obvious that the trivial solution is unstable. □

These definitions of stability can be immediately generalized to RDEs with infinite delay. In this case, however, the space of initial disturbances ϕ (see subs. 2.2.5) must be specified. In this connection we note that a solution which is stable under a certain class of disturbances is also stable under a narrower class of disturbances. On the other hand, a solution which is unstable for a certain class of disturbances may well be stable under given a wider class of disturbances. We consider by a simple example the dependence of stability on the choice of the initial function space.

EXAMPLE 1.3. Given the Cauchy problem

$$\dot{x}(t) = \int_{-t}^{t} x(\tau)\,d\tau, \qquad 0 \le t < \infty;$$
$$x(t) = \phi(t) \qquad (-\infty < t \le 0), \qquad \phi \in C(-\infty, 0].$$

Its solution has the form

$$x(t) = -\int_0^t \sin(t-\tau)\phi(-\tau)\,d\tau + \phi(0)\cos t, \qquad 0 \le t < \infty.$$

This implies that the trivial solution is stable (unstable) if the Banach space c functions $\phi \in C(-\infty, 0]$ with finite norm $\|\phi\| = \sup_{-\infty<t\le 0}(|\phi(t)|e^{-t})$ (respectivel $\sup_{-\infty<t\le 0}|\phi(t)|$) is taken as initial set of functions. □

1.2. Stability theorems for equations with bounded delay. We will conside the stability problem for the following equation in $\mathbf{R}^n$ with bounded delay $h > 0$:

$$\left.\begin{aligned} \dot{x}(t) &= f(t, x_t) & (t &\ge t_0), \\ x_t(\theta) &= x(t+\theta), & -h &\le \theta \le 0; \\ f(t,0) &\equiv 0; & x_{t_0} &= \phi. \end{aligned}\right\} \tag{1.2}$$

We assume that $\phi \in C[-h, 0]$, and that for some $H > 0$ the functional $((t,\psi) \mapsto f(t,\psi)) \colon \mathbf{R}\times Q_H \to \mathbf{R}^n$ is continuous and bounded, and locally Lipschitz with respec to

$$\psi \in Q_H := \{\psi \in C[-h, 0] : \|\psi\| < H\}.$$

From the stability theory of ODEs the profound efficiency of Liapunov's direc method is well known, based as it is on deep energy relations. N. Krasovskii was th first to generalize this method to RDEs. In addition, since to each solution of an RDI there is naturally related an integral curve in the space $\mathbf{R}\times C[-h, 0]$ (see subs. 2.2.3) it is natural to use functionals in this space instead of Liapunov functions. Sucl functionals are often called *Liapunov–Krasovskii functionals*.

Let $V \colon \mathbf{R}\times Q_H \to \mathbf{R}$ be a continuous functional such that $V(t, 0) \equiv 0$. Substitutin in the second argument of V the function $t \mapsto x_t$ related with a solution $x \colon J_x \to \mathbf{R}^n$ of (1.2), we obtain a composite function of t, $[t \mapsto V(t) := V(t, x_t)] \colon J_x \to \mathbf{R}$. B definition, $\dot{V}(t_0)$ is the right upper derivative:

$$\dot{V}(t_0) := \limsup_{\Delta t\to 0+} \frac{1}{\Delta t}[V(t_0+\Delta t) - V(t_0)]; \tag{1.3}$$

it is called the *derivative of* $V(\cdot,\cdot)$ *along the solution* of (1.2).

It is obvious that $\dot{V}(t_0)$ depends on ϕ. Since we can proceed similarly for arbitrar $t \in \mathbf{R}$, $\psi \in Q_H$, we obtain a generalized functional $\dot{V} \colon \mathbf{R}\times Q_H \to [-\infty, \infty]$, witl values

$$\dot{V}(t,\psi) = \limsup_{\Delta t\to 0+} \frac{1}{\Delta t}[V(t+\Delta t, x_{t+\Delta t}(\cdot; t, \psi)) - V(t,\psi)]. \tag{1.4}$$

The functional $\dot{V}$ is called generalized because it can take infinite values.

We give some examples. Let $n = 1$ and $V(t,\psi) = q(\psi(\alpha))$, where $q \in C^1(\mathbf{R})$, $-h \le \alpha < 0$. Then

$$\dot V(t,\psi) = \begin{cases} \dot q(\psi(\alpha))\dot\psi^{ur}(\alpha) & \text{if } \dot q(\psi(\alpha)) > 0,\\ \dot q(\psi(\alpha))\dot\psi^{lr}(\alpha) & \text{if } \dot q(\psi(\alpha)) < 0,\end{cases}$$

where ur means 'upper right' and lr 'lower right'; the case $\dot q(\psi(\alpha)) = 0$ is not considered here. We can see that in this example $\dot V(t,\psi)$ does not take finite values on all of $\mathbf{R} \times Q_H$, but only on a subspace of it, since $\dot\psi^{ur}$ and $\dot\psi^{lr}$ may become infinite.

As another example we take $V(t,\psi) = q(t,\psi(0))$ for any n, where $q \in C^1(\mathbf{R}\times Q_H)$. Then

$$\dot V(t,\psi) = q_t(t,\psi(0)) + [\operatorname{grad} q(t,\psi(0))]' f(t,\psi),$$

where r_t denotes partial derivative with respect to t and the gradient is calculated for t fixed. As a result we obtain a functional that is finite and continuous on $\mathbf{R} \times Q_H$.

Finally, for $n = 1$ we consider the functional

$$V(t,\psi) = \int_{-h}^{0} K(t,\theta)\psi(\theta)\, d\theta, \tag{1.5}$$

where $K \in C^1(\mathbf{R} \times [-h,0])$. Using (1.4) we find

$$\dot V(t,\psi) = \int_{-h}^{0} K_t(t,\theta)\psi(\theta)\,d\theta + \left[\frac{d}{d\tau}\int_{-h}^{0} K(t,\theta)x_\tau(\theta;t,\psi)\,d\theta\right]_{\tau=t} =$$
$$= \int_{-h}^{0} K_t(t,\theta)\psi(\theta)\,d\theta + \left[\frac{d}{d\tau}\int_{-h}^{0} K(t,\theta)x(\tau+\theta;t,\psi)\,d\theta\right]_{\tau=t}.$$

Here we cannot directly use Leibnitz's rule, since in general $\dot x(t+\theta;t,\psi)$ need not exist. Therefore we first integrate by parts:

$$\int_{-h}^{0} K(t,\theta)x(t+\theta;t,\psi)\,d\theta = K(t,0)\int_0^{\tau} x(s;t,\psi)\,ds +$$
$$-K(t,-h)\int_0^{\tau-h} x(s;t,\psi)\,ds - \int_{-h}^{0} K_\theta(t,\theta)\left[\int_0^{\tau+\theta} x(s;t,\psi)\,ds\right] d\theta.$$

Differentiation with respect to τ and substituting $\tau = t$ leads to

$$\dot V(t,\psi) = K(t,0)\psi(0) - K(t,-h)\psi(-h) +$$
$$+ \int_{-h}^{0} [K_t(t,\theta) - K_\theta(t,\theta)]\psi(\theta)\,d\theta. \tag{1.6}$$

(The same result is obtained by first using Leibnitz's rule and then integrating by parts; a justification of this can be given using distribution theory.) So, in this example the functional $\dot V(t,\psi)$ is finite on the whole space $\mathbf{R} \times Q_H$, and is independent of the righthand side of (1.2).

We can similarly prove the following general rule, which holds in many actual examples. Assume that for arbitrary n the functional $V\colon \mathbf{R} \times C[-h,0] \to \mathbf{R}$ can be represented in the form $V(t,\psi) = W(t,\psi(0),\psi)$, where the functional $W\colon \mathbf{R} \times \mathbf{R}^n \times C[-h,0]$ has continuous derivative in the first argument, continuous gradient in the second (vectorial) argument, while in the third (functional) argument the following linearization holds:

$$W(t,\psi^0,\psi+\delta\psi)-W(t,\psi^0,\psi)=$$
$$=\int_{-h}^{0} K'(t,\psi^0,\theta,\psi(\theta))\delta\psi(\theta)\,d\theta+o(\|\delta\psi\|) \qquad (\|\delta\psi\|\to 0).$$

Here, $K'=(K_1,\dots,K_n)$ is a continuous vector function on $\mathbf{R}\times\mathbf{R}^n\times[-h,0]\times\mathbf{R}^n$. Assume that there is a scalar continuous function S, depending on the same arguments, which is of class C^1 in the last two arguments and such that $\partial S/\partial\psi_i\equiv K_i$ $(i=1,\dots,n)$. Then

$$\dot V(t,\psi)=W_t(t,\psi(0),\psi)+[\mathrm{grad}_{\psi(0)}W(t,\psi(0),\psi)]'f(t,\psi)+$$
$$+S(t,\psi(0),0,\psi(0))-S(t,\psi(0),-h,\psi(-h))+$$
$$-\int_{-h}^{0} S_\theta(t,\psi(0),\theta,\psi(\theta))\,d\theta.$$

For functionals V of this kind we can construct $\dot V$ as follows. Replace ψ by x_t in $V(t,\psi)$, pass to x, differentiate in t (so that $\dot x$ appears only for the current (not delayed) value of t), substitute $\dot x(t)$ from the RDE, pass from x to x_t, and replace x_t by ψ. We illustrate this procedure for the functional (1.4):

$$V(t,x_t)=\int_{-h}^{0}K(t,\theta)x_t(\theta)\,d\theta=\int_{-h}^{0}K(t,\theta)x(t+\theta)\,d\theta=$$
$$=\int_{t-h}^{t}K(t,s-t)x(s)\,ds$$

(the last transformation is done to avoid the appearance of $\dot x(t+\theta)$ after differentiation). This implies

$$\dot V(t,x_t)=$$
$$=K(t,0)x(t)-K(t,-h)x(t-h)+\int_{t-h}^{t}[K_I(t,s-t)-K_{II}(t,s-t)]x(s)\,ds=$$
$$=K(t,0)x_t(0)-K(t,-h)x_t(-h)+\int_{-h}^{0}[K_t(t,\theta)-K_\theta(t,\theta)]x_t(\theta)\,d\theta,$$

where K_I and K_{II} are the partial derivatives of K in the first and second argument, respectively. As a result we obtain (1.6). Having in mind this procedure, the expressions for V and $\dot V$ are (in practice) often immediately written in terms of x, keeping in mind the transition from x to ψ. E.g., in this example, $V=\int_{-h}^{0}K(t,\theta)x(t+\theta)\,d\theta$, $\dot V=K(t,o)x(t)-\dots$.

Let Ω be the set of scalar nondecreasing functions $\omega\in C[0,\infty)$ such that $\omega(r)>0$ for $r>0$ and $\omega(0)=0$.

A functional $(t,u)\mapsto V(t,u)$ is called *positive definite* (*negative definite*, or is said to have *infinitesimal upper bound*) if there is a function $\omega\in\Omega$ such that $V(t,\psi)\geq\omega(|\psi(0)|)$ (respectively, $V(t,\psi)\leq-\omega(\psi(0))$, $|V(t,\psi)|\leq\omega(\|\psi\|)$) for all $t\in\mathbf{R}$, $\psi\in Q_H$.

THEOREM 1.1. *Assume that for some $H > 0$ there is a positive definite continuous functional $((t,\psi) \mapsto V(t,\psi))\colon \mathbf{R} \times Q_H \to \mathbf{R}$ which has infinitesimal upper bound and whose derivative $\dot V$ is a negative definite functional on $\mathbf{R} \times Q_H$. Then the trivial solution of* (1.2) *is asymptotically stable.*

PROOF. Let $\omega_1(|\psi(0)|) \le V(t,\psi) \le \omega_2(\|\psi\|)$ for all $t \in \mathbf{R}$, $\psi \in Q_H$, where $\omega_1, \omega_2 \in \Omega$. Take an $\epsilon \in (0,H)$, and choose $\delta > 0$ satisfying $\omega_2(\delta) < \omega_1(\epsilon)$. Then $\delta < H$, and for an arbitrary function $\phi \in Q_\delta$ and corresponding solution of (1.2), we find that for $t \ge t_0$ and $x(\tau) \in Q_H$, $\forall \tau \in [t_0, t]$:

$$\omega_1(|x(t)|) \le V(t,x_t) \le V(t_0,\phi) \le \omega_2(\|\phi\|) \le \omega_2(\delta) < \omega_1(\epsilon).$$

Hence $|x(t)| < \epsilon$ for all $t \in [t_0,\infty)$, since ω_1 is monotone. This implies stability of the trivial solution.

We now prove that $x(t) \to 0$ as $t \to \infty$ for any $\phi \in Q_l$, where $l > 0$ is such that $\omega_1(H) > \omega_2(l)$. Thus, for all $t \ge t_0$, $\phi \in Q_l$ we have $|x(t)| < H$. By (1.2) we now have $|\dot x(t)| \le C \in (0,\infty)$. Suppose that the solution $x(t)$ corresponding to some $\phi \in Q_l$ does not tend to zero. Then there are $\epsilon > 0$ and a sequence $\{t_i\}$ such that $t_0 \le t_i \underset{i\to\infty}{\longrightarrow} \infty$ and $|x(t_i)| \ge \epsilon$. Without loss of generality we may take

$$t_{i+1} - t_i \ge 2\Delta, \qquad i = 0,1,\ldots; \qquad \Delta := \frac{\epsilon}{2c}.$$

Therefore $|x(t_i+\tau)| \ge \epsilon/2$ for all τ, $|\tau| \le \Delta$. The negative definiteness of $\dot V$ implies $\dot V(t_i+\tau, x_{t_i+\tau}) \le -\alpha = \text{const} < 0$ for all such τ. Let $N(t)$ be the number of t_i's for which $t_0 + \Delta \le t_i \le t - \Delta$. Then

$$V(t) - V(t_0) \le \sum_{t_0-\Delta \le t_i \le t-\Delta} [V(t_i+\Delta) - V(t_i - \Delta)] \le -2\Delta N(t)\alpha.$$

Since $N(t) \to \infty$ as $t \to \infty$, we have $V(t) \to -\infty$. This contradicts the positive definiteness of V. So, $x(t,t_0,\phi) \to 0$ for $\phi \in Q_l$ and $t \to \infty$. □

We can similarly state other stability conditions. Some of these are as follows.

THEOREM 1.2. *If there is a continuous positive definite functional $V\colon \mathbf{R} \times Q_H \to \mathbf{R}$ with $\dot V \le 0$, then the trivial solution of* (1.2) *is stable.*

THEOREM 1.3. *A necessary and sufficient condition for the exponential stability of the trivial solution of* (1.2) *is that there exists a continuous functional $V\colon \mathbf{R} \times Q_H \to \mathbf{R}$ such that*

$$C_1\|\psi\| \le V(t,\psi) \le C_2\|\psi\|, \qquad \dot V(t,\psi) \le -C_3\|\psi\|,$$
$$|V(t,\psi) - V(t,\xi)| \le C_4\|\psi - \xi\|.$$

Here, C_i are positive constants, and $\psi, \xi \in Q_H$.

Certain stability conditions can be derived by using Liapunov functions. The main idea of this way is due to B. Razumikhin. We will consider some assertions of this kind.

THEOREM 1.4. *Assume that there is a positive definite continuous function* $((t,\psi)\mapsto V(t,\psi))\colon \mathbf{R}\times\mathbf{R}^n\to\mathbf{R}$ *such that* $\dot V(t,\psi)\le 0$ *for any function* $\psi\in Q_H$ *satisfying the inequality* $V(t+\theta,\psi(\theta))<V(t,\psi(0))$, $-h\le\theta<0$. *Then the trivial solution of* (1.2) *is stable.*

For proofs of thms. 1.2–1.4 see [286, 291, 304, 470].

EXAMPLE 1.4. Consider the following scalar equation with continuous coefficients a and b:

$$\dot x(t)=-a(t)x(t)-b(t)x(t-h),\qquad t\ge t_0,\quad h>0. \tag{1.7}$$

To derive stability conditions for (1.7), we introduce a functional $V\colon \mathbf{R}\times C[-h,0]\to\mathbf{R}$ by

$$V(t,\psi)=\psi^2(0)+\int_{-h}^{0}|b(t+\theta+h)|\psi^2(\theta)\,d\theta.$$

To compute $\dot V$ we apply (1.4), putting $x(\tau)=x(t;t,\psi)$ $(\tau\ge t)$. Changing $\tau+\theta=s$ under the integral sign, we obtain

$$\begin{gathered}
\dot V(t,\psi)=\left[\frac{d}{d\tau}V(\theta,x_\tau)\right]_{\theta=t}=\\
\left[\frac{d}{d\tau}\left(x^2(s)+\int_{\tau-h}^{\tau}|b(s+h)|x^2(s)\,ds\right)\right]_{\tau=t}=\\
x^2(t)[-2a(t)+|b(t+h)|]-2b(t)x(t)x(t-h)-|b(t)|x^2(t-h)=\\
=\psi^2(0)[-2a(t)+|b(t+h)|]-2b(t)\psi(0)\psi(-h)-|b(t)|\psi^2(-h)\le\\
\le-\gamma(t)\psi^2(0),\\
\gamma(t):=2a(t)-|b(t)|-|b(t+h)|.
\end{gathered}$$

Therefore the trivial solution of (1.7) is stable if $\gamma(t)\ge 0$, by thm. 1.2, and asymptotically stable if $\sup_{t\ge t_0}|b(t)|<\infty$, $\inf_{t\ge t_0}\gamma(t)>0$, by thm. 1.1. Note that these stability conditions are due to the negative instantaneous feedback $-a(t)x(t)$. The following sufficient condition for asymptotic stability was obtained in [377.3] for the case $a(t)\ge 0$, $b(t)\ge 0$:

$$h\sup_{t\ge t_0}[a(t)+b(t)]<1.5. \qquad\square$$

EXAMPLE 1.5. Consider the problem of the stationary values V_a^0 and V_b^0 of arterial and venous blood pressure. By (1.6.1) and (1.6.2), these are equal to

$$\begin{gathered}
V_a^0=(Q_i+k_5)(k_4k_a)^{-1}=x_0,\\
V_b^0=[Q_i(k_2-k_3x_0)+k_ax_0][k_b(1+k_1(k_2-k_3x_0))]^{-1}=y_0.
\end{gathered}$$

The stability conditions for the stationary solutions have the form

$$\left.\begin{aligned}
-2(k_1k_b+k_4k_a)+\tfrac12\alpha_0k_b+3\alpha_0k_3(k_ax_0-k_by_0)(k_2-k_3x_0)<0,\\
-k_1k_b+\tfrac12\alpha_0k_a+\alpha_0k_3(k_ax_0-k_by_0)(k_2-k_3x_0)^{-1}<0.
\end{aligned}\right\} \tag{1.8}$$

Here $\alpha_0 = (k_2 - k_3\delta)^{-1}$, and $\delta \in (0, k_2/k_3)$ is a constant. To prove this it suffices to use thm. 1.2, applied to the functional [202]

$$V(x_t, y_t) = x_t^2(0) + y_t^0(0) + C_0C_1^{-1}[x_t(0) + y_t(0)]^2 + 2\alpha_0 C_2 \int_{-\tau}^{0} x_t^2(\theta)\, d\theta,$$

where $x = V_a - x_0$, $y = V_b - y_0$, $C_0 = k_1k_b$, $C_1 = k_4k_a$, $C_2 = k_3(k_ax_0 - k_by_0)(k_2 - k_3x_0)^{-1}$.

Let us examine the equations (1.8). We note that the stationary value V_a^0 increases as Q_i and k_5 increase while k_4 decreases. By (1.8) the rate with which any solution tends to a stationary value decreases if k_3, x_0 increase, while k_1, k_2, k_a, k_b decrease. These conclusions have been experimentally confirmed [202]. □

1.3. Stability of equations with unbounded delay. Stability problems for RDEs with unbounded delay can be stated in much the same way as for RDEs with bounded delay.

Let $\mathcal{K}$ be a metric space of continuous functions $\phi\colon (-\infty, 0] \to \mathbf{R}^n$, with metric ρ, satisfying the conditions in subs. 2.2.5. Consider the following Cauchy problem for $x\colon J_x \to \mathbf{R}^n$:

$$\left.\begin{array}{ll} \dot{x}(t) = f(t, x_t), & t \geq t_0, \\ x_t(\theta) = x(t+\theta), & -\infty < \theta \leq 0, \\ x_{t_0} = \phi \in \mathcal{K}. & \end{array}\right\} \tag{1.9}$$

Here, $f\colon \mathbf{R} \times \mathcal{K} \to \mathbf{R}^n$ is a continuous function which is locally Lipschitz in the second argument and $f(t, 0) \equiv 0$.

The definitions of stability of the trivial solution of (1.9) are the same as in the case of equation (1.1) with bounded delay, with $C[-h, 0]$ replaced by $\mathcal{K}$ and $|\psi|$ replaced by $\rho(\psi, 0)$, so that

$$Q_H := \{\psi \in \mathcal{K}:\ \rho(\psi, 0) < H\}.$$

We state some conditions for the stability of the trivial solution of (1.9); these can be proved similar to thm. 1.1.

THEOREM 1.5. *Assume that for some $H > 0$ and $\omega_i \in \Omega$ there is a continuous functional $((t, \psi) \mapsto V(t, \psi))\colon \mathbf{R} \times Q_H \to \mathbf{R}$ such that*

$$\omega_1(|\psi(0)|) \leq V(t, \psi) \leq \omega_2(\rho(\psi, 0)),$$
$$\dot{V}(t, \psi) \leq -\omega_3(|\psi(0)|).$$

Then the trivial solution of (1.9) *is asymptotically stable.*

EXAMPLE 1.6. We will derive stability conditions for the equilibrium $\psi = 0$, $\phi = 0$ ($\psi, \phi\colon \mathbf{R} \to \mathbf{R}^n$) of equation (1.2.1). This equation describes the motion of a rigid body with PID regulator:

$$u = -A_1\phi - A_2\dot{\phi} - \int_0^t g(t-s)\phi(s)\, ds, \qquad A_1, A_2 > 0,$$

where $A_1, A_2 \in \mathbf{R}^{n\times n}$ are symmetric matrices, and the function $g\colon (-\infty, 0] \to \mathbf{R}$ is integrable. In the sequel we put $H = H(\phi(t))$, $\dot{\phi} = \dot{\phi}(t)$, $\dot{\psi} = \dot{\psi}(t)$, $L = L(\phi, \psi)$

($H \in \mathbf{R}^{n\times n}$, $L \in \mathbf{R}$), and also $\phi(s) = 0$, $\psi(s) = 0$ for $s < 0$. Then we can write (1.2.1) as

$$\left.\begin{aligned} \dot\psi &= -L_\phi - A_1\phi - A_2H^{-1}\psi - \textstyle\int_0^\infty g(s)\phi(t-s)\,ds, \\ \dot\phi &= H^{-1}\psi, \qquad t \geq 0, \\ 2L &= \psi' H^{-1}(\phi)\psi, \qquad L_\phi = \tfrac{\partial L}{\partial \phi}. \end{aligned}\right\} \tag{1.10}$$

Put

$$\rho_0 = \int_0^\infty |g(s)|\,ds.$$

Since the operator on the righthand side of (1.10) satisfies the local Lipschitz condition with respect to ϕ, ψ for any $\phi(0)$, $\psi(0)$, there is a unique solution on a certain interval $[0, \delta)$, $\delta > 0$ (see the end of subs. 2.2.4). Let $\mathcal{K}$ be the space $C(-\infty, 0]$ with $\|f\| := \sup_{-\infty<t\leq 0} |f(t)| < \infty$. This space does not satisfy all the conditions of subs. 2.2.5, but this is not essential because of the remarks in subs. 2.5.2, the condition $\rho_0 < \infty$, an the form of (1.10).

To study the possibility of prolongation of a solution and the stability of the trivial solution, we use the functional (where ϵ is a positive constant)

$$\left.\begin{aligned} V_1(v, w; \epsilon) &= V(v(0), w(0); \epsilon) + \tfrac{1+\epsilon}{2} \textstyle\int_0^\infty ds \int_{-s}^0 |g(s)| \cdot |v(\theta)|^2\,d\theta, \\ V(v(0), w(0); \epsilon) &= L(v(0), w(0)) + \tfrac{1}{2}v'(0)A_1w(0) + \epsilon v'(0)w(0), \\ & v, w \in \mathcal{K}. \end{aligned}\right\} \tag{1.11}$$

Note that for sufficiently small $\epsilon > 0$ the function $(\phi, \psi) \mapsto V(\phi, \psi; \epsilon)$ is positive definite in ϕ, ψ and has an infinitesimal upper bound.

First we consider the case $g(s) \equiv 0$. We compute the derivative $\dot V = \dot V(\phi, \psi; \epsilon)$ of the function V along the trajectories of (1.10). We then find

$$\begin{aligned} \dot V &= \dot\phi' L_\phi + \dot\psi' L_\psi + \\ &+ \dot\phi' A_1\phi + \epsilon\dot\psi'\phi + \epsilon\psi'\dot\phi = \\ &= \psi' H^{-1} L_\phi - (L_\phi + A_2H^{-1}\psi + A_1\phi)' H^{-1}\psi + \\ &+ \psi' L^{-1} A_1\phi - \epsilon(L_\phi + A_2H^{-1}\psi + A_1\phi)'\phi + \epsilon\psi' H^{-1}\psi = \\ &= \psi'(H^{-1}A_2H^{-1} - \epsilon H^{-1})\psi - \epsilon\phi' L_\phi - \epsilon\phi' A_1\phi - \epsilon\phi' A_2H^{-1}\psi. \end{aligned}$$

Using the estimates

$$\epsilon\phi' A_2 H^{-1}\psi \leq \frac{1}{2}\epsilon^2\phi' A_2\phi + \frac{1}{2}\psi' H^{-1}A_2H^{-1}\psi,$$

$$\epsilon\phi' L_\phi = -\frac{1}{2}\epsilon\psi' H^{-1}\left(\sum_{i=1}^n \phi_i \frac{\partial H}{\partial \phi_i}\right) H^{-1}\psi$$

we obtain

$$\dot V \leq W,$$

$$W := -\epsilon\phi'\left(A_1 - \frac{1}{2}\epsilon A_2\right)\phi - \psi' H^{-1}\left[\frac{1}{2}A_2 - \epsilon H - \frac{1}{2}\epsilon\sum \phi_i \frac{\partial H}{\partial \phi_i}\right] H^{-1}\psi.$$

Together with (1.11) this implies that the function $(\phi,\psi)\mapsto V(\phi(0),\psi(0);\epsilon)$ satisfies the conditions of the theorem on global asymptotic stability [291].

For an arbitrary scalar function $s\mapsto g(s)$ which is absolutely integrable on $[0,\infty)$ the derivative of the functional (1.11) along the trajectories of (1.10) satisfies the inequality

$$\begin{aligned}\dot V_1 \le W - \psi' H^{-1}\int_0^\infty g(s)\phi(t-s)\,ds + \\ -\epsilon\phi'\int_0^\infty g(s)\phi(t-s)\,ds + \frac{1+\epsilon}{2}\rho_0|\phi(t)|^2 + \\ -\frac{1+\epsilon}{2}\int_0^\infty |g(s)|\cdot|\phi(t-s)|^2\,ds.\end{aligned}$$

Note that

$$\begin{aligned}\epsilon\left|\phi'(t)\int_0^\infty g(s)\phi(t-s)\,ds\right| \le \\ \le \frac{\epsilon}{2}\rho_0|\phi(t)|^2 + \frac{\epsilon}{2}\int_0^\infty |g(s)|\cdot|\phi(t-s)|^2\,ds, \\ \left|\psi'(t)H^{-1}\int_0^\infty g(s)\phi(t-s)\,ds\right| \le \\ \le \frac{\rho_0}{2}\psi'(t)H^{-2}\psi(t) + \frac{1}{2}\int_0^\infty |g(s)|\cdot|\phi(t-s)|^2\,ds.\end{aligned}$$

This implies the inequality

$$\begin{aligned}\dot V_1 \le -\phi'\left[\epsilon A_1 - \frac{1}{2}\epsilon^2 A_2 - \left(\epsilon+\frac{1}{2}\right)\rho_0 I\right]\phi + \\ -\psi' H^{-1}\left[\frac{1}{2}A_2 - \epsilon H - \frac{1}{2}\epsilon\sum \phi_i\frac{\partial H}{\partial \phi_i} - \frac{1}{2}\rho_0 I\right]H^{-1}\psi,\end{aligned} \tag{1.12}$$

where I is the identity matrix.

Let D_0 be a bounded domain in the (ϕ,ψ)-space. We show that we can choose the function g such that the equilibrium position of (1.10) will be asymptotically stable, with domain of attraction containing D_0. Let $M=\sup_{(\phi,\psi)\in D_0} V(\phi,\psi)$. There is a ball D outside which the inequality $V(\phi,\psi)>M$ is valid for all sufficiently small $\epsilon>0$. By the choice of g we can make ρ_0 arbitrarily small. Let ϵ and ρ_0 be so small that the matrices between square brackets in (1.11) are positive definite. Then $\dot V_1$ is negative definite in D. We verify that the trajectories of (1.6) starting in D_0 do not leave D. In fact, for $t=0$ we have $V_1=V\le M$. Further, the functional V_1 does not increase in D, and on the boundary the inequality $V_1\ge V\ge M$ holds. This and (1.11) imply the asymptotic stability of the trivial solution, with domain of attraction containing D_0. □

EXAMPLE 1.7. Suppose we are given the scalar equation

$$\dot x(t) = -a(t)x(t) + b(t)\int_0^\infty \alpha(s)x(t-s)\,ds, \qquad t\ge 0, \tag{1.13}$$

with functions $a, b, \alpha \in L_1[0,\infty)$ and b, α integrable on $[0,\infty)$. Let $\mathcal{K}$ be the space o bounded uniformly continuous functions on $(-\infty, 0]$, with norm $\|\phi\| = \sup_{-\infty<t\leq 0} |\phi(t$ To obtain a stability condition we consider the functional

$$V(t,\psi) = \psi^2(0) + \int_0^\infty |\alpha(s)| \left[\int_{t-s}^t |b(\tau+s)|\psi^2(\tau - t)\, d\tau\right] ds.$$

Using the procedure of subs. 1.2 (which can be used for unbounded delay as well), we find

$$V(t,x_t) = x^2(t) + \int_0^\infty |\alpha(s)| \left[\int_{t-s}^t |b(\tau+s)| x^2(\tau)\, d\tau\right] ds,$$

$$\begin{aligned}\dot V &= 2x(t)\dot x(t) + \int_0^\infty |\alpha(s)|[|b(t+s)|x^2(t) - |b(t)|x^2(t-s)]\, ds = \\ &= 2x(t)\left[-a(t)x(t) + b(t)\int_0^\infty \alpha(s)x(t-s)\, ds\right] + \\ &\quad + \int_0^\infty |\alpha(s)|[|b(t+s)|x^2(t) - |b(t)|x^2(t-s)]\, ds \leq \\ &\leq -2a(s)x_t^2(0) + \int_0^\infty |\alpha(s)|[|b(s)| + |b(t+s)|]\, ds \cdot x_t^2(0) + \\ &\quad -|b(t)|\int_0^\infty |\alpha(s)|[x_t(0) - x_t(-s)]^2\, ds.\end{aligned}$$

Replacing x_t by ψ we obtain an estimate for $\dot V(t,\psi)$. One can easily see that the trivial solution of (1.13) is stable (hence all its solutions are stable), provided that

$$2a(t) \geq \int_0^\infty |\alpha(s)|[|b(s)| + |b(t+s)|]\, ds, \qquad 0 \leq t < \infty. \qquad \square$$

Note that in this example the function $t \mapsto \dot V(t,x_t)$ is, for a given initial condition, only defined for almost all $t \in [0,\infty)$, in general. However, this does not impair the use of the stability theorem, since $t \mapsto V(t,x_t)$ is absolutely continuous.

1.4. Stability of linear nonautonomous equations. Here we will consider the Cauchy problem for the linear equation

$$\dot x(t) = \int_{-\infty}^0 [d_\theta R(t,\theta)]x(t+\theta), \qquad t \geq t_0, \qquad x(t) \in \mathbf{R}^n, \tag{1.14}$$

$$x(t_0+\theta) = \phi(t_0+\theta), \qquad -\infty < \theta \leq 0. \tag{1.15}$$

We assume that the matrix-kernel R satisfies the condition of subs. 2.5.2 related to (2.5.8). For the space $\mathcal{K}$ of initial functions ϕ we take $C(-\infty, t_0]$, with norm $\|\phi\| := \sup |\phi(t)| < \infty$.

It is clear that if the solution of (1.14) is (asymptotically) stable for some initial function, then the same holds for any initial function, and the same is true for the inhomogeneous equation (2.5.8). Therefore, for linear equations it is possible to talk of the stability of not only an individual solution, but also of the (*asymptotic*) *stability of the equation* itself.

The following simple, but useful result is due to A. Zverkin.

THEOREM 1.6. *Equation* (1.14) *is stable if and only if for every initial function* $\phi \in \mathcal{K}$ *the corresponding solution* $x(t; t_0, \phi)$ *is bounded.*

PROOF. Necessity is obvious.

To prove sufficiency we define for each $t \geq t_0$ a functional U_t on the Banach space $\mathcal{K}$ by $U_t\phi = x(t; t_0, \phi)$. Since the solution depends continuously on the initial data (cf. the end of subs. 2.5.2), this U_t is continuous.

The assumptions of thm. 1.6 imply that also $\sup_{t \geq t_0} |U_t\phi| < \infty$ for all $\phi \in \mathcal{K}$. Using the Banach–Steinhaus theorem we find that the norms of the functionals U_t are bounded in totality. This implies stability. □

COROLLARY 1.1. *If* $x(t; t_0, \phi) \to 0$, $t \to \infty$, *for any initial function* $\phi \in \mathcal{K}$, *then* (2.1) *is asymptotically stable.*

1.5. Stability of linear periodic differential equations. Here we consider the problem (1.2), where $f(t, \psi)$ is, in addition, linear in ψ and periodic in t, with period $\omega > 0$. We introduce the *monodromy operator* (also called *shift operator*) $U(t)\colon C[-h, 0] \to C[-h, 0]$ by $U(t)\phi = x_{t+\omega}(t + \omega; t, \phi)$. Here, $C[-h, 0]$ must be regarded as a space of complex-valued vector functions.

THEOREM 1.7. ([238]). *The problem* (1.2) *with functional* f *linear in* x_t *and periodic in* t *is:*

1) *asymptotically stable if and only if the spectrum of the operator* $U(t)$ *lies inside the unit circle;*
2) *stable if the eigenvalues* ρ_i *of* $U(t)$ *are such that* $|\rho_i| \leq 1$, *while the eigenvalues* ρ_j *with* $|\rho_j| = 1$ *have simple elementary divisors.*

In addition, the spectrum of $U(t)$ does not depend on t.

1.6. Application of comparison theorems. In certain cases we can obtain stability conditions by comparing the solutions of the initial problem with solutions of auxiliary problems. We give an example of this using the linear equation

$$\dot{x}(t) = A(t)x(t) + \int_{-\infty}^{0} [d_\theta R(t, \theta)]x(t + \theta), \qquad t \geq t_0. \tag{1.16}$$

Here, $A = (a_{ij})$ is a continuous matrix function with $a_{ij} \geq 0$ for $i \neq j$. The kernel $R(t, \theta)$ and initial function ϕ satisfy the conditions A and B of subs. 2.5.2.

Along with (1.16) we consider the inequality

$$\dot{z}(t) \geq A(t)z(t) + \int_{-\infty}^{0} [d_\theta V(t, \theta)]z(t + \theta), \qquad t \geq t_0 \qquad z(t) \in \mathbf{R}^n \tag{1.17}$$

(we write $u \geq v$ for $u, v \in \mathbf{R}^n$ if $u_1 \geq v_1, \dots, u_n \geq v_n$). Here, the entry $v_{ij}(t, \theta)$ of the matrix V is equal to the variation of the function $r_{ij}(t, \cdot)$ on $[\theta, 0]$.

THEOREM 1.8. *If there is a function* $z \in C(\mathbf{R}, \mathbf{R}^n)$ *such that* $z|_{[t_0,\infty)} \in C^1[t_0, \infty)$, $z(t) \geq 0$ $(t \in \mathbf{R})$, $z_i(t) \geq |\phi_i(t)|$ $(t \leq t_0)$, $z_i(t) > 0$ $(t \geq t_0)$, $i = 1, \dots, n$, *and satisfying* (1.17) *for all* $t \geq t_0$, *then* $|x_i(t)| \leq z_i(t)$ $(t \geq t_0)$, $i = 1, \dots, n$, *where* x *is the solution of* (1.16)–(1.15).

PROOF. Let $\epsilon > 0$ be arbitrary. Let I be the identity matrix, and x^ϵ the solutior of the equation

$$\dot{x}^\epsilon(t) = [A(t) - \epsilon I]x^\epsilon(t) + \int_{-\infty}^{0} [d_\theta R(t,\theta)]x^\epsilon(t+\theta), \qquad t \geq t_0,$$

with initial condition $x^\epsilon(t_0+\theta) = (1-\epsilon)\phi(t_0+\theta)$ $(-\infty < \theta \leq 0)$, where $\epsilon \in (0,2)$ We check that $|x^\epsilon(t)| < z(t)$, $t \geq t_0$.

This assertion holds for $t = t_0$. Suppose, on the contrary, that it does not hold fo all $t > t_0$, and let $\tau > t_0$ be the first moment such that $|x^\epsilon(\tau)| = z_i(\tau)$ for some i Without loss of generality we may assume that $x^\epsilon(\tau) > 0$. Then

$$\dot{x}^\epsilon(\tau) - \dot{z}_i(\tau) \leq$$
$$\leq -\epsilon x^\epsilon(\tau) + \sum_{j=1}^{n}\left[\int_{-\infty}^{0} x_j^\epsilon(\tau+\theta)d_\theta r_{ij}(\tau,\theta) - \int_{-\infty}^{0} z_j(\tau+\theta)|d_\theta r_{ij}(\tau,\theta)|\right] \leq$$
$$\leq -\epsilon x_i^\epsilon(\tau) < 0.$$

this contradicts the definition of the moment τ. Now thm. 2.2.3 implies $x(t) = \lim x^\epsilon(t)$, $\epsilon \to 0^+$. Hence $|x_i(t)| \leq z_i(t)$, $i = 1,\dots,n$. □

EXAMPLE 1.8. A sufficient condition for stability of (1.16) with $A \in C([t_0,\infty),\mathbf{R}^{n\times}$ $a_{ij}(t) \geq 0$ $(i \neq j)$ is that

$$a_{ii}(t) \leq -\sum_{j\neq i} a_{ij}(t) - \sum_{j=1}^{n} V_{(-\infty,0]}r_{ij}(t,\cdot), \qquad i = 1,\dots,n. \tag{1.18}$$

(It suffices to note that under the conditions (1.18) the assumptions of thm. 1.8 holc for the function $z(t) = c > 0$.) □

1.7. Stability in the first approximation. Here we consider the equation (1.1 with righthand side disturbed by a functional $G(t,x_t)$:

$$\dot{x}(t) = f(t,x_t) + G(t,x_t), \qquad t \geq t_0. \tag{1.19}$$

Here f is the same as in (1.1), and $G \in C([t_0,\infty)\times C[-h,0],\mathbf{R}^n)$ is such that

$$|G(t,\psi)| \leq \beta(\psi)\|\psi\|. \tag{1.20}$$

THEOREM 1.9. ([291]). *Assume that $\beta(\psi) \to 0$ as $\|\psi\| \to 0$ in (1.20), and that th trivial solution of (1.1) is asymptotically stable. Then the trivial solution of (1.19) i asymptotically stable.*

EXAMPLE 1.9. We investigate the stability of the nonzero stationary solution $x_0 \equiv K$ of the logistic equation (1.5.1):

$$\dot{x}(t) = \gamma[1 - K^{-1}x(t-h)]x(t), \qquad \gamma > 0, \quad K > 0, \quad h \geq 0. \tag{1.21}$$

We define a new variable $y(t)$ by

$$x(t) = K[1 + y(t)].$$

For $y(t)$ we obtain

$$\dot{y}(t) = -\gamma y(t-h) - \gamma y(t)y(t-h).$$

We will study the first approximation $\dot{y}(t) = -\gamma y(t-h)$ in subs. 3.2.1 (see fig. 3.2.1). There we will show that this equation is exponentially stable, if only $2\gamma h < \pi$. By thm. 1.9 this implies that the stationary solution $x_0 \equiv K$ of (1.21) is asymptotically stable if $\gamma h < \pi/2$, A more refined analysis [260] has shown that the solution $x_0 \equiv K$ is globally asymptotically stable under all admissible disturbances of the initial position satisfying $x(0) > 0$, if only $\gamma h < 37/24$. For $\gamma h > \pi/2$ there is a nonconstant periodic solution of (1.21). □

1.8. L_2-stability

DEFINITION 1.1. The trivial solution of (1.1) is called *L_2-stable* if there is a constant $H > 0$ such that for any initial function $\phi \in C[-h, 0]$ with $\|\phi\| \le H$ we have

$$\int_{t_0}^{\infty} |x(t; t_0, \phi)|^2 \, dt < \infty. \qquad \square$$

Certain conditions for L_2-stability can be obtained by using Liapunov's direct method (see, e.g., [282, 286.I]). Here we will consider another condition for L_2-stability, derived in [199] for equations with bounded delay h:

$$\left.\begin{aligned} \dot{x}(t) &= \textstyle\int_{-h}^{0} [dR(\theta)] x(t+\theta) + G(t, x_t), \qquad t \ge t_0, \\ x_{t_0} &= \phi \in C[-h, 0]. \end{aligned}\right\} \tag{1.22}$$

Here, G is as in (1.19), and the entries of R are functions of bounded variation on $[-h, 0]$. We also assume that for any continuous square-integrable function ψ: $[t_0 - h, \infty) \to \mathbf{R}^n$ the following relation holds:

$$\left[\int_{t_0}^{\infty} |G(t, \psi_t)|^2 \, dt\right]^{1/2} \le q \left[\int_{t_0}^{\infty} |\psi(t)|^2 \, dt\right]^{1/2} + \alpha(\psi_{t_0}). \tag{1.23}$$

Here, $q = \text{const} > 0$ and α: $C[-h, 0] \to \mathbf{R}^n$ is a bounded continuous functional. Let z be a complex number, and let $A(z)$ be the $(n \times n)$-matrix

$$A(z) = zI - \int_{-h}^{0} e^{z\theta} \, dR(\theta),$$

where I is the $(n \times n)$ identity matrix. Let $\lambda_j(A(z))$ be the eigenvalues of $A(z)$, $j = 1, \dots, n$, and let $g(A)$ be defined as

$$g(A) = \left[\operatorname{Tr}(AA^*) - \sum_{j=1}^{n} |\lambda_j(A)|^2\right]^{1/2}.$$

Here, A^* is the adjoint matrix and Tr denotes trace. Finally we introduce the quantity

$$q_1(\omega) = \sum_{l=0}^{n-1} [g(A(i\omega))]^l (l!)^{-1/2} [d(i\omega)]^{-l-1},$$

$$i^2 = -1, \qquad \omega \in \mathbf{R}, \qquad d(z) = \min_{j=1,\dots,n} |\lambda_j(A(z))|.$$

THEOREM 1.10. ([199]). *Assume that* (1.23) *holds, that* $\det A(z)$ *is zero-free in th halfplane* $\operatorname{re} z \geq 0$, *and also that*

$$q^{-1} > \sup_{\omega \in \mathbf{R}} q_1(\omega).$$

Then the trivial solution of (1.20) *is* L_2*-stable.*

REMARK 1.3. Liapunov's direct method is used not only in stability theory, bu also in other areas, such as solvability theorems, existence of stationary and periodi measures induced by a solution, asymptotic properties of solutions, etc. [285]. In thes cases the properties of Liapunov functionals and the manners of their applicatio differs essentially from those described above.

E.g., in [98, 346] a Liapunov functional with discrete values is introduced for study ing the exponential decay rate of the solution x of the scalar RDE

$$\dot{x}(t) = f(t, x(t), x(t-1)), \qquad t \geq 0, \qquad f(t, 0, 0) \equiv 0.$$

The exponential decay rate of x is defined as

$$\alpha(x) = \inf \left\{ \alpha \in \mathbf{R} : \lim_{t \to \infty} e^{-\alpha t} x(t) = 0 \right\}.$$

The solution x is called a *superexponential solution* if $\alpha(x) = -\infty$ and $\|x|_{[t,\infty)}\| >$ for all $t \in [0, \infty)$.

In this case the Liapunov functional V is defined as the number of zeros of x i some unit time interval. It has been proved [98] that a solution x is superexponentia if and only if V tends to infinity as $t \to \infty$. □

2. Linear autonomous equations

2.1. General stability conditions. In this section we derive stability condition for equations

$$\left.\begin{aligned} \dot{x}(t) &= \textstyle\int_{-h}^{0} [dR(\theta)] x(t+\theta), \qquad x(t) \in \mathbf{R}^n, \qquad t \geq 0, \\ x(\theta) &= \phi(\theta), \qquad -\infty < \theta \leq 0. \end{aligned}\right\} \tag{2.1}$$

We assume that the matrix entries r_{ij} of R are functions of bounded variation, and

$$\int_{-\infty}^{0} |\theta| \cdot |dr_{ij}(\theta)| < \infty, \qquad i, j = 1, \ldots, n. \tag{2.2}$$

The initial functions are chosen to belong to the linear space of continuous function ϕ with finite norm $\|\phi\|_0$ given by

$$\|\phi\|_0 = |\phi(0)| + \left(\int_0^{\infty} |F_\phi(t)|^2 \, dt \right)^{1/2},$$

where

$$F_\phi(t) := \int_{-\infty}^{-t} [dR(\theta)] \phi(t+\theta).$$

The definitions of stability for (2.1) coincide with the definitions A–D in subs. 1.1 and 1.3, with $\rho(\phi, 0)$ replaced by $\|\phi\|_0$. Let $\tilde{x}(z)$, $\tilde{F}_\phi(z)$ and $\tilde{R}(z)$ be the Laplace transforms of the functions x, F_ϕ and R:

$$\tilde{x}(z) = \int_0^\infty e^{-zt} x(t)\, dt,$$
$$\tilde{F}_\phi(z) = \int_0^\infty e^{-zt} F_\phi(t)\, dt,$$
$$\tilde{R}(z) = \int_{-\infty}^0 e^{z\theta}\, dR(\theta).$$

By (2.1),

$$[zI - \tilde{R}(z)]\tilde{x}(z) = \phi(0) + \tilde{F}_\phi(z),$$

with I the identity matrix.

The characteristic function $\Delta(z)$ of (2.1) is equal to

$$\Delta(z) := \det[zI - \tilde{R}(z)]. \tag{2.3}$$

REMARK 2.1. Usually, for retarded systems the equation $\Delta(z) = 0$ has infinitely many roots. However, there do exist RDEs (2.1) (called *degenerate RDEs*) whose characteristic function has only finitely many roots (see subs. 2.6.1). It can be proved that in this case the function Δ is a polynomial of degree n. E.g., consider the equation $\dot{x}(t) = Ax(t) + Bx(t-h)$, $n = 3$. If

$$A = \begin{pmatrix} 0 & 1 & 0 \\ -0.5 & 0 & 1 \\ 0 & -0.5 & 0 \end{pmatrix}, \qquad B = \begin{pmatrix} 0 & 1 & 0 \\ 0 & 0 & -1 \\ 0 & 0 & 0 \end{pmatrix},$$

then $\Delta(z) = z(z^2 + 1)$. If

$$A = \begin{pmatrix} 0 & 1 & 0 \\ 0 & 0 & 0 \\ 0 & 0 & 1 \end{pmatrix}, \qquad B = \begin{pmatrix} -e & -e & -e^2 \\ 0 & e & e^2 \\ 1 & 0 & 0 \end{pmatrix},$$

then $\Delta(z) = z(z^2 - 1)$.

If the RDE is degenerate, then for sufficiently large t its solution space is finite-dimensional, and we can give an equivalent ODE for each component of a solution. E.g., for the first example above such equations have the form

$$\ddot{x}_1 + x_1 = 0, \qquad \ddot{x}_2 + x_2 = 0, \qquad \dddot{x}_3 + \dot{x}_3 = 0. \qquad \square$$

In the case of a finite integration interval in (2.1), the following theorem would follow immediately from subs. 2.6.2.

THEOREM 2.1. *Suppose the characteristic function* $\Delta(z)$ *is zero-free in the halfplane* $\operatorname{re} z \geq 0$, *and let* (2.2) *hold. Then* (2.1) *is asymptotically stable. Moreover,* $|x(t, \phi)| \leq c_\phi \exp(-\alpha t)$, $\alpha > 0$, *if*

$$\int_{-\infty}^0 e^{-\alpha\theta} |dr_{ij}(\theta)| < \infty$$

and the equation $\Delta(z) = 0$ *has no roots in the halfplane* $\operatorname{re} z \geq -\alpha$.

Note that the roots of (2.3) are such that each halfplane $\operatorname{re} z \geq \text{const}$ contains only finitely many of them.

Many stability criteria for linear autonomous equations can be derived by methods of complex function theory (e.g., from Rouché's theorem, the argument principle, the residue theorem, etc.). Below we consider some of these.

Let C_R be a contour consisting of the interval $[-iR, iR]$ and the halfcircle $z = R\exp(i\psi)$, $-\pi/2 \leq \psi \leq \pi/2$, $R > 0$, $i^2 = -1$. Let L be the limit of the images of the contours C_R as $R \to \infty$ under the map $z \mapsto \Delta(z)$. The argument principle implies:

THEOREM 2.2. *Suppose L does not encircle the point $z = 0$. Then* (2.1) *is asymptotically stable.*

Put $z = i\omega$, and represent (2.3) as

$$\Delta(i\omega) = u(\omega) + iv(\omega).$$

The graph of the function $\omega \mapsto \Delta(i\omega)$ in the complex plane is usually called the *Mikhailov hodograph.*

THEOREM 2.3. *Equation* (2.1) *is asymptotically stable if and only if $|\Delta(i\omega)|^2$ $(= u^2(\omega) + v^2(\omega)) > 0$ $(\omega \in \mathbf{R})$, while the variation of* $\arg \Delta(i\omega)$ *when ω varies from 0 to ∞ is equal to $n\pi/2$ ($n = \dim(2.1)$). The last condition is equivalent to the equality*

$$\int_{\omega=0}^{\infty} \frac{u(\omega)\, dv(\omega) - v(\omega)\, du(\omega)}{u^2(\omega) + v^2(\omega)} = \frac{n\pi}{2}.$$

For proofs of thms. 2.1–3 see [286, 303, 304, 470]. In theses references one can also find various methods for a detailed study of the roots of the equation $\Delta(z) = 0$.

2.2. Scalar nth order equations. Here we consider the problem of the stability of a scalar linear equation

$$\left.\begin{aligned} x^{(n)}(t) = \textstyle\sum_{j=0}^{n-1} \int_{-\infty}^{0} x^{(j)}(t+\theta)\, dr_j(\theta), \qquad t \geq 0, \\ x^{(j)}(t) := \tfrac{d^j x}{dt^j}, \qquad x(t) \in \mathbf{R}. \end{aligned}\right\} \tag{2.4}$$

Put

$$\alpha_{lj} = \int_{-\infty}^{0} |\theta|^l |dr_j(\theta)| (\leq \infty),$$

$$\beta_{lj} = \int_{-\infty}^{0} |\theta|^l \, dr_j(\theta) \quad (\text{if } \alpha_{lj} < \infty).$$

Since (2.4) can be reduced to a system of the form (2.1), here we will only consider the problem of when the assumptions of thm. 2.1 are valid for this system. According to (2.3), the characteristic function $\Delta_n(z)$ of (2.4) has the form

$$\Delta_n(z) = z^n - \sum_{j=0}^{n-1} z^j \int_{-\infty}^{0} e^{z\theta}\, dr_j(\theta).$$

The zeros of $\Delta_n(z)$ with $\operatorname{re} z \geq 0$ will be called *p-zeros.*

Case $n = 1$. (a) We prove that if $\beta_{00} < 0$, $\alpha_{10} < 1$, then $\Delta_1(z)$ is p-zero-free. Let $z = \alpha + i\beta$, and write $\Delta_1(z) = 0$ as

$$\operatorname{im} \Delta_1(z) = \beta - \int_{-\infty}^{0} e^{\alpha\theta} \sin \beta\theta \, dr_0(\theta) = 0, \tag{2.5}$$

$$\operatorname{re} \Delta_1(z) = \alpha - \int_{-\infty}^{0} e^{\alpha\theta} \cos \beta\theta \, dr_0(\theta) = 0. \tag{2.6}$$

If $\beta \neq 0$, then (2.5) implies

$$1 - \int_{-\infty}^{0} e^{\alpha\theta} \frac{\sin \beta\theta}{\beta\theta} \theta \, dr_0(\theta) = 0.$$

This equation is root-free for $\alpha \geq 0$, since its lefthand side is positive for $\alpha \geq 0$. However, for $\beta = 0$, $\alpha \geq 0$ equation (2.6) is root-free because the function

$$\Phi(\alpha) := \alpha - \int_{-\infty}^{0} e^{\alpha\theta} \, dr_0(\theta) = \operatorname{re} \Delta_1(z)|_{\beta=0}$$

satisfies the estimates

$$\Phi(0) = -\beta_{00} > 0,$$

$$\frac{d\Phi(\alpha)}{d\alpha} = 1 - \int_{-\infty}^{0} \theta e^{\alpha\theta} \, dr_0(\theta) \geq 1 - \alpha_{10} > 0.$$

Hence, under the conditions $\beta_{00} < 0$, $\alpha_{10} < 1$ the equations (2.5) and (2.6) cannot be simultaneously satisfied in the domain $\operatorname{re} z \geq 0$. □

(b) Suppose r_0 is nonincreasing, $r_0(\theta) \equiv \text{const}$ for $\theta < -h \leq 0$ and $\beta_{00} < 0$, $h\alpha_{00} < \pi/2$. Then the function $\Delta_1(z)$ is p-zero-free.

PROOF. Put $z = \alpha + i\beta$, and write the equation $\Delta_1(z) = 0$ in the equivalent form (2.5), (2.6). We show that these equations cannot be simultaneously satisfied. If $|\beta| < \pi/(2h)$, $\alpha \geq 0$, then (2.6) has no roots, since

$$\alpha - \int_{-\infty}^{0} e^{\alpha\theta} \cos \beta\theta \, dr_0(\theta) \geq - \int_{-h}^{0} e^{\alpha\theta} \cos \beta\theta \, dr_0(\theta) \geq$$
$$\geq -\beta_{00} \cos \beta h > 0.$$

If $|\beta| \geq \pi/(2h)$, then (2.5) has no roots, since

$$\left| \int_{-\infty}^{0} e^{\alpha\theta} \sin \beta\theta \, dr_0(\theta) \right| \leq \left| \int_{-h}^{0} dr_0(\theta) \right| < \frac{\pi}{2h}.$$

Hence (2.5) and (2.6) are incompatible. Note that the bound $h\alpha_{00} < \pi/2$ is sharp (see subs. 2.6.3). □

Case $n = 2$. (a) If the function r_1 has a jump $a_0 < 0$ at the point $\theta = 0$, and $\alpha_{11} < 2$, $\beta_{00} < 0$,

$$\gamma := a_0 + \alpha_{10} + \int_{-\infty}^{0} |dr_1(\theta)| < 0,$$

then Δ_2 is p-zero-free.

PROOF. Put $z = \alpha + i\beta$. If $\beta = 0$, $\alpha \geq 0$, then

$$\operatorname{re}\Delta_2(z)(=\Delta_2(z)) = \alpha^2 - \alpha\int_{-\infty}^{0} e^{\alpha\theta}\,dr_1(\theta) - \int_{-\infty}^{0} e^{\alpha\theta}\,dr_0(\theta) =$$

$$= \alpha^2 - \alpha a_0 - \alpha\int_{-\infty}^{0-} e^{\alpha\theta}\,dr_1(\theta) - \int_{-\infty}^{0}(e^{\alpha\theta}-1)\,dr_0(\theta) - \int_{-\infty}^{0} dr_0(\theta) \geq$$

$$\geq \alpha^2 - \alpha\left(a_0 + \int_{-\infty}^{0-}|dr_1(\theta)| + \alpha_{10}\right) - \beta_{00} > 0.$$

On the other hand, for $\beta \neq 0$, $\alpha \geq 0$,

$$\frac{\operatorname{im}\Delta_2(z)}{\beta} =$$

$$= 2\alpha - \alpha\int_{-\infty}^{0} e^{\alpha\theta}\frac{\sin\beta\theta}{\beta}\,dr_1(\theta) - \int_{-\infty}^{0}(\cos\beta\theta)e^{\alpha\theta}\,dr_1(\theta) - \int_{-\infty}^{0} e^{\alpha\theta}\frac{\sin\beta\theta}{\beta}\,dr_0(\theta) \geq$$

$$\geq \alpha\left(2 - \int_{-\infty}^{0}|\theta|\cdot|dr_1(\theta)|\right) - \gamma > 0.$$

So the equations $\operatorname{re}\Delta_2(z) = 0$ and $\operatorname{im}\Delta_2(z) = 0$ are incompatible. □

(b) Assume that

$$\beta_{00} < 0, \qquad 2\alpha_{11} + \alpha_{20} < 2, \qquad \beta_{10} + \beta_{01} < 0. \tag{2.7}$$

Then the function $\Delta_2(z)$ is p-zero-free.

PROOF. Put $\Delta_2(z) = \psi_1(z) + \psi_2(z)$, where

$$\psi_1(z) = z^2 - z(\beta_{01} + \beta_{10}) - \beta_{00},$$

$$\psi_2(z) = -z\int_{-\infty}^{0}(e^{z\theta}-1)\,dr_1(\theta) - \int_{-\infty}^{0}(e^{z\theta} - 1 - z\theta)\,dr_0(\theta).$$

Under the conditions (2.7) the functions ψ_1 is p-zero-free, and

$$|\psi_2(z)| \leq |z|^2\left(\alpha_{11} + \frac{1}{2}\alpha_{20}\right).$$

Consequently, on the boundary of the domain $\operatorname{re} z \geq 0$ the inequality $|\psi_1(z)| > |\psi_2(z)|$ holds. Rouché's theorem now implies that the conditions (2.7) are sufficient for Δ_2 to be p-zero-free. □

2.3. Equations with discrete delays. Here we will consider equation (2.1) with a piecewise constant kernel $R(\theta)$ which may have jumps at finitely many points. In this case the characteristic function (2.3), now also called a quasipolynomial, can be represented as

$$\Delta(z) = z^n + \sum_{l=0}^{n-1}\sum_{j=1}^{m(l)} a_{lj}z^l e^{-zh_{lj}}, \qquad h_{lj} \geq 0. \tag{2.8}$$

Numerous studies have been devoted to the investigation of the zeros of the quasipolynomial (2.8) [47, 282, 286.2, 339].

As has been pointed out in subs. 2.6.1, all zeros of the function (2.8) ly in the halfplane $\operatorname{re} z \leq \alpha$, for a certain α. In general, only finitely many zeros, located near the origin, can be found either approximately or numerically (these are called *nonasymptotic zeros*). The *asymptotic zeros* of the function (2.8) are such that $\operatorname{re} z_k \to -\infty$ as $k \to \infty$, and are distributed along a finite number of chains. Any sector containing the imaginary axis also contains all zeros of (2.8), except finitely many.

Note that the distance between two asymptotic roots of the equation $\Delta(z) = 0$ with a single delay is always larger than a positive number. In general, however, the distance between two zeros of (2.8) may become arbitrarily small.

We will describe some methods for determining conditions under which (2.8) is p-zero-free.

The method of D-subdivision. This method has already been described in subs. 2.6.1. Recall that it requires one to consider in the parameter (a_{lj}, h_{lj})-space of the function (2.8) all hypersurfaces whose points correspond to a quasipolynomial with at least one pure imaginary root. This procedure was called D-subdivision. Now assume that the parameters a_{lj}, h_{lj} vary. Since the zeros of (2.8) depend continuously on the parameters, the number of p-zeros can only change when certain zeros cross the imaginary axis. Hence, each domain in the D-subdivision corresponds to quasipolynomials with the same number of p-zeros, counted with their multiplicity. In particular, the domain without p-zeros is the stability region.

We will consider some concrete examples.

1) Let us find the stability domain for the scalar equation

$$\dot{x}(t) + ax(t) + bx(t-h) = 0.$$

The characteristic quasipolynomial of this equation is

$$\Delta(z) = z + a + be^{-hz}.$$

The equality $\Delta(i\omega) = 0$ implies that for fixed $h > 0$ the boundaries of the domains in a D-subdivision of the (a, b)-plane are formed by the straight line $a + b = 0$ and the parametric curve $a = -\omega \cot h\omega$, $b = \omega/\sin(h\omega)$, where $\omega \in \mathbf{R}$, which has infinitely many branches (see Fig. 3.2.1, in which one branch is depicted and the domain of asymptotic stability is shaded; in Figs. 3.2.1–4 the number p indicates the quantity of p-zeros in the corresponding domain of the D-subdivision).

For given coefficients a and b, (2.9) is unstable if $a+b < 0$ and stable if $a+b > 0$, $a > b$. Suppose $a + b > 0$, $b > a$, i.e. $b > |a|$. Put

$$h_0 := q^{-1} \arccos -\frac{a}{b}, \qquad q := (b^2 - a^2)^{1/2},$$

where $\arccos(-a/b) \in (0, \pi)$. Then, for $h < h_0$ all roots of the equation $\Delta(z) = 0$ have negative real parts; for $h = h_0$ this equation has two purely imaginary roots, $\pm iq$; and the equation acquires two additional roots with positive real parts if h passes through the values $h_0 + 2\pi j$, $j = 0, 1, \ldots$, in increasing direction.

Note that by replacing t by ht we can take $h = 1$ in (2.9).

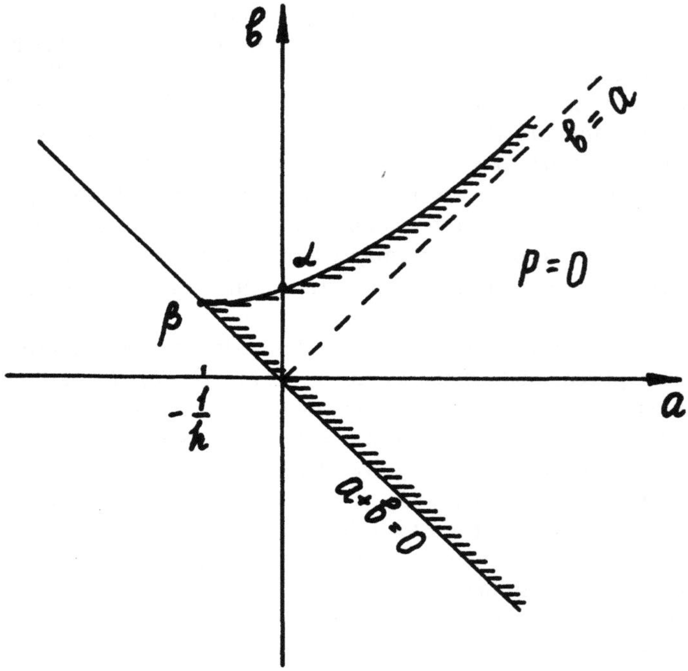

FIGURE 3.2.1. $\beta = \left(-\frac{1}{h}, \frac{1}{h}\right)$, $\alpha = \left(0, \frac{1}{2h}\right)$. Stability domain for the equation $\dot{x}(t) + ax(t) + bx(t-h) = 0$

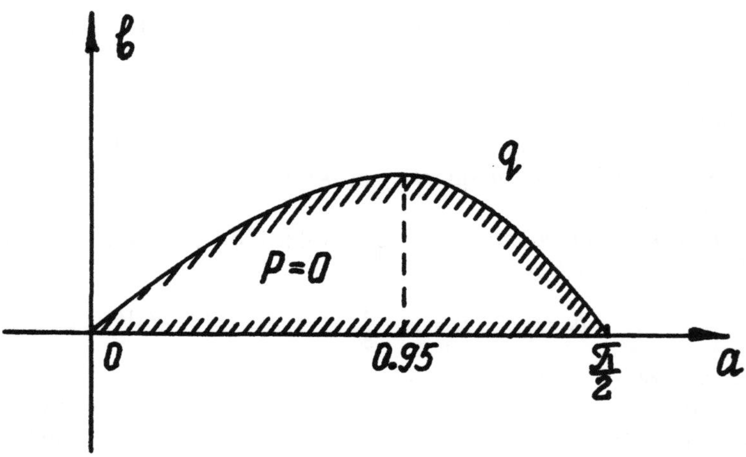

FIGURE 3.2.2. Stability domain for the equation $\ddot{x}(t) + a\dot{x}(t-1) + bx(t-1) = 0$

2) The stability domain of the equation

$$\ddot{x}(t) + a\dot{x}(t-1) + bx(t-1) = 0$$

is depicted in Fig. 3.2.2, in which the upper bound (the curve q) is given by the relations $a = \omega \sin \omega$, $b = \omega^2 \cos \omega$, $0 \le \omega \le \pi/2$. Note that the slope of the curve q (i.e. db/da) is equal to 1 for $a = 0$ and to $-(\pi/2)^2$ for $a = \pi/2$.

3) The stability domain of the equation

$$\ddot{x}(t) + a\dot{x}(t-1) + bx(t) = 0$$

is depicted in Fig. 3.2.3. It consists of a sequence of sides of triangles, formed by parts of the a- and the b-axis and the line segments

$$b = (-1)^{l+1}\frac{2l+1}{2}\pi a + \left(\frac{2l+1}{2}\pi\right)^2, \qquad l = 0, 1, \ldots.$$

The points β_l have coordinates

$$\beta_l = \left((-1)^{l+1}\pi, \frac{(2l+1)(2l+3)}{4}\pi^2\right), \qquad l = 0, 1, \ldots.$$

4) The stability domain of the equation

$$\ddot{x}(t) + a\dot{x}(t) + bx(t-1) = 0$$

is depicted in Fig. 3.2.4. The boundary curve q is given by the parametric equations

$$a = \omega \tan \omega, \qquad b = \frac{\omega^2}{\cos \omega}, \qquad 0 \le \omega < \infty.$$

The curve q is downwards convex, has slope $db/da = 1$ for $a = 0$, and tends to the straight line with slope $\pi/2$ as $\omega \to \infty$ (the dashed line in Fig. 3.2.4).

5) The equation

$$\ddot{x}(t) + 2\dot{x}(t-1) + x(t-2) + \alpha x(t) = 0$$

is stable for $-1 < \alpha < \alpha_0$, $\alpha_{2l-1} < \alpha < \alpha_{2l}$, $l = 1, 2, \ldots$, where

$$\alpha_j = \left(\frac{\pi}{2} + j\pi - (-1)^j\right)^2, \qquad j = 0, 1, \ldots.$$

Generalization of the Routh–Hurwitz conditions. Chebotarev's theorem. We write the quasipolynomial (2.8) as the product of a quasipolynomial $\Delta_1(z)$ and the function $z \mapsto \exp(-zh_{pq})$, where $h_{pq} = \max_{l,j} h_{lj}$, i.e.

$$\Delta(z) = e^{-zh_{pq}}\Delta_1(z).$$

By expanding all exponents in Taylor series we obtain

$$\Delta_1(z) = \sum_{j=0}^{\infty} a_j z^j.$$

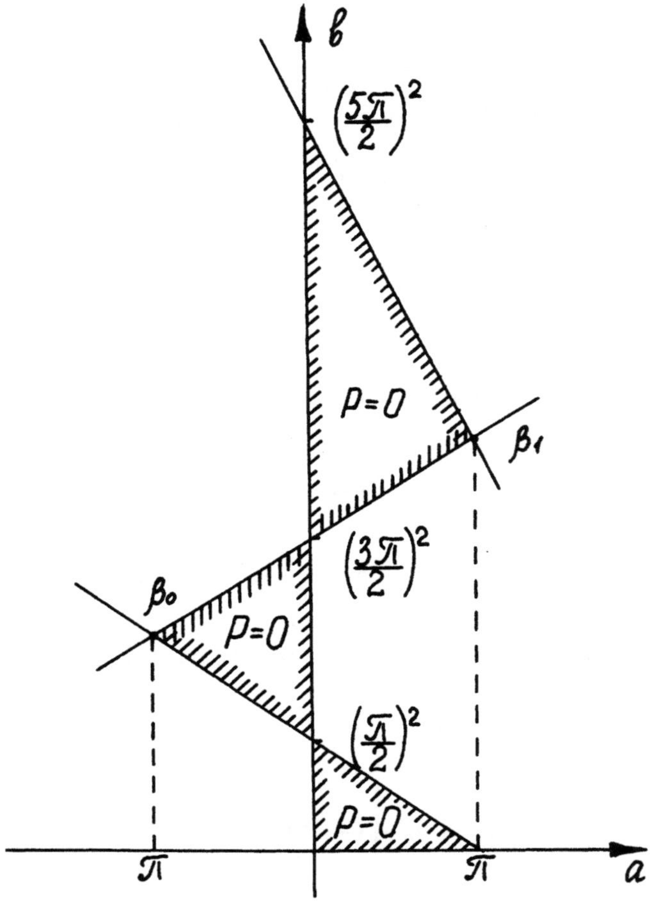

FIGURE 3.2.3. Stability domain for the equation $\ddot{x}(t) + a\dot{x}(t-1) + bx(t) = 0$

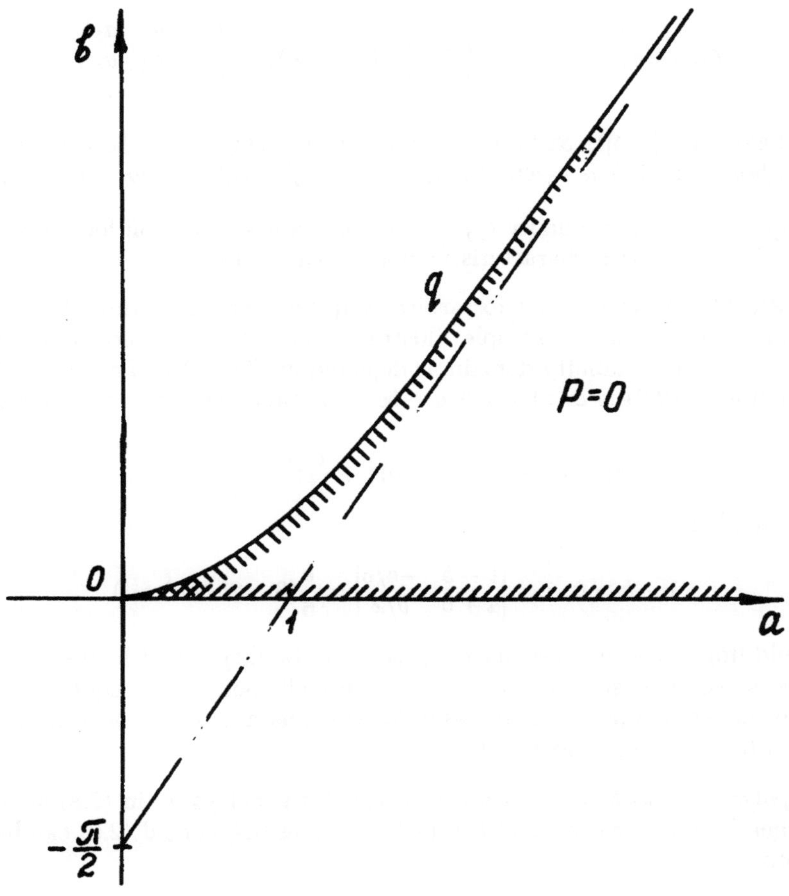

FIGURE 3.2.4. Stability domain for the equation $\ddot{x}(t) + a\dot{x}(t) + bx(t-1) = 0$

It is obvious that the zeros of $\Delta(z)$ coincide with those of $\Delta_1(z)$. We further consider functions $y \mapsto u(y)$ and $y \mapsto v(y)$, $y \in \mathbf{R}$, such that $\Delta_1(iy) = u(y) + iv(y)$,

$$u(y) = \sum_{j=0}^{\infty}(-1)^j a_{2j} y^{2j}, \qquad v(y) = \sum_{j=0}^{\infty}(-1)^j a_{2j+1} y^{2j+1},$$

and define the matrices $Q_j = (a_{2k-i})_{i,k=1}^{j}$, $j = 1, 2, \ldots$, and put $a_{2k-i} := 0$ for $2k - i < 0$. E.g.,

$$Q_1 = (a_1), \qquad Q_2 = \begin{pmatrix} a_1 & a_3 \\ a_0 & a_2 \end{pmatrix}, \qquad Q_3 = \begin{pmatrix} a_1 & a_3 & a_5 \\ a_0 & a_2 & a_4 \\ 0 & a_1 & a_3 \end{pmatrix}.$$

Theorem 2.4. ([114]). *Suppose the functions v and u do not have common zeros. Then the function* (2.8) *is p-zero-free if and only if* $\det Q_j > 0$ *for all* $j = 1, 2, \ldots$.

Consequently every inequality $Q_j > 0$ is a necessary condition for the asymptotic stability of an RDE with characteristic quasipolynomial (2.8).

Remark 2.2. We stress that for arbitrary quasipolynomials thm. 2.4 is not true, in general. The following example illustrates this. Consider equation (2.9) with $h = 1$, $a + b > 0$ and stability domain as depicted in Fig. 3.2.1. The quasipolynomial corresponding to (2.9) is $\Delta(z) = z + a + be^{-z}$. Expanding e^{-z} in a Taylor series, we find

$$\Delta(z) = (a + b) + (1 - b)z + \frac{b}{2!}z^2 - \frac{b}{3!}z^3 + \ldots.$$

Thm. 2.4 implies

$$1 - b > 0, \qquad \begin{vmatrix} 1 - b & -b/6 \\ a + b & b/2 \end{vmatrix} = \frac{b}{6}(3 - 2b + a) > 0.$$

This would imply that a necessary condition for stability would be $0 < b < 1$. However, this is not true, since for $a = 1$, $b = -0.1$ all roots of the equation $\Delta(z) = 0$ belong to the left halfplane. The reason for this mistake is that we have not passed from $\Delta(z)$ to $\Delta_1(z)$ (see above). □

Quasipolynomial with commensurate delays. If the delays h_{lj} in (2.8) are commensurate, then by a linear change of t in (2.1) the quasipolynomial (2.8) can be written in the form

$$\Delta(z) = z^n + \sum_{l=0}^{n-1}\sum_{j=1}^{m(l)} a_{lj} z^l e^{jz}.$$

Again we introduce functions u and v by $\Delta_1(i\omega) = u(\omega) + iv(\omega)$, where $\omega \in \mathbf{R}$. The following theorem is a generalization of the Hermite–Biler theorem, and has been proved in [114].

Theorem 2.5. *The function $\Delta(z)$ is zero-free in the halfplane* $\operatorname{re} z \geq 0$ *if one of the following conditions holds:*

1) *all zeros of the functions u or v, analytically continued to the complex plane, are real and simple, and for each zero ω the relation*

$$\dot{u}(\omega)v(\omega) - \dot{v}(\omega)u(\omega) > 0, \qquad \dot{} = \frac{d}{d\omega}, \tag{2.9}$$

holds;

2) *all zeros of the functions u and v are real, simple, alternate, and for at least one ω the relation* (2.9) *holds.*

Stability for arbitrary delays. Few papers have been devoted to conditions under which there is stability for arbitrary delays [14, 38, 68, 189, 277, 286.3]. We will consider such conditions for a scalar equation

$$x^{(n)}(t) + \sum_{l=1}^{n} \left[a_l x^{(n-l)}(t) + \sum_{j=1}^{m} b_{lj} x^{(n-l)}(t - h_j) \right] = 0, \tag{2.10}$$

where a_l, b_{lj} and $h_j \geq 0$ are certain constants. The quasipolynomial $\Delta(z)$ corresponding to this equation is

$$\Delta(z) = P(z) + \sum_{j=1}^{m} Q_j(z) e^{-h_j z}, \tag{2.11}$$

$$P(z) = z^z + \sum_{l=1}^{n} a_l z^{n-l}, \qquad Q_j(z) = \sum_{l=1}^{n} b_{lj} z^{n-l}.$$

THEOREM 2.6. ([286.3]). *Assume that $|b_{n1}| + |b_{n2}| + \cdots + |b_{nm}| < |a_n|$. Then the system* (2.10) *is asymptotically stable for all delays $h_j \geq 0$ if and only if the polynomial $P(z)$ is asymptotically stable and*

$$\sum_{j=1}^{m} |Q_j(i\omega)| < |P(i\omega)|, \qquad i^2 = -1, \qquad \omega > 0.$$

EXAMPLE 2.1. Consider the equation

$$\ddot{x}(t) + a_1 \dot{x}(t) + a_2 x(t) = b\dot{x}(t - h).$$

Here $P(z) = z^2 + a_1 z + a_2$, $Q(z) = bz$. By thm. 2.6, this equation is stable for arbitrary delays $h \geq 0$ if and only if $a_1 > |b|$, $a_2 > 0$. □

Robust stability. In [14, 38, 277] the following *robust stability problems* for a class of *uncertain delay equations* were considered. It was assumed that the coefficients of the equations are known only within certain bounds. In general, stability of uncertain FDEs is not implied by checking their stability for extremal values of coefficients. However, some boundary theorems are valid for equations (2.10) with

$$\underline{a}_l \leq a_l \leq \overline{a}_l, \qquad \underline{b}_{lj} \leq b_{lj} \leq \overline{b}_{lj}, \tag{2.12}$$

where $\underline{a}_l$, $\overline{a}_l$, $\underline{b}_{lj}$, and $\overline{b}_{lj}$ are given constants.

We introduce four polynomials $P_k(z)$, $k = 1, 2, 3, 4$, associated with the polynomial $P(z)$ in (2.11). The coefficients of $P_1(z)$ are:

$$\left.\begin{aligned} a_{n-2l} &= \begin{cases} \overline{a}_{n-2l} & \text{if } l \text{ is even,} \\ \underline{a}_{n-2l} & \text{if } l \text{ is odd,} \end{cases} \\ a_{n-2l-1} &= \begin{cases} \overline{a}_{n-2l-1} & \text{if } l \text{ is even,} \\ \underline{a}_{n-2l-1} & \text{if } l \text{ is odd.} \end{cases} \end{aligned}\right\} \tag{2.13}$$

The coefficients of $P_2(z)$ are:

$$\left.\begin{aligned} a_{n-2l} &= \begin{cases} \underline{a}_{n-2l} & \text{if } l \text{ is even,} \\ \overline{a}_{n-2l} & \text{if } l \text{ is odd,} \end{cases} \\ a_{n-2l-1} &= \begin{cases} \underline{a}_{n-2l-1} & \text{if } l \text{ is even,} \\ \overline{a}_{n-2l-1} & \text{if } l \text{ is odd.} \end{cases} \end{aligned}\right\} \tag{2.14}$$

The coefficients of $P_3(z)$ are:

$$\left.\begin{aligned} a_{n-2l} &= \begin{cases} \underline{a}_{n-2l} & \text{if } l \text{ is even,} \\ \overline{a}_{n-2l} & \text{if } l \text{ is odd,} \end{cases} \\ a_{n-2l-1} &= \begin{cases} \overline{a}_{n-2l-1} & \text{if } l \text{ is even,} \\ \underline{a}_{n-2l-1} & \text{if } l \text{ is odd.} \end{cases} \end{aligned}\right\} \tag{2.15}$$

The coefficients of $P_4(z)$ are:

$$\left.\begin{aligned} a_{n-2l} &= \begin{cases} \overline{a}_{n-2l} & \text{if } l \text{ is even,} \\ \underline{a}_{n-2l} & \text{if } l \text{ is odd,} \end{cases} \\ a_{n-2l-1} &= \begin{cases} \underline{a}_{n-2l-1} & \text{if } l \text{ is even,} \\ \overline{a}_{n-2l-1} & \text{if } l \text{ is odd.} \end{cases} \end{aligned}\right\} \tag{2.16}$$

We similarly define four polynomials $Q_{jk}(z)$ for each $Q_j(z)$.

Theorem 2.7. ([277]). *Assume that*

$$\sum_{j=1}^{m} \max\{|\underline{b}_{nj}|, |\overline{b}_{nj}|\} < \min\{|\underline{a}_n|, |\overline{a}_n|\}.$$

Then the systems (2.10), (2.12) *are asymptotically stable if and only if each of the* 4^{m+1} *quasipolynomials*

$$P_k(z) + \sum_{j=1}^{m} Q_{jl_j}(z) e^{-h_j z}, \qquad k, l_j = 1, 2, 3, 4,$$

is asymptotically stable.

CHAPTER 4

Stability of neutral type functional differential equations

In this chapter we will give some methods for investigating the stability of various classes of NDEs. The definitions of stability for NDEs and the notation are the same as for RDEs, given in §3.1.

1. Direct Liapunov's method

1.1. Degenerate Liapunov functionals. Consider the problem

$$[x(t) - G(t, x_t)]^{\bullet} = F(t, x_t), \qquad t \geq t_0, \tag{1.1}$$

$$x_{t_0} = \phi \in C[-h, 0], \qquad x_t(\theta) = x(t+\theta), \quad -h \leq \theta \leq 0. \tag{1.2}$$

Here $F, G\colon [t_0, \infty) \times Q_H \to \mathbf{R}^n$ are continuous maps, $Q_H = \{u \in C[-h, 0]\colon \|u\| < H\}$, $0 < H < \infty$, satisfying all the general conditions in subs. 2.3.5, as well as

$$F(t, 0) \equiv G(t, 0) \equiv 0, \qquad |F(t, \psi)| \leq C, \qquad t \geq t_0, \quad \psi \in Q_H. \tag{1.3}$$

The method given below for studying stability is based on the use of positive semidefinite functionals and the study of the stability of the following type of functional inequalities:

$$|z(t, y_t)| \leq f(t), \qquad t \geq t_0; \qquad y_{t_0} = \phi, \tag{1.4}$$

where $z(t, y_t) := y(t) - G(t, y_t)$, and $f\colon [t_0, \infty) \to [0, \infty)$ is a continuous function.

DEFINITION 1.1. The trivial solution of inequality (1.4) is said to be:

1) *stable* if for any $\epsilon > 0$ there is a $\delta(\epsilon) > 0$ such that for any solution y: : $[t_0, T) \to \mathbf{R}^n$ $(t_0 < T \leq \infty)$ of (1.4) we have $|y(t)| \leq \epsilon$, $t_0 \leq t < T$, for all ϕ and f, $\|\phi\| \leq \delta(\epsilon)$, $\|f\|$ $(:= \sup_{t_0 \leq t < T} |f(t)|) \leq \delta(\epsilon)$;
2) *asymptotically stable* if it is stable and $\lim_{t\to\infty} y(t) = 0$ for all ϕ, f, $\|\phi\| < \delta_0$, $\|f\| < \delta_0$ (δ_0 fixed), where $f(t) \to 0$ as $t \to \infty$. □

THEOREM 1.1. *Suppose that the trivial solution of* (1.4) *is stable and that there exists a continuous functional* $V : [t_0, \infty) \times Q_H \times \mathbf{R}^n \to \mathbf{R}$ *such that*

$$\omega_1(|z(t, \psi)|) \leq V(t, \psi, z(t, \psi)) \leq \omega_2(\psi), \tag{1.5}$$

$$\dot{V} := \frac{d}{dt} V(t, x_t, z(t, x_t)) \leq 0 \tag{1.6}$$

for all solutions x of equation (1.1), *where ω_i and d/dt have the same meaning as in subs.* 3.1.2. *Then the trivial solution of* (1.1) *is stable. If the trivial solution of* (1.4) *is asymptotically stable, V satisfies* (1.5), *and*

$$\dot{V} \leq -\omega_3(|z(t, x_t)|),$$

then the trivial solution of (1.1) *is asymptotically stable.*

PROOF. Take $\epsilon \in (0, H)$ arbitrary, and choose $\delta_1 = \delta_1(\epsilon) > 0$ such that $|y(t)| \leq \epsilon$, $t \geq t_0$, if $f(t) \leq \delta_1$, $\|\phi\| \leq \delta_1$. Define $\delta_2 \in (0, \delta_1)$ by $\omega_2(\delta_2) \leq \omega_1(\delta_1)$. Consider an arbitrary initial function ϕ with $\|\phi\| \leq \delta_2$, and suppose the solution x of (1.1)–(1.2) is defined and satisfies the inequality $|x(t)| < H$ on an interval $[t_0, T)$. Then, taking into account (1.5)–(1.6), we find that for $t_0 \leq t < T$,

$$\begin{aligned} &\omega_1(|z(t, x_t)|) \leq V(t, x_t, z(t, x_t)) \leq \\ &\leq V(t_0, \phi, z(t_0, \phi)) \leq \omega_2(\delta_2) \leq \omega_1(\delta_1). \end{aligned}$$

Putting $f(t) := |z(t, x_t)|$ and using the definition of δ_1 we obtain $|x(t)| \leq \epsilon$, $t_0 \leq t \leq T$. Hence this inequality holds for $t_0 \leq t < \infty$ and arbitrary ϕ with $\|\phi\| \leq \delta_2$.

This proves stability.

To prove asymptotic stability we take $\delta > 0$ such that $|x(t; t_0, \phi)| \leq H$, $t \geq t_0$, for any $\phi \in Q_\delta$. It suffices to verify that

$$z(t, x_t(t; t_0, \phi)) \to 0, \qquad \phi \in Q_\delta, \qquad t \to \infty. \tag{1.7}$$

Suppose (1.7) is not true. Then we can find a $\gamma \in (0, H)$ and a sequence $t_i \to \infty$ such that $|z(t_i, x_{t_i})| \geq \gamma$. The conditions (1.1)–(1.3) give

$$|\dot{z}(t, x_t)| = |F(t, x_t)| \leq C, \qquad \phi \in Q_\delta.$$

Using now the same arguments as in the proof of thm. 2.1.1, we obtain

$$V(t, x_t, z(t, x_t)) - V(t_0, \phi, z(t_0, \phi)) \to -\infty, \qquad t \to \infty.$$

This relation contradicts (1.5), hence (1.7) holds. Relation (1.7) and the asymptotic stability of (1.4) imply the asymptotic stability of the trivial solution of (1.1). □

We can similarly prove

THEOREM 1.2. *Suppose the conditions of thm.* 1.1 *hold with $C[-h, 0]$ instead of Q_H*, $\sup_{t_0 \leq t < \infty, \|\psi\| < H} |F(t, \psi)| < \infty$ *for all $H > 0$, $\omega_1(r) \to \infty$ as $r \to \infty$, and suppose that all solutions of* (1.4) *are bounded for any bounded t. Then $x(t; t_0, \phi) \to 0$, $t \to \infty$, for any initial function $\phi \in C[-h, 0]$.*

We give some stability conditions for inequality (1.4) and equations (1.1).

THEOREM 1.3. *Suppose that the continuous functional G: $[t_0, \infty) \times C[-h, 0] \to \mathbf{R}^n$ satisfies the condition*

$$|G(t, \psi) - G(t, \chi)| \leq \gamma \|\psi - \chi\|, \qquad \gamma < 1,$$

if $\|\psi\| \leq \delta_0$, $\|\chi\| \leq \delta_0$ ($\delta_0 > 0$ fixed). Then the trivial solution of (1.4) *is stable.*

PROOF. Put $\alpha(t) = \max_{t_0 \le s \le t} |y(s)|$. Now (1.4) implies that if $|y(t)| \le \delta_0$ $(t_0 - h \le t < T)$, then for $t_0 \le t < T$,

$$|y(t)| \le |G(t, y_t)| + f(t) \le \gamma \|y_t\| + f(t),$$
$$\alpha(t) \le \gamma\alpha(t) + \gamma\|\phi\| + \max_{t_0 \le s \le t} f(s),$$
$$\alpha(t) \le (1-\gamma)^{-1}(\gamma\|\phi\| + \|f\|_{[t_0,\infty)}) \qquad (t_0 \le t < T).$$

This immediately implies that one can take $T = \infty$ if $\max\{\|\phi\|, \|f\|_{[t_0,\infty)}\} < (1-\gamma)(1+\gamma)^{-1}$. Hence inequality (1.4) is stable.

Further, we assume that the last inequality is valid and that $f(t) \to 0$ as $t \to \infty$, but $\limsup_{t\to\infty} |y(t)| = y_0 > 0$. Put $\beta := (1-\gamma)(2+\gamma)^{-1} y_0$, and let $t_1 \in [t_0, \infty)$ be such that $|y(t_1)| > y_1 - \beta$, $\|y_{t_1}\| < y_0 + \beta$, $f(t_1) < \beta$. Then (1.4) implies

$$y_0 - \beta < |y(t_1)| \le |G(t_1, y_{t_1})| + f(t_1) < \gamma(y_0 + \beta) + \beta.$$

Therefore $y_0 < (2+\gamma)(1-\gamma)^{-1}\beta = y_0$, which is impossible. □

With natural modifications, thms. 1.1–3 can be immediately generalized to systems with infinite delay (in particular, thm. 1.3 then only concerns stability).

THEOREM 1.4. *Suppose the functionals $F, G\colon [t_0, \infty) \times C[-h, 0] \to \mathbf{R}^n$ are continuous, and*

$$|G(t+s, \psi) - G(t, \chi)| \le \omega_5(s) + \gamma\|\psi - \chi\|,$$
$$0 < \gamma < 1, \qquad \forall \psi, \chi \in Q_H, \qquad t \ge t_0, \qquad s \ge 0.$$

Assume that there is a continuous functional V satisfying condition (1.5) *and*

$$\dot V \le -\omega_6(|x(t)|).$$

Then the trivial solution of the problem (1.1)–(1.2) *is asymptotically stable.*

(Here, $\omega_{5,6} \in \Omega$, see subs. 3.1.2.)

EXAMPLE 1.1. Consider the system of scalar equations

$$\left.\begin{aligned} &\left[x_1(t) + \textstyle\sum_{j=1}^m \alpha_j x_1(t - h_j)\right]^{\bullet} = x_2(t), \\ &\dot x_2(t) = -g\left(x_1(t) + \textstyle\sum_{j=1}^m \alpha_j x_1(t - h_j)\right). \end{aligned}\right\} \tag{1.8}$$

Here $0 < h_1 < \cdots < h_m$, $g \in C(\mathbf{R})$, $yg(y) > 0$ for $y \ne 0$,

$$\sum_{j=1}^m |\alpha_j| < 1.$$

We will prove that under these conditions the trivial solution of (1.8) is stable. Define a functional V for arbitrary $x_1, x_2 \in C[-h_m, 0]$ by:

$$V(x_{1t}, x_{2t}) = \frac{1}{2}[x_2(t)]^2 + \int_0^{z(t)} g(s)\, ds,$$
$$z(t) := x_1(t) + \sum_{j=1}^m \alpha_j x_1(t - h_j).$$

Then $\dot{V} = x_2\dot{x}_2 + g(z)\dot{z} = 0$. Therefore, by thms. 1.1, 1.3 the system (1.8) is stable. □

1.2. Stability in a first approximation. Here we consider an equation of the type (1.1) with finite delay:

$$\left.\begin{array}{c}[x(t) - G(t, x_t) - G_1(t, x_t)]^\bullet = F(t, x_t) + F_1(t, x_t), \qquad t \geq t_0, \\ x_{t_0} = \phi \in C[-h, 0].\end{array}\right\} \tag{1.9}$$

Along with the general assumptions of subs. 1.1, the functionals G_1 and F_1 are assumed to satisfy the following condition uniformly for $t \in [t_0, \infty)$:

$$|F_1(t, \psi)| + |G_1(t, \psi)| = o(\|\psi\|) \qquad (\|\psi\| \to 0).$$

The functionals $(t, \psi) \mapsto G(t, \psi)$ and $(t, \psi) \mapsto F(t, \psi)$ are assumed to be linear in ψ, and the linear system

$$[x(t) - G(t, x_t)]^\bullet = F(t, x_t)$$

(called the *system of first approximation*) is assumed to be uniformly exponentially stable.

THEOREM 1.5. ([238]). *Under the above assumptions the trivial solution of* (1.9) *is asymptotically stable.*

EXAMPLE 1.2. Consider the NDE

$$\dot{x}(t) = \gamma x(t)(1 - K^{-1}(x(t-h)) + \gamma c\dot{x}(t-h)$$
$$(\gamma, K, h, c = \text{const} > 0).$$

This equation has steady state equilibrium $x(t) \equiv K$; we study its stability. We write $x(t) = K(y(t) + 1)$. Then the equation for $y(t)$ takes the form

$$\dot{y}(t) = -\gamma(y(t) + 1)y(t-h) + \gamma c\dot{y}(t-h).$$

So the equation of first approximation is

$$\dot{y}(t) = -\gamma y(t-h) + \gamma c\dot{y}(t-h).$$

Sufficient conditions for the exponential stability of this equation are [203]:

$$0 < \gamma < 1, \qquad 1 < \gamma h < \beta(1 - \gamma^2 c^2),$$

where $\beta \in (\pi/2.\pi)$ is the root of the equation $\beta \tan \beta = -h/c$. By thm. 1.5, these conditions give us those for asymptotic stability of the initial logistic equation. □

1.3. The use of functionals depending on derivatives. We consider the stability problem for NDEs of the form

$$\left.\begin{aligned} \dot{x}(t) = f(t, x_t, \dot{x}_t), \qquad & x\colon [t_0 - h, \infty) \to \mathbf{R}^n, \qquad t \geq t_0, \\ & x_{t_0} = \phi. \end{aligned}\right\} \tag{1.10}$$

Let W be the Banach space of absolutely continuous functions $[-h, 0] \to \mathbf{R}^n$ with square-integrable derivative, with norm

$$\|\psi\|_W = \left[|\psi(0)|^2 + \int_{-h}^{0} |\dot{\psi}(\theta)|^2\, d\theta\right]^{1/2}.$$

Let $Q_H \subset W$ be a ball, $Q_H := \{\psi\colon \|\psi\|_W \leq H\}$, $H > 0$. Let $f\colon [t_0, \infty) \times Q_H \times L_2[-h, 0] \to \mathbf{R}^n$ be a continuous function which satisfies a Lipschitz condition in the second and third arguments, with Lipschitz constant for the third argument less than 1, and $f(t, 0, 0) \equiv 0$.

THEOREM 1.6. *If there is a continuous functional* $V\colon [t_0, \infty) \times Q_H \times L_2[-h, 0] \to \mathbf{R}$ *satisfying the inequalities*

$$\left.\begin{aligned} \omega_1(|\psi(0)|) \leq V(t, \psi, \chi) \leq \omega_2(\|\psi\|_W), \\ \dot{V} \leq 0, \end{aligned}\right\}$$

then the solution $x(t) \equiv 0$ *is stable. If, moreover, we have* $\dot{V} \leq -\omega_3(|x(t)|)$, *then the trivial solution of* (1.10) *is asymptotically stable.*

(Here all $\omega_i \in \Omega$.)

EXAMPLE 1.3. Consider the scalar equation

$$\dot{x}(t) = -a(t)x(t) + b(t)\dot{x}(t - h), \quad t \geq 0, \qquad h > 0, \tag{1.11}$$

where a, b are bounded continuous functions, and $a(t) \geq c > 0$. Consider also the functional

$$V(t, x_t, \dot{x}_t) = x^2(t) + \int_{-h}^{0} \frac{\dot{x}^2(t + \theta)}{a(t + \theta)}\, d\theta.$$

We obtain

$$\dot{V} = 2x(t)\dot{x}(t) + \frac{\dot{x}^2(t)}{a(t)} - \frac{\dot{x}^2(t - h)}{a(t - h)}.$$

Eliminating $\dot{x}(t)$ according to (1.11) we find

$$\dot{V} = -a(t)x^2(t) - \left[\frac{1}{a(t - h)} - \frac{b^2(t)}{a(t)}\right] \dot{x}^2(t - h).$$

Hence, by thm. 1.6 the trivial solution of (1.11) is asymptotically stable if

$$a(t) - b^2(t)a(t - h) \geq 0. \qquad \square$$

2. Stability of linear autonomous equations

2.1. General case. Suppose we are given the NDE

$$\left.\begin{aligned} \dot{x}(t) &= \textstyle\int_{-\infty}^{0}[dR_0(\theta)]x(t+\theta) + \int_{-\infty}^{0}[dR_1(\theta)]\dot{x}(t+\theta), \qquad t \geq 0, \\ x(\theta) &= \phi(\theta), \quad -\infty < \theta \leq 0, \qquad x\colon \mathbf{R} \to \mathbf{R}^n, \qquad \phi \in C^1(-\infty, 0]. \end{aligned}\right\} \tag{2.1}$$

Here the entries of the matrix R_0 are functions of bounded variation and with bounded first moments. Also, by definition,

$$\int_{-\infty}^{0}[dR_1(\theta)]\dot{x}(t+\theta) = \sum_{j=1}^{\infty} L_j\dot{x}(t-h_j) + \int_{-\infty}^{0}\lambda(\theta)\dot{x}(t+\theta)\,d\theta,$$

where the constants $h_j \geq 0$, the matrices L_j, and the absolutely integrable matrix $\lambda(\theta)$ with entries λ_{ij} satisfy

$$\sum_{j=1}^{\infty}\|L_j\| + \int_{-\infty}^{0}\|\lambda(\theta)\|\,d\theta < 1. \tag{2.2}$$

(Here $\|\cdot\|$ is the euclidean matrix norm.) Assume that the initial function ϕ satisfies

$$\|\phi\|_{(-\infty,o]} < \infty, \qquad \|\dot{\phi}\|_{(-\infty,0]} < \infty,$$
$$\|\phi\|_1 :=$$
$$:= |\phi(0)| + |\dot{\phi}(0)| + \left[\int_0^{\infty}|F_0(t)|^2\,dt\right]^{1/2} + \left[\int_0^{\infty}|F_1(t)|^2\,dt\right]^{1/2} < \infty,$$

where

$$F_j(t) = \int_{-\infty}^{-t}[dR_j(\theta)]\phi^{(j)}(t+\theta), \qquad j = 0,1, \qquad 0 \leq t < \infty.$$

The definitions of stability of the trivial solution of (2.1) coincide with those in subs. 3.1.3, with $\|\phi\|_1$ instead of $\|\phi\|$. Put

$$\tilde{R}_j(z) = \int_{-\infty}^{0} e^{z\theta}\,dR_j(\theta), \qquad \tilde{F}_j(z) = \int_0^{\infty} e^{-zt}F_j(t)\,dt,$$
$$\tilde{x}(z) = \int_0^{\infty} e^{-zt}x(t)\,dt.$$

Then

$$[Iz - \tilde{R}_1(z)z - \tilde{R}_0(z)]\tilde{x}(z) = [I - \tilde{R}_1(z)]\phi(0) + \tilde{F}_0(z) + \tilde{F}_1(z).$$

(Here I is the identity matrix.) Let

$$\Delta(z) = \det[Iz - \tilde{R}_1(z)z - \tilde{R}_0(z)]$$

be the so-called characteristic function of this equation.

THEOREM 2.1. ([281.5]). *Suppose the above stated conditions on the kernels R_0, R_1 hold, and suppose also the function $z \mapsto \Delta(z)$ is zero-free in the domain* $\operatorname{re} z \geq 0$. *Then (2.1) is asymptotically stable.*

REMARK 2.1. If condition (2.2) does not hold, then (2.1) may be unstable even if all zeros of the function $\Delta(z)$ belong to $\operatorname{re} z < 0$ [79, 286.3]. □

2.2. Scalar equations. Here we consider the Cauchy problem for a scalar equation of type (2.1):

$$x^{(n)}(t) = \sum_{j=0}^{n} \int_{-\infty}^{0} x^{(j)}(t+\theta)\, dr_j(\theta), \qquad t \geq 0,$$
$$x^{(j)}(\theta) = \phi^{(j)}(\theta), \qquad \theta \leq 0, \qquad j = 0, \dots, n.$$

Put

$$\alpha_{lj} = \int_{-\infty}^{0} |\theta|^l |dr_j(\theta)|,$$
$$\beta_{lj} = \int_{-\infty}^{0} |\theta|^l \, dr_j(\theta) \qquad \text{if } \alpha_{lj} < \infty,$$
$$\Delta_n(z) = z^n[1 - \tilde{r}_n(z)] - \sum_{j=0}^{n-1} z^j \tilde{r}_j(z).$$

Here we assume that all functions r_j are of bounded variation on $(-\infty, 0]$, $\alpha_{0n} < 1$, and r_n does not contain a singular component (see subs. 2.5.2). The initial function $\phi \in C^n(-\infty, 0]$, and

$$\|\phi\|_n := \sum_{j=0}^{n} \left[|\phi^{(j)}(0)| + \left(\int_0^{\infty} |F_j(t)|^2 \, dt \right)^{1/2} \right] < \infty,$$

where

$$F_j(t) = \int_{-\infty}^{-t} \phi^{(j)}(t+\theta)\, dr_j(\theta).$$

Thm. 2.1 implies that if the function $z \mapsto \Delta(z)$ is zero-free in the half-plane $\operatorname{re} z \geq 0$, then the trivial solution of the equation under consideration is asymptotically stable. We will consider some concrete cases.

Case $n = 1$. For $n = 1$ we obtain

$$\Delta_1(z) = z\left[1 - \int_{-\infty}^{0} e^{z\theta}\, dr_1(\theta)\right] - \int_{-\infty}^{0} e^{z\theta}\, dr_0(\theta).$$

A. Assume that

$$\beta_{00} < 0, \qquad \alpha_{10} + \alpha_{01} < 1.$$

Then the function Δ_1 is zero-free in $\operatorname{re} z \geq 0$.

For the proof we choose functions $z \mapsto \psi_1(z)$ and $z \mapsto \psi_2(z)$ as follows

$$\psi_1(z) = z - \beta_{00}, \qquad \psi_2(z) = \Delta_1(z) - \psi_1(z).$$

The function ψ_1 is zero-free in $\operatorname{re} z \geq 0$, and $\psi_2(z)| \leq (\alpha_{10} + \alpha_{01})|z|$ for $\operatorname{re} z \geq 0$. So, on the boundary of the halfplane $\operatorname{re} z \geq 0$ we have $|\psi_2(z)| < |\psi_1(z)|$. Using Rouché's theorem this implies that Δ_1 is zero-free in $\operatorname{re} z \geq 0$.

B. Assume that the kernel $r_0(\theta)$ has a jump $a_0 < 0$ at the point $\theta = 0$, and that

$$\gamma := a_0(1 - \alpha_{01}) + (1 + 2\alpha_{01})(\alpha_{00} + a_0) < 0. \tag{2.3}$$

Then the function Δ_1 is zero-free in $\operatorname{re} z \geq 0$.

PROOF. Assume the contrary, i.e. the equation

$$\Delta_1(z) = 0 \tag{2.4}$$

has a root $z_0 = \alpha + i\beta$ with real part $\alpha \geq 0$. Then

$$\operatorname{re} z_0 \left|1 - \int_{-\infty}^{0} e^{z_0\theta}\, dr_1(\theta)\right|^2 \geq 0. \tag{2.5}$$

We prove that (2.5) is impossible. Equation (2.4) implies

$$z_0 \left[1 - \int_{-\infty}^{0} e^{z_0\theta}\, dr_1(\theta)\right] = a_0 + \int_{-\infty}^{0^-} e^{z_0\theta}\, dr_0(\theta). \tag{2.6}$$

Multiplying (2.6) by the complex conjugate of $1 - \int_{-\infty}^{0} e^{z_0\theta}\, dr_1(\theta)$, we find by (2.3)

$$\operatorname{re} z_0 \left|1 - \int_{-\infty}^{0} e^{z\theta}\, dr_1(\theta)\right|^2 =$$

$$= \left[a_0 + \int_{-\infty}^{0^-} e^{\alpha\theta} \cos\beta\theta\, dr_0(\theta)\right] \cdot \left[1 - \int_{-\infty}^{0} e^{\alpha\theta} \cos\beta\theta\, dr_1(\theta)\right] +$$

$$- \int_{-\infty}^{0^-} e^{\alpha\theta} \sin\beta\theta\, dr_0(\theta) \cdot \int_{-\infty}^{0} e^{\alpha\theta} \sin\beta\theta\, dr_1(\theta) \leq \gamma < 0,$$

contradicting (2.5). □

C. Assume that the kernels $r_0(\theta)$, $r_1(\theta)$ are constant for $\theta \leq -h < 0$, nonincreasing, and

$$\beta_{00} < 0, \qquad \alpha_{00} < \frac{\pi}{2h}\left[1 - \left(1 + \frac{1}{e}\right)\alpha_{01}\right]. \tag{2.7}$$

Then equation (2.4) does not have roots in $\operatorname{re} z \geq 0$.

PROOF. For $z = \alpha + i\beta$, (2.4) implies the two equations

$$q_1(\alpha,\beta) := \alpha \left[1 - \int_{-h}^{0} e^{\alpha\theta} \cos\beta\theta\, dr_1(\theta)\right] +$$

$$+\beta \int_{-h}^{0} e^{\alpha\theta} \sin\beta\theta\, dr_1(\theta) - \int_{-h}^{0} e^{\alpha\theta} \cos\beta\theta\, dr_0(\theta) = 0, \tag{2.8}$$

$$q_2(\alpha,\beta) := \beta \left[1 - \int_{-h}^{0} e^{\alpha\theta} \cos\beta\theta\, dr_1(\theta)\right] +$$

$$-\alpha \int_{-h}^{0} e^{\alpha\theta} \sin\beta\theta\, dr_1(\theta) - \int_{-h}^{0} e^{\alpha\theta} \sin\beta\theta\, dr_0(\theta) = 0. \tag{2.9}$$

Equation (2.8) does not have a solution in the domain $2h|\beta| < \pi$, $\alpha \geq 0$, since under the conditions (2.7) we have $q_1(\alpha,\beta) > 0$ in this domain. We verify that for $2h|\beta| \geq \pi$, $\alpha \geq 0$ the equation (2.9) also does not have a solution. In fact, we obtain

$$\left|\alpha \int_{-h}^{0} e^{\alpha\theta} \sin\beta\theta\, dr_1(\theta)\right| =$$

$$= \left|\beta \int_{-h}^{0} \alpha\theta e^{-\alpha\theta} \frac{\sin\beta\theta}{\beta\theta}\, dr_1(\theta)\right| \leq \frac{|\beta|}{e}\alpha_{01}.$$

This and (2.9) imply that for $2h\beta \geq \pi$, $\alpha \geq 0$ we have $q_2(\alpha, \beta) \geq \beta[1 - \alpha_{01}(1 + 1/e)] - \alpha_{00} > 0$, while if $2h\beta \leq -\pi$, $\alpha \geq 0$ we have $q_2(\alpha, \beta) < 0$. Hence equations (2.8) and (2.9) are not compatible for $\alpha \geq 0$. □

Case $n = 2$. The function Δ_2 is equal to

$$\Delta_2(z) = z^2 \left[1 - \int_{-\infty}^{0} e^{z\theta}\, dr_2(\theta)\right] - z \int_{-\infty}^{0} e^{z\theta}\, dr_1(\theta) - \int_{-\infty}^{0} e^{z\theta}\, dr_0(\theta).$$

Asume that

$$\alpha_{02} + \alpha_{11} + \frac{\alpha_{20}}{2} < 1, \qquad \beta_{10} - \beta_{01} > 0, \qquad \beta_{00} < 0.$$

Then the function Δ_2 is zero-free in the halfplane $\operatorname{re} z \geq 0$.

To prove this we write Δ_2 as a sum $\Delta_2(z) = \psi_1(z) + \psi_2(z)$, where

$$\psi_1(z) = z^2 - (\beta_{01} + \beta_{10})z - \beta_{00},$$

$$\psi_2(z) =$$

$$= -z^2 \int_{-\infty}^{0} e^{z\theta}\, dr_2(\theta) - z \int_{-\infty}^{0} (e^{z\theta} - 1)\, dr_1(\theta) - \int_{-\infty}^{0} (e^{z\theta} - 1 - z\theta)\, dr_0(\theta).$$

Note that under the conditions imposed the function ψ_1 is zero-free in the domain $\operatorname{re} z \geq 0$, and also $|\psi_1(z)| > |\psi_2(z)|$ on the boundary of this domain. Hence, by Rouché's theorem the function Δ_2 is zero-free in this halfplane as well.

2.3. Stability of NFDEs with discrete delays. Consider the following scalar equation with discrete delays $h_j \geq 0$ and constant coefficients a_l and b_{lj}:

$$x^{(n)}(t) + \sum_{l=1}^{n} a_l x^{(n-l)}(t) + \sum_{l=0}^{n} \sum_{j=1}^{m} b_{lj} x^{(n-l)}(t - h_j) = 0. \tag{2.10}$$

Assume that

$$\sum_{j=1}^{m} |b_{0j}| < 1, \qquad \sum_{j=1}^{m} |b_{nj}| < |a_n|.$$

Then for (2.10) to be asymptotically stable it is necessary and sufficient [286.3] that the polynomial $P(z)$ be asymptotically stable and that

$$\sum_{j=1}^{m} |Q_j(i\omega)| < |P(i\omega)|, \qquad i^2 = -1, \qquad \omega > 0,$$

where

$$P(z) = z^n + \sum_{l=1}^{n} a_l z^{n-l}, \qquad Q_j(z) = \sum_{l=0}^{n} b_{lj} z^{n-l}. \tag{2.11}$$

Robust stability of NDEs. Consider (2.10) with coefficients a_l and b_{lj} unknown but satisfying the conditions (3.2.12). Define the four polynomials $P_k(z)$, $k = 1, 2, 3, 4$, using the relation (2.11), by (3.2.13)–(3.2.16). Define for each Q_j the $Q_{jk}(z)$ in a similar manner.

THEOREM 2.2. ([277]). *Assume that*

$$\sum_{j=1}^{m} \beta_{0j} < 1, \qquad \sum_{j=1}^{m} \beta_{nj} < \alpha_n.$$

Then all systems (2.10), (3.2.12) *are asymptotically stable if and only if each of the* 4^{m+1} *quasipolynomials*

$$P_k(z) + \sum_{j=1}^{m} Q_{jl}(z)e^{-h_j z}$$

is asymptotically stable.

Some remarks about the roots of quasipolynomials of NDEs with discrete delays. Numerous investigations have been devoted to the zeros of the characteristic function [47, 286.1–3]

$$\Delta(z) = \sum_{l=0}^{n} \sum_{j=1}^{r(l)} a_{lj} z^l e^{-h_{lj} z}.$$

It has been proved that if the delays h_{nj} are commensurate, then the asymptotic roots z_k of the equation $\Delta(z) = 0$ belong to certain chains, and $\lim \operatorname{im} z_k = \pm\infty$, $k \to \infty$. The real parts of the roots z_k on each chain are either uniformly bounded, or tend to $-\infty$ as $k \to \infty$.

If the delays h_j are incommensurate, then there are, in addition to these chains, finitely many strips parallel to the imaginary axis containing certain zeros of the function Δ. The real parts of these zeros densely fill the intervals on the real axis corresponding to these strips. The imaginary parts of these zeros tend to an almost asymptotically periodic point set. The maximal multiplicity of a zero of Δ is at most $n + (n+1)\max_{0 \le l \le n} r(l)$.

Suppose we have only a single delay $h > 0$, i.e. $m = 1$ in (2.10), and $b_{01} > 0$. By the substitution $x(t) = b_{01}^{t/h} y(t)$ we get

$$\Delta(z) = P(z) + Q(z)e^{-hz},$$

$$P(z) = z^n + \sum_{l=1}^{n} a_l z^{n-l}, \qquad Q(z) = z^n + \sum_{l=1}^{n} b_l z^{n-l}.$$

Assume that the equation $\Delta(z) = 0$ has a root z_0 of maximal possible multiplicity $2n+1$. Then all roots z_k of this equation are such that $\operatorname{re} z_k = \operatorname{re} z_0$ [400]. From this we can derive that certain NDEs have infinitely many fundamental frequencies (defined as the positive roots of $\Delta(i\omega) = 0$). A necessary and sufficient condition for this is that $|P(i\omega)| = |Q(i\omega)|$ for all $\omega \in \mathbf{R}$. If, in addition, $P(z) \not\equiv Q(z)$, then the roots ω_k of the equation $\Delta(i\omega) = 0$ satisfy

$$\omega_k = -\frac{2k+1}{h}\pi - \frac{(b_1 + a_1)h}{(2k+1)\pi} + \mathcal{O}\left(\frac{1}{k^2}\right), \qquad k \to \pm\infty.$$

On the other hand, if $P(z) \equiv Q(z)$, then

$$\omega_k = -\pi\frac{2k+1}{h}i, \qquad k = 0, \pm 1, \ldots.$$

Note that the case when the number of fundamental frequencies is infinite is exceptional. In general, the number of distinct fundamental frequencies ω does not exceed n, since ω is a root of the algebraic equation $|P(i\omega)|^2 = |Q(i\omega)|^2$, which is of order $2n$ and involves only even powers of ω.

2.4. The influence of small delays on stability. Many actual phenomena involve small delays which are often neglected in the process of mathematical modeling. Sometimes this leads to false conclusions. As an example we will consider a scalar NFDE which is asymptotically stable for $h = 0$ but unstable for arbitrary $h > 0$:

$$\ddot{x}(t) + \dot{x}(t) + x(t) = a[\ddot{x}(t-h) + \dot{x}(t-h) + x(t-h)], \tag{2.12}$$

where $a > 1$. If $h = 0$, the system (2.12) is asymptotically stable. The characteristic quasipolynomial of (2.12) is $\Delta(z) = (z^2 + z + 1)(1 - ae^{-hz})$, and has the following zeros with positive real part if $h > 0$:

$$z_k = \frac{1}{h}(\ln a + 2k\pi i), \qquad i^2 = -1, \qquad k = 0, \pm 1, \ldots.$$

So, for any $h > 0$ the trivial solution of (2.12) is unstable.

CHAPTER 5

Stability of stochastic functional differential equations

1. Statement of the problem

1.1. Definitions of stability. Here we will consider the Itô type SRDE

$$\left.\begin{aligned} dx(t) &= a_1(t, x_t)\,dt + a_2(t, x_t)\,d\xi(t), \qquad t \geq t_0, \\ x_t(\theta) &= x(t+\theta), \qquad -h \leq \theta \leq 0, \qquad x\colon J_x \to \mathbf{R}^n. \end{aligned}\right\} \tag{1.1}$$

Here $\xi\colon [t_0, \infty) \to \mathbf{R}^l$ is the standard Wiener process, and the continuous functionals a_1, a_2 are defined on $[t_0, \infty) \times C[-h, 0]$. The initial condition for (1.1) is

$$x_{t_0} = \phi. \tag{1.2}$$

The initial function ϕ is a stochastic process with continuous trajectories, independent of the σ-algebra $B_{t_0\infty}(d\xi)$, where $B_{t_1t_2}(d\xi)$ is the minimal σ-algebra generated by the random variables $\xi(s) - \xi(t)$ for arbitrary s, t, $t_1 \leq t \leq s \leq t_2$. A solution of the problem (1.1)–(1.2) is a process $t \mapsto x(t)$ such that $B_{t\infty}(d\xi)$ is independent of $B_{-\infty t}(x) \cup B_{t_0t}(d\xi)$, $t \geq t_0$, and such that with probability 1,

$$x(t) = \phi(t_0) + \int_{t_0}^{t} a_1(s, x_s)\,ds + \int_{t_0}^{t} a_2(s, x_s)\,d\xi(s).$$

Here, the last term is an Itô stochastic integral.

Similarly, the inequality

$$dx(t) \leq a_1(t, x_t)\,dt + a_2(t, x_t)\,d\xi(t)$$

means that for arbitrary t, t_1, $t \geq t_1 \geq t_0$, we have with probability 1:

$$x(t) \leq x(t_1) + \int_{t_1}^{t} a_1(s, x_s)\,ds + \int_{t_1}^{t} a_2(s, x_s)\,d\xi(s),$$

where this inequality is interpreted componentwise.

As usual in stability theory one may assume that $a_{1,2}(t, 0) \equiv 0$. Sometimes we will write $t \mapsto x(t, \phi)$ for the solution of (1.1)–(1.2).

DEFINITION 1.1. The trivial solution of (1.1)–(1.2) is said to be

1) *p-stable* ($0 < p < \infty$) if for each $\epsilon > 0$ there is a $\delta(\epsilon) > 0$ such that $x(t, \phi)$ exists for $t_0 \le t < \infty$ and $\mathcal{E}|x(t, \phi)|^p < \epsilon$ ($t_0 \le t < \infty$), if only

$$\sup_{-h\le\theta\le 0} \mathcal{E}|\phi(\theta)|^p < \delta(\epsilon), \qquad \sup_{-h\le\theta\le 0} \mathcal{E}|\phi(\theta)|^{2p} < \infty; \tag{1.3}$$

2) *asymptotically p-stable* if it is stable and there is an $\epsilon > 0$ such that for all initial functions satisfying (1.3) we have

$$\lim_{t\to\infty} \mathcal{E}|x(t, \phi)|^p = 0 \tag{1.4}$$

(the set of all ϕ for which these conditions hold is called the *domain of attraction* of the trivial solution);

3) *exponentially p-stable* if there are positive constants C_1, C_2 such that

$$\mathcal{E}|x(t, \phi)|^p \le C_1 \sup_{-h\le\theta\le 0} \mathcal{E}|\phi(\theta)|^p e^{-C_2 t}, \qquad t_0 \le t < \infty,$$

for any ϕ satisfying the second inequality in (1.3);

4) *stable in probability* if for arbitrary $\epsilon_1 > 0$, $\epsilon_2 > 0$ there is a $\delta > 0$ such that

$$\mathcal{P}\left\{ \sup_{t_0\le t<\infty} |x(t, \phi)| \le \epsilon_1 \right\} \ge 1 - \epsilon_2$$

for any initial function ϕ satisfying with probability 1 the inequality

$$\sup_{-h\le\theta\le 0} |\phi(\theta)| \le \delta.$$

For $p = 2$ we obtain the definition of *mean-square stability*. □

Without especially saying so, we assume in this chapter that the functionals $a_i(t, \psi)$ satisfy the following conditions:

$$|a_i(t, \psi) - a_i(t, \chi)|^2 \le \int_{-h}^{0} |\psi(\theta) - \chi(\theta)|^2 \, dr_i(\theta),$$

where the $r_i(\theta)$ are nondecreasing bounded functions.

All definitions above can be immediately generalized to equations with unbounded delay.

1.2. Itô's formula. In many studies related to stability and control problems for SRDEs it is useful to have the following Itô formula [331], which gives an expression for the stochastic differential of the process $t \mapsto \eta(t) = u(t, x(t))$. Here, $t \mapsto x(t)$ is a solution of (1.1)–(1.2), and the function $u\colon [t_0, \infty) \times \mathbf{R}^n \to \mathbf{R}$ has continuous partial derivatives u_t, u_x, u_{xx}. Then

$$\begin{aligned} d\eta(t) = \Big[& u_t(t, x(t)) + u_x'(t, x(t)) a_1(t, x_t) + \\ & + \frac{1}{2} \operatorname{Tr}(a_2(t, x_t) a_2'(t, x_t) u_{xx}(t, x(t))) \Big] \, dt + \\ & + u_x'(t, x(t)) a_2(t, x_t) \, d\xi(t), \end{aligned} \tag{1.5}$$

where Tr denotes the matrix trace and the prime denotes transposition.

2. Liapunov's direct method

2.1. Asymptotic stability. Certain stability conditions for SRDEs can be stated in terms of Liapunov functionals, similar to the appropriate theorems for deterministic FDEs. In the sequel C_i are positive numbers, not alway the same on each occurrence.

THEOREM 2.1. *Suppose there is a continuous functional* $V\colon [t_0,\infty)\times C[-h,0]\to \mathbf{R}$ *such that for any solution of* (1.1)–(1.2) *the following inequalities hold:*

$$V(t,x_t)\ge C_1|x(t)|^2, \tag{2.1}$$

$$\mathcal{E}V(t,x_t)\le C_2 \sup_{-h\le\theta\le 0}\mathcal{E}|x(t+\theta)|^2, \tag{2.2}$$

while for arbitrary $t\ge t_0$, $s\ge t$,

$$\mathcal{E}[V(s,x_s)-V(t,x_t)]\le -C_3\int_t^s \mathcal{E}|x(\tau)|^2\,d\tau. \tag{2.3}$$

Then the trivial solution of (1.1) *is asymptotically mean-square stable.*

PROOF. By (2.3) we have

$$\mathcal{E}V(t,x_t)\le \mathcal{E}V(t_0,\phi)-C_3\int_{t_0}^t \mathcal{E}|x(s)|^2\,ds.$$

It follows from (2.1)–(2.2) that

$$C_1\mathcal{E}|x(t)|^2\le C_2\sup_{-h\le\theta\le 0}\mathcal{E}|\phi(\theta)|^2.$$

This proves stability.

We verify (1.4). Note that by the previous two inequalities,

$$\int_{t_0}^{\infty}\mathcal{E}|x(t)|^2\,dt<\infty,\qquad \sup_{t\ge t_0}\mathcal{E}|x(t)|^2<\infty. \tag{2.4}$$

Applying Itô's formula to the function $|x(t)|^2$ we find

$$d|x(t)|^2=2x'(t)\,dx(t)+\operatorname{Tr}[a_2(t,x_t)a_2'(t,x_t)]\,dt.$$

This implies that for arbitrary points t_1,t_2, $t_0\le t_1\le t_2$,

$$\mathcal{E}[|x(t_2)|^2-|x(t_1)|^2]=$$

$$=2\mathcal{E}\int_{t_1}^{t_2}[x'(t)a_1(t,x_t)+\operatorname{Tr}(a_2(t,x_t)a_2'(t,x_t))]\,dt.$$

By (2.4) there is a constant C_3 such that

$$2\mathcal{E}|x'(t)a_1(t,x_t)|\le \mathcal{E}[|x(t)|^2+|a_1(t,x_t)|^2]\le$$

$$\le \mathcal{E}\left[|x(t)|^2+\int_{-h}^0|x(t+\theta)|^2\,dr_1(\theta)\right]\le C_3,$$

$$\mathcal{E}|\operatorname{Tr}[a_2(t,x_t)a_2'(t,x_t)]|\le$$

$$\le\int_{-h}^0|x(t+\theta)|^2\,dr_2(\theta)\le C_3.$$

Hence,

$$\left|\mathcal{E}|x(t_2)|^2 - \mathcal{E}|x(t_1)|^2\right| \leq C_3(t_2 - t_1),$$

i.e. the function $\mathcal{E}|x(t)|^2$ satisfies the Lipschitz condition. This and (2.4) imply (1.4) with $p = 2$. □

2.2. Examples.

EXAMPLE 2.1. Consider the scalar equation

$$dx(t) = \left[-a(t)x(t) + \int_{-h}^{0} x(t+\theta)\, dr(\theta)\right] dt + \\ +b(t)x(t)\, d\xi(t), \tag{2.5}$$

where $\theta \mapsto r(\theta)$ is a function of bounded variation on $[-h, 0]$ and a, b are continuous functions. Consider the functional

$$\psi \mapsto V(\psi) = \psi^2(0) + \int_{-h}^{0} \left[\int_{\theta}^{0} \psi^2(\tau)\, d\tau\right] |dr(\theta)|.$$

For the stochastic differential of the functional V we find

$$dV(x_t) \leq -2x^2(t)\alpha(t)\, dt + 2b(t)x^2(t)\, d\xi(t),$$

where

$$\alpha(t) := a(t) - V_{[-h,0]}r - \frac{1}{2}b^2(t).$$

So, by thm. 2.1 the trivial solution of (2.5) is asymptotically mean-square stable if $\inf_{t \geq t_0} \alpha(t) > 0$. □

In this example the stability is due to the instantaneous negative feedback $-a(t)x(t)$. However, in certain cases one can use negative feedback with delay to stabilize a system, even if the system has only positive instantaneous feedback.

EXAMPLE 2.2. Consider the stochastic equation

$$dx(t) = -bx(t-h)\, dt + cx(t)\, d\xi(t), \qquad t > t_0,$$

with constants $b, c, h > 0$.

Sufficient conditions for asymptotic mean-square stability of this equation are:

$$0 < bh < 1, \qquad b(1 - bh) > \frac{c^2}{2}.$$

To prove this one can use the functional

$$V(\psi) = \left[\psi(0) - b\int_{-h}^{0} \psi(\theta)\, d\theta\right]^2 + b^2 \int_{-h}^{0} ds \int_{s}^{0} \psi^2(\theta)\, d\theta.$$

Using Itô's formula we find

$$2V(x_t) =$$
$$= 2\left[x(t) - b\int_{t-h}^{t} x(\theta)\,d\theta\right][dx(t) - bx(t)\,dt + bx(t-h)\,dt] +$$
$$+ \left[c^2x^2(t) + b^2hx^2(t) - b^2\int_{t-h}^{t} x^2(\theta)\,d\theta\right]dt =$$
$$= 2\left[x(t) - b\int_{t-h}^{t} x(\theta)\,d\theta\right][cx(t)\,d\xi(t) - bx(t)\,dt] +$$
$$+ \left[c^2x^2(t) + b^2hx^2(t) - b^2\int_{t-h}^{t} x^2(\theta)\,d\theta\right]dt.$$

However,

$$2b^2x(t)\int_{t-h}^{t} x(\theta)\,d\theta \le b^2\left[hx^2(t) + \int_{t-h}^{t} x^2(\theta)\,d\theta\right].$$

Therefore,

$$dV(x_t) \le 2c\left[x(t) - b\int_{t-h}^{t} x(\theta)\,d\theta\right]x(t)\,d\xi(t) +$$
$$-[2b(1-bh) - c^2]x^2(t).$$

Integration of both parts of this inequality from $s \in [t_0, t]$ to t and taking the expectation yields

$$\mathcal{E}[V(x_t) - V(x_s)] \le -[2b(1-bh) - c^2]\int_{s}^{t} \mathcal{E}x^2(\tau)\,d\tau.$$

This inequality means that

$$\mathcal{E}V(x_t) \le \mathcal{E}V(x_{t_0}), \qquad t \ge t_0.$$

Therefore

$$\mathcal{E}\left[x(t) - b\int_{t-h}^{t} x(\theta)\,d\theta\right]^2 \le \mathcal{E}V(x_{t_0}), \qquad \int_{t_0}^{\infty} \mathcal{E}x^2(s)\,ds < \infty.$$

This and the condition $bh < 1$ implies mean-square stability, since

$$\sup_{t \ge t_0} \mathcal{E}x^2(t) \le C_1 \sup_{-h \le \theta \le 0} \mathcal{E}\phi^2(\theta).$$

Note also that the function $t \mapsto \mathcal{E}x^2(t)$ satisfies the Lipschitz condition (see the proof of thm. 2.1). Together with the relation

$$\int_{t_0}^{\infty} \mathcal{E}x^2(s)\,ds < \infty$$

this implies asymptotic mean-square stability, since

$$\lim_{t \to \infty} \mathcal{E}x^2(t) = 0. \qquad \square$$

EXAMPLE 2.3. We will derive mean-square stability conditions for the equation

$$\left.\begin{aligned} dx(t) = \left[\textstyle\sum_{j=0}^{r} K_j x(t-h_j) + a(t,x_t)\right] dt + b(t,x_t)\, d\xi(t), \\ t \le t_0, \qquad x\colon J_x \to \mathbf{R}^n. \end{aligned}\right\} \tag{2.6}$$

Here the h_j are nonnegative constants, $h_0 = 0$; $a(t,x_t) = a(t,x(t),x(t-\tau_1(t)),\dots,x(t-\tau_m(t)))$; $b(t,x_t) = b(t,x(t),x(t-\theta_1(t)),\dots,x(t-\theta_l(t)))$. The matrices K_j are assume to be constant, and the nonnegative delays $\tau_i(t)$, $\theta_j(t)$ are continuously differentiabl and such that

$$0 \le \dot{\tau}_i(t) \le \alpha_i < 1, \qquad i = 1,\dots,m,$$
$$0 \le \dot{\theta}_j(t) \le \beta_j < 1, \qquad j = 1,\dots,l.$$

The functionals a, b are assumed to be continuous and to satisfy the inequalities

$$|a(t,z_0,\dots,z_m)|^2 \le \sum_{i=0}^{m} a_i^2 |z_i|^2,$$
$$|b(t,z_0,\dots,z_l)|^2 \le \sum_{j=0}^{l} b_j^2 |z_j|^2,$$

where a_i and b_j are constants and z_i are arbitrary vectors in $\mathbf{R}^n$.

We introduce the following notations:

$$k_j = \|K_j\|, \qquad K = \sum_{j=1}^{r} K_j,$$
$$(kh) = \sum_{j=1}^{r} k_j h_j, \qquad k = \sum_{j=1}^{r} k_j,$$
$$\gamma_j = \|(K_0 + K)' K_j\|, \qquad (\gamma h) = \sum_{j=1}^{r} \gamma_j h_j,,$$

where $\|\cdot\|$ is the euclidean matrix norm. The constants $\bar{a}$ and $\bar{b}$ are defined by th relations

$$\bar{a}^2 = a_0^2 + \sum_{i=1}^{m} a_i^2 (1-\alpha_i)^{-1}, \qquad \bar{b}^2 = b_0^2 + \sum_{j=1}^{l} b_j^2 (1-\beta_j)^{-1}.$$

THEOREM 2.2. *The trivial solution of* (2.6) *is asymptotically mean-square stable there is an* $\epsilon \ge 0$ *such that the matrix*

$$L := K_0 + K + (1+\epsilon)^{-1}[kI - K + (\gamma h)\epsilon I] + \frac{1}{2} I \bar{b}^2 +$$
$$+ \bar{a}[1 + (1+\epsilon)^{-1}(kh)\epsilon] I$$

is negative definite. Here I is the identity matrix.

To prove this it suffices to consider the functional

$$V(t,\psi) = |\psi(0)|^2 + \epsilon \left|\psi(0) + \sum_{j=1}^{r} \int_{-h_j}^{0} K_j\psi(\theta)\,d\theta\right|^2 +$$
$$+\epsilon\sum_{j=1}^{r}(\gamma_j + \bar{a}k_j)\int_{-h_j}^{0}(\theta - h_j - t)|\psi(\theta)|^2\,d\theta +$$
$$+\sum_{j=1}^{r} k_j \int_{-h_j}^{0} |\psi(\theta)|^2\,d\theta +$$
$$+\bar{a}^{-1}\left[1 + \epsilon(1+(kh))\sum_{i=1}^{m} a_i^2(1-\alpha_i)^{-1}\int_{-\tau_i(t)}^{0}|\psi(\theta)|^2\,d\theta\right] +$$
$$+(1+\epsilon)\sum_{j=1}^{l} b_j^2(1-\beta_j)^{-1}\int_{-\theta_j(t)}^{0}|\psi(\theta)|^2\,d\theta.$$

By computation we find for any $t_1 \in [t_0,\infty)$, $t_2 \geq t_1$,

$$\mathcal{E}[V(t_2, x_{t_2}) - V(t_1, x_{t_1})] \leq 2(1+\epsilon)\int_{t_1}^{t_2} x'(s)Lx(s)\,ds.$$

Hence, by thm. 2.1, if the conditions of thm. 2.2 hold, then the trivial solution of (2.6) is asymptotically mean-square stable. □

2.3. Exponential stability. Similar to thm. 2.1 we can state sufficient conditions for other types of stability, e.g., exponential stability, stability in probability, etc. Here we will consider sufficient conditions for the trivial solution of (1.1) to be exponentially stable. Our general assumptions are the same as in subs. 1.1.

THEOREM 2.3. *Suppose there is a continuous functional* $V\colon [t_0,\infty)\times C[-h,0] \to \mathbf{R}$ *such that*

$$\left.\begin{array}{ll} C_1\|\psi\|^2 \leq V(t,\psi) \leq C_2\|\psi\|^2, & t_0 \leq t < \infty, \\ \limsup_{\Delta t\to 0^+} \frac{1}{\Delta t}[V(t+\Delta t, y_{t+\Delta t}) - V(t,\psi)] \leq -C_3\|\psi\|^2, & \end{array}\right\} \tag{2.7}$$

where y is a solution of equation (1.1) *with initial condition $y_t = \psi \in C[-h,0]$. Then the trivial solution of* (1.1) *is exponentially mean-square stable.*

Note that for an equation (1.1) with stationary coefficients (i.e. $a_i(t,\psi) \equiv a(\psi)$), the conditions (2.7) in thm. 2.3 are not only sufficient, but also necessary for exponential stability [486].

2.4. Stability in the first approximation. The method of first approximation for investigating stability is based on the linearized initial equation. Under certain additional assumptions, stability of the initial equation follows from that of the linearized initial equation. To illustrate this method we consider the following equation of the form (1.1):

$$dx(t) = Ax(t)\,dt + a_2(t,x_t)\,d\xi(t), \qquad t \geq t_0, \qquad x\colon J_x \to \mathbf{R}^n. \tag{2.8}$$

Here, A is a constant, and a_2 is a continuous matrix function satisfying

$$|a_2(t,\psi)|^2 \le \int_{-h}^{0} |\psi(\theta)|^2\, dr(\theta), \tag{2.9}$$

where r is a nondecreasing function. As first approximation we take the deterministic equation

$$\dot{x}(t) = Ax(t). \tag{2.10}$$

We assume that (2.10) is uniformly asymptotically stable, and show that the trivial solution of (2.8) is asymptotically mean-square stable if $r_0 := r(0) - r(-h)$ is sufficiently small.

Our assumption implies the existence of two positive definite symmetric matrices P and Q such that

$$A'P + PA = -Q. \tag{2.11}$$

Now we take the following functional:

$$V(t,\psi) = \psi'(0)P\psi(0) + \|P\| \int_{-h}^{0} dr(\theta) \int_{\theta}^{0} |\psi(\tau)|^2\, d\tau,$$

where $\|\cdot\|$ is the euclidean matrix norm. The conditions (2.1)–(2.2) in thm. 2.1 clearly hold. Using Itô's formula and (2.11) we find

$$\begin{aligned} dV(t,x_t) &= [dx'(t)]Px(t) + x'(t)Px(t) + \\ &+ \left[\operatorname{Tr}(Pa_2(t,x_t)a_2'(t,x_t)) - \|P\| \int_{-h}^{0} |x(t+\theta)|^2\, dr(\theta) + r_0\|P\|\,|x(t)|^2\right] dt \le \\ &\le d\xi'(t)a_2'(t,x_t)Px(t) + x'(t)Pa_2(t,x_t)\, d\xi(t) + \\ &+ [r_0\|P\|\,|x(t)|^2 - x'(t)Qx(t)]\, dt. \end{aligned} \tag{2.12}$$

Since all moments of the solution of (2.8) are bounded on any finite time interval (2.12) implies that (2.3) holds. Hence, by thm. 2.1 the system (2.8) is asymptotically mean-square stable.

This assertion about stability in the first approximation can be generalized to equation with infinite delay, provided that

$$\int_{-\infty}^{0} |\theta|\, dr(\theta) < \infty.$$

2.5. Stability under persistent disturbances. The idea of this method is similar to the method of first approximation. Namely, a certain part of the equation is interpreted as being undisturbed, and the other part as the disturbance. Assuming that the undisturbed equation is stable and majorizes, in a certain sense, the disturbed equation, this method makes it possible to obtain stability conditions.

Using this approach we will derive conditions for the mean-square stability of a scalar second-order system with constant coefficients a_0, a_1:

$$d\dot{x}(t) = -[a_0\dot{x}(t) + a_1x(t)]\, dt + a_2(t,\dot{x}_t)\, d\xi(t). \tag{2.13}$$

In this case the undisturbed system is

$$\ddot{x}(t) + a_0\dot{x}(t) + a_1x(t) = 0,$$

and necessary and sufficient conditions for its asymptotic stability are $a_0 > 0$, $a_1 > 0$. The intensity $a_2(t, \dot{x}_t)$ of the disturbances is supposed to satisfy a condition like (2.9):

$$[a_2(t,\psi)]^2 \leq \int_{-h}^{0} [\psi(\theta)]^2 \, dr(\theta), \tag{2.14}$$

where r is a nondecreasing function. Conditions for stability of the disturbed system take the form

$$a_0 > r_0 := r(0 - r(-h), \qquad a_1 > 0. \tag{2.15}$$

To prove this one can use thm. 2.1, applied to the functional $V\colon C^1[-h,0] \to \mathbf{R}$:

$$V(\psi) = 2a_1\psi^2(0) + \dot{\psi}^2(0) +$$
$$+2\int_{-h}^{0} dr(\theta) \int_{\theta}^{0} \dot{\psi}^2(\tau)\, d\tau + [\dot{\psi}(0) + a_0\psi(0)]^2.$$

By Itô's formula, (2.13) and (2.14) we obtain

$$dV(x_t) = 2\dot{x}(t)d\dot{x}(t) +$$
$$+\left[4a_1x(t)\dot{x}(t) + 2a_2^2(t,x_t) + 2r_0\dot{x}^2(t) - 2\int_{-h}^{0} \dot{x}^2(t+\theta)\, dr(\theta)\right] dt +$$
$$+2[\dot{x}(t) + a_0x(t)][-a_1x(t)\, dt + a_2(t,x_t)\, d\xi(t)] \leq$$
$$\leq -2[(a_0 - r_0)\dot{x}^2(t) + a_1a_0x^2(t)]\, dt +$$
$$+2[2\dot{x}(t) + a_0x(t)]a_2(t,x_t)\, d\xi(t).$$

This, (2.15) and (2.16) imply that the conditions of thm. 2.1 hold. Hence the system (2.13) is asymptotically mean-square stable.

3. Boundedness of moments of solutions

3.1. General conditions for boundedness of moments. By slightly modifying the proof of thm. 2.1 we can justify the following conditions for boundedness of moments of solutions of (1.1) ([281.13]).

THEOREM 3.1. *Suppose there is a continuous functional $V\colon [t_0,\infty) \times C[-h,0] \to \mathbf{R}^n$ such that for any solution x of (1.1)–(1.2) the following inequalities hold:*

$$\left.\begin{aligned} V(t,x_t) &\geq C_1|x(t)|^2, \\ \mathcal{E}V(t,x_t) &\leq C_2\left[1 + \sup\nolimits_{-h\leq\theta\leq 0} \mathcal{E}|x(t+\theta)|^2\right], \end{aligned}\right\} \tag{3.1}$$

and for any $t \geq t_0$, $s \geq t$,

$$\mathcal{E}[V(s,x_s) - V(t,x_t)] \leq -C_3\int_t^s \mathcal{E}|x(\tau)|^p\, d\tau + C_4(s-t). \tag{3.2}$$

(Here $p \geq 2$, $C_i > 0$.) Then

$$\mathcal{E}|x(t,\phi)|^p \leq \left[1 + C_4 C_3^{-1} + \sup_{-h\leq\theta\leq 0} \mathcal{E}|\phi(\theta)|^p\right] C_2 C_1^{-1}.$$

We will consider some applications of thm. 3.1 to concrete systems.

3.2. Scalar equations. We will obtain a condition for the boundedness of the second moment of solutions of the scalar equation

$$dx(t) = [-a_0(t)x(t) + a_1(t,x_t)]\,dt + a_2(t,x_t)\,d\xi(t). \tag{3.3}$$

THEOREM 3.2. *Suppose that a_0 is a bounded measurable function, and*

$$[a_i(t,\psi)]^2 \leq q_i + \int_{-h}^{0} \psi^2(\theta)\,dr_i(\theta), \qquad i = 1, 2, \tag{3.4}$$

where q_i are constants and r_i nondecreasing functions. If the inequality

$$\inf_{t_0\leq t<\infty}\left[a_0(t) - \sqrt{\alpha_{01}} - \frac{1}{2}\alpha_{02}\right] =: q_3 > 0, \qquad \alpha_{ij} := \int_{-h}^{0} |\theta|^i |dr_j(\theta)|, \tag{3.5}$$

holds, then the solution x of (3.3), (1.2) satisfies

$$\mathcal{E}x^2(t) \leq C_1\left[1 + \sup_{-h\leq\theta\leq 0} \mathcal{E}\phi^2(\theta)\right], \qquad t_0 \leq t < \infty, \tag{3.6}$$

where the constant C_1 is independent of ϕ.

PROOF. Consider the functional

$$V(\psi) = \psi^2(0) + \int_{-h}^{0} dr_3(\theta)\int_{\theta}^{0}\psi^2(\tau)\,d\tau, \qquad r_3 := r_2 + \alpha_{01}^{-1/2} r_1.$$

Itô's formula and (3.3) imply that the stochastic differential dV satisfies

$$\begin{aligned} dV(x_t) &= \\ &= \Big[a_2^2(t,x_t) + x^2(t)(\sqrt{\alpha_{01}} + \alpha_{02} - 2a_0(t)) + \\ &\quad - \int_{-h}^{0} x^2(t+\theta)\,dr_3(\theta) + 2x(t)a_1(t,x_t)\Big]\,dt + \\ &\quad + 2x(t)a_2(t,x_t)\,d\xi(t). \end{aligned} \tag{3.7}$$

We estimate the individual summands in (3.7). By (3.4) we have

$$\begin{aligned} |2x(t)a_1(t,x_t)| &\leq \sqrt{\alpha_{01}}x^2(t) + \alpha_{01}^{-1/2}a_1^2(t,x_t) \leq \\ &\leq \sqrt{\alpha_{01}}x^2(t) + \alpha_{01}^{-1/2}\left[q_1 + \int_{-h}^{0} x^2(t+\theta)\,dr_1(\theta)\right]. \end{aligned}$$

Using for a_1 the estimate (3.4), we find

$$\begin{aligned} dV(x_t) &\leq \\ &\leq [-2q_3x^2(t) + q_4]\,dt + 2x(t)a_2(t,x_t)\,d\xi(t), \\ q_4 &:= q_2 + q_1\alpha_{01}^{-1/2}. \end{aligned} \tag{3.8}$$

Integrating both sides of this inequality from $t \geq t_0$ to $s \geq t$ and computing the expectation, by (3.5) we find that (3.2) is valid with $p = 2$. The inequalities (2.1) and (3.1) with $p = 2$ also hold. Hence, by thm. 3.1, the estimate (3.6) holds. □

Now we will obtain conditions for boundedness of moments without the assumption of an instantaneous negative feedback. Consider the scalar equation

$$dx(t) = \left[-\int_{-h}^{0} x(t+\theta)\, dr_0(\theta) + a_1(t, x_t)\right] dt + a_2(t, x_t)\, d\xi(t).$$

The functionals a_i in this expression are the same as in (3.3), and r_0 is a function of bounded variation. In the sequel we put

$$\beta_{ij} = \int_{-h}^{0} |\theta|^i\, dr_j(\theta).$$

THEOREM 3.3. *The estimate* (3.6) *holds if*

$$C_1 := (1-\alpha_{10})\beta_{00} - (1+\alpha_{10})\sqrt{\alpha_{01}} - \frac{1}{2}\alpha_{02} > 0, \qquad \alpha_{10} < \frac{1}{2}. \tag{3.9}$$

PROOF. Consider the functional $V\colon C[-h,0] \to \mathbf{R}$:

$$\begin{aligned} V(\psi) = &\left[\psi(0) - \int_{-h}^{0} dr_0(\theta) \int_{\theta}^{0} \psi(\tau)\, d\tau\right]^2 + \\ &+(\beta_{00} + \sqrt{\alpha_{01}} \int_{-h}^{0} |dr_0(\theta)| \int_{\theta}^{0} d\tau \int_{\tau}^{0} \psi^2(\tau_1)\, d\tau_1 + \\ &+ \int_{-h}^{0} [dr_2(\theta) + (1+\alpha_{10})\alpha_{01}^{-1/2}\, dr_1(\theta)] \int_{\theta}^{0} \psi^2(\tau)\, d\tau. \end{aligned}$$

Using Itô's formula we obtain, similarly to (3.8),

$$\begin{aligned} dV(x_t) \leq &\, [-C_1 x^2(t) + C_2]\, dt + \\ +2 &\left[x(t) - \int_{-h}^{0} dr_0(\theta) \int_{t+\theta}^{t} x(\tau)\, d\tau\right] a_2(t, x_t)\, d\xi(t), \\ C_2 = &\, q_2 + q_1(\alpha_{01}^{-1/2} + \alpha_{10}). \end{aligned} \tag{3.10}$$

The relations (3.9), (3.10) show that

$$\mathcal{E}V(x_t) \leq C\left(1 + \sup_{-h\leq\theta\leq 0} [\phi(\theta)]^2\right) = C_3.$$

Hence

$$\mathcal{E}\left[x(t) - \int_{-h}^{0} dr_0(\theta) \int_{t+\theta}^{t} x(\tau)\, d\tau\right]^2 \leq C_3.$$

This implies that

$$(1-\alpha_{10})\mathcal{E}x^2(t) \leq C_3 + \int_{-h}^{0} |dr_0(\theta)| \int_{t+\theta}^{t} \mathcal{E}x^2(\tau)\, d\tau.$$

Putting $\rho(t) = \max_{t_0 \leq \tau \leq t} \mathcal{E}x^2(\tau)$ we obtain

$$(1-2\alpha_{10})\rho(t) \leq C_3 + \alpha_{10} \sup_{-h\leq\theta\leq 0} \mathcal{E}[\phi(\theta)]^2.$$

This proves thm. 3.3. □

3.3. Second order equations. Consider the system of equations

$$\dot{x}(t) = y(t),$$
$$dy(t) = \left[-a_0(t)y(t) - \int_{-h}^{0} y(t+\theta)\,dr_0(\theta) - \int_{-h}^{0} x(t+\theta)\,dr_1(\theta)\right] dt +$$
$$+a_2(t, y_t)\,d\xi(t).$$

Here, $a_0(t)$ is a bounded measurable function, r_i are functions of bounded variation, and the functional a_2 is the same as in (3.3).

THEOREM 3.4. *Suppose that*

$$\beta_{01} > 0, \qquad \beta := \inf_{t \geq t_0} [a_0(t) - \alpha_{00} - \alpha_{11} - \alpha_{02}] > 0.$$

Then

$$\mathcal{E}[x^2(t) + y^2(t)] \leq C\left[1 + \sup_{-h \leq \theta \leq 0} \mathcal{E}(\phi^2(\theta) + \dot{\phi}^2(\theta))\right]. \tag{3.11}$$

PROOF. By thm. 3.1 it suffices to give a functional V satisfying (2.1), (3.1) and (3.2). Take

$$V(\psi, \chi) = 2\beta_{01}\psi^2(0) + \chi^2(0) + [V_0(\psi, \chi)]^2 +$$
$$+2\int_{-h}^{0} dr_2(\theta) \int_{\theta}^{0} \chi^2(\tau)\,d\tau + \int_{-h}^{0} |dr_0(\theta)| \int_{\theta}^{0} [\beta_{01}\psi^2(\tau) + \chi^2(\tau)]\,d\tau +$$
$$+\int_{-h}^{0} |dr_1(\theta)| \int_{\theta}^{0} d\tau_1 \int_{\tau_1}^{0} [\beta_{01}\psi^2(\tau) + \chi^2(\tau)]\,d\tau,$$

where

$$V_0(\psi, \chi) = \chi(0) + \int_{-h}^{0} \psi(\theta)\,dr_0(\theta) + \int_{-h}^{0} dr_1(\theta) \int_{\theta}^{0} \psi(\tau)\,d\tau.$$

Using Itô's formula, the same calculation as in the proof of thm. 3.2 gives

$$dV(x_t, y_t) \leq -2\beta_{01}x^2(t)[a_0(t) - \alpha_{00} - \alpha_{11}]\,dt +$$
$$+2[q_2 - \beta y^2(t)] + 2a_2(t, y_t)[V_0(x_t, y_t) + y(t)]\,d\xi(t).$$

Hence, by thm. 3.1, the estimate (3.11) holds. □

REMARK 3.1. The methods for constructing the above functionals can also be used in other situations. E.g., (3.11) holds under the conditions [281.13]

$$a_0(t) \geq 0, \qquad q_0 := \alpha_{10} + \frac{1}{2}\alpha_{21} < 1, \qquad \beta_{01} > 0,$$
$$\beta_{00} - \beta_{11} > \max\{q_0, (\beta_{01}q_0 + \alpha_{02})(1 - q_0)^{-1}\}. \qquad \square$$

We also give certain conditions for the boundedness of the fourth moments of solutions of (3.3). Assume that

$$\inf_{t \geq t_0} a_0(t) > (\alpha_{01})^{1/4} + \frac{\sqrt{3}}{2}(\alpha_{02})^{1/2}.$$

Then

$$\mathcal{E}x^4(t) \leq C\left[1 + \sup_{-h \leq \theta \leq 0} \mathcal{E}\phi^4(\theta)\right].$$

CHAPTER 6

Problems of control for deterministic FDEs

Many papers have been devoted to control problems for FDEs (see, e.g., [3, 6, 19, 29, 39, 80, 99, 190, 191, 226, 283, 284, 287, 325, 348, 353, 354, 386, 389, 391, 405, 418, 425, 426, 443, 450, 476, 499]).

Control and estimation problems for FDEs are classified according to the type of equation and cost functional, restrictions on the trajectory and control, initial and boundary conditions, etc. A distinguishing feature of these problems is the dependence of the state of the system on the previous values of the trajectory. Using this one can give definitions of controllability in various senses: conditional, relative, full, pointwise, zero-controllability, quasicontrollability, etc.

In this chapter we will give certain optimality conditions, and methods and algorithms for solving control problems.

1. The dynamic programming method for deterministic equations. Bellman's equation

In this section we consider a general approach to control problems for RDEs using the dynamic programming method. This method allows us to build optimal closed-loop controls for certain problems.

1.1. Statement of the problem. Consider the equation

$$\dot{x}(t) = f(t, x_t, u(t)), \qquad t_0 \le t \le T. \tag{1.1}$$

Here, $x(t) \in \mathbf{R}^n$, $u(t) \in U \subseteq \mathbf{R}^m$, $x_t(\theta) = x(t+\theta)$, $-h \le \theta \le 0$, $f\colon [t_0, T] \times C[-h, 0] \times U \to \mathbf{R}^n$. The function f, time moments t_0, $T > t_0$, and set U are prescribed. It is assumed that the functional $(t, \psi, u) \mapsto f(t, \psi, u)$ satisfies the *modified Carathéodory condition* (see subs. 2.2.4): for almost all t this functional is continuous in (ψ, u), for all (ψ, u) it is measurable in t, and for some neighborhood of an arbitrary point in $[t_0, T] \times C[-h, 0]$ there is a function $\lambda \in L_1([t_0, T], [0, \infty))$ such that for any $q, q_1, q_2 \in U$ we have in this neighborhood:

$$|f(t, \psi, q)| \le \lambda(t),$$
$$|f(t, \psi_1, q_1) - f(t, \psi_2, q_2)| \le \lambda(t)(\|\psi_1 - \psi_2\| + |q_1 - q_2|).$$

The initial condition is

$$x_{t_0} = \phi \in C[-h,0], \tag{1.2}$$

where ϕ is a given function.

Suppose we are also given a cost functional

$$J(u) = F_0(x_T) + \int_{t_0}^{T} F_1(t, x_t, u(t))\,dt. \tag{1.3}$$

Here, $F_0\colon C[-h,0] \to \mathbf{R}$ and $F_1\colon [t_0,T]\times C[-h,0]\times U \to \mathbf{R}$ are continuous functionals. Note that the righthand side depends on u directly and also via x (which depends on u).

A piecewise continuous function $u\colon [t_0,T] \to U$ is called an *admissible control* if for an arbitrary $(t_1,\psi) \in [t_0,T) \times C[-h,0]$ there is a solution of (1.1) on $[t_1,T]$ satisfying the initial condition $x_{t_1} = \psi$ under this control u. The set of all admissible controls is denoted by W. It is assumed that $W \neq \emptyset$.

The *optimal control problem* under consideration is to find an admissible control minimizing the cost functional $J(u)$.

We define the *Bellman functional* $V\colon [t_0,T] \times C[-h,0] \to \mathbf{R}$ as follows. Let $t \mapsto x(t;t_1,\psi;u)\colon [t_1,T] \to \mathbf{R}^n$ ($t_1 \in [t_0,T]$, $\psi \in C[-h,0]$, $u \in W$) be the solution of (1.1) with control u and initial condition $x_{t_1} = \psi$. Put $x_\tau(t,\psi;u) = x(\tau + \cdot\,; t,\psi;u)$. Then

$$V(t,\psi) :=$$

$$:= \inf_{u\in W}\left[F_0(x_T(t,\psi;u)) + \int_t^T F_1(s, x_s(t,\psi;u), u(s))\,ds\right].$$

Define the operator $L_uV\colon [t_0,T) \times C[-h,0] \to \mathbf{R}$ by

$$L_uV(t,\psi) = \limsup\nolimits_{\Delta t\to 0^+} \frac{1}{\Delta t}[V(t+\Delta t, x_{t+\Delta t}(t,\psi;u)) - V(t,\psi)].$$

Note that L_uV is the total derivative of V along the trajectories of the system (1.1) with control u.

If, in addition, the functional $(t,\psi) \mapsto V(t,\psi)$ is locally Lipschitz in ψ, then the value $L_uV(t,\psi)$ at any point (t,ψ) does not depend on the complete control u but only on the value $u(t^+)$.

REMARK 1.1. A control u defined as a function of t is called an *open-loop control*. Next to it one can consider controls of the form $t \mapsto u(x(t))$, or, more generally, $t \mapsto u(t,x_t)$, which are called *closed-loop controls*. Transition from an open-loop control to a closed-loop control cannot improve the solution of a deterministic control problem, since each function $t \mapsto u(t,x_t)$ is at the same time a function of t. However, such a transition is in certain concrete situations useful in realizing optimal or 'quasi-optimal' control. (In particular, such a transition is made in the dynamic programming method, described in subs. 1.2.) Therefore, in papers on optimal control the initial equation (in our case (1.1)) is usually written with u instead of $u(t)$ or $u(t,x_t)$, assuming that the arguments will be determined more precisely during the further investigation. This representation is useful, but not quite consistent. □

1.2. Optimality conditions

THEOREM 1.1. *Suppose there are a continuous functional* $V_0\colon [t_0,T]\times C[-h,0]\to \mathbf{R}$ *which is locally Lipschitz in the second argument and a functional* $u_0\colon [t_0,T]\times C[-h,0]\to \mathbf{R}^m$ *satisfying the Carathéodory condition, such that*

$$\inf_{u\in U}[L_u V_0(t,\psi)+F_1(t,\psi,u)]=$$

$$=L_{u_0(t,\psi)}V_0(t,\psi)+F_1(t,\psi,u_0(t,\psi))=0, \tag{1.4}$$

$$V_0(T,\psi)=F_0(\psi), \tag{1.5}$$

for all $t\in[t_0,T)$, $\psi\in C[-h,0]$. *Then* u_0 *is the optimal control direction at* (t,ψ) (i.e. the optimal solution and optimal control in the problem (1.1)–(1.3) are defined by (1.1) and (1.2) with $u(t)=u_0(t,x_t)$) *and* V_0 *is the Bellman functional in this problem.*

REMARK 1.2. We stress that the infimum in (1.4) must be computed with respect to the vector parameter $u\in U$. □

PROOF. Substituting in the middle part of (1.4) the trajectory $x(\cdot;t,\psi;u)$ with $u(\tau)=u_0(\tau,x_\tau)$ $(t\le\tau\le T)$, we find

$$L_{u_0}V_0(\tau,x_\tau(t,\psi;u))+F_1(\tau,x_\tau(t,\psi;u),u_0)=0,\qquad t\le\tau\le T, \tag{1.6}$$

where $u_0=u_0(\tau,x_\tau(t,\psi;u_0))$. Here $\tau\mapsto V_0(\tau,x_\tau(t,\psi;u))$ is an absolutely continuous function, and hence, by (1.5)–(1.6),

$$\left.\begin{aligned} V_0(t,\psi)=F_0(x_T(t,\psi);u)+\textstyle\int_t^T F_1(\tau,x_\tau(t,\psi;u),u(\tau))\,d\tau,\\ u(\tau)=u_0(\tau,x_\tau). \end{aligned}\right\} \tag{1.7}$$

Consider an arbitrary admissible control u_1. By (1.4) we find

$$L_{u_1}V_0(\tau,x_\tau(t,\psi;u_1))+F_1(\tau,x_\tau(t,\psi;u_1),u_1(\tau))\ge 0.$$

This implies that

$$V_0(t,\psi)\le F_0(x_T(t,\psi;u_1))+\int_t^T F_1(\tau,x_\tau(t,\psi;u_1),u_1(\tau))\,d\tau.$$

Comparison of this inequality with (1.7) proves thm. 1.1. □

2. Linear quadratic problems

2.1. Optimal control synthesis. Here we will determine the solution of the optimal control problem with quadratic cost functional (1.3) and linear equations

$$\dot x(t)=A(t)x(t)+A_1(t)x(t-h)+$$

$$+\int_{-h}^0 G(t,\theta)x(t+\theta)\,d\theta+B(t)u(t,x_t). \tag{2.1}$$

The initial conditions have the form (1.2). Here $t\in[t_0,T]$, $t_0<T$ is a given time moment, $x(t)\in\mathbf{R}^n$, $u\colon[t_0,T]\times C[-h,0]\to\mathbf{R}^m$, the matrix functions A,A_1,G,B are

piecewise continuous, and $U = \mathbf{R}^m$. It is required to choose u so that the quadratic cost functional

$$J(u) := x'(T)N_0x(T) + \\ + \int_{t_0}^{T} [u'(t,x_t)N_1(t)u(t,x_t) + x'(t)N_2(t)x(t)]\,dt \to \min_u . \tag{2.2}$$

Here, prime denotes transposition, all matrices $N_0, N_{1,2}(t)$ are symmetric nonnegative definite, $N_1(t)$ is positive definite, $N_{1,2}(\cdot)$ are continuous, $N_0, N_2(t) \in \mathbf{R}^{n\times n}$, $N_1(t) \in \mathbf{R}^{m\times m}$.

To solve this problem (2.1)–(2.2) we will use thm. 1.1. The relations (1.4), (1.5) become

$$\left.\begin{array}{c} \inf_{u\in\mathbf{R}^m}[L_uV_0(t,\psi) + u'N_1(t)u + \psi'(0)N_2\psi(0)] = 0, \\ V_0(T,\psi) = \psi'(0)N_0\psi(0). \end{array}\right\} \tag{2.3}$$

We assume that the Bellman functional V_0 has the form

$$V_0(t,\psi) = \psi'(0)P(t)\psi(0) + \psi'(0)\int_{-h}^{0} Q(t,\theta)\psi(\theta)\,d\theta + \\ + \int_{-h}^{0}\psi'(\theta)Q'(t,\theta)\,d\theta\,\psi(0) + \int_{-h}^{0}\int_{-h}^{0}\psi'(\theta)R(t,\theta,\rho)\psi(\rho)\,d\theta\,d\rho, \tag{2.4}$$

where $P(t)$ is a symmetric nonnegative definite matrix, $R(t,\theta,\rho) = R'(t,\rho,\theta)$, and P, Q, R are Lipschitz matrix functions with piecewise continuous derivatives. Substituting (2.4) into (2.3) and equating to zero the quadratic form in $\psi(t+\theta)$, $-h \le \theta \le 0$, we find that the matrices P, Q, R satisfy the matrix equations

$$\left.\begin{array}{c} \dot P(t) + A'(t)P(t) + P(t)A(t) + Q(t,0) + Q'(t,0) + N_2(t) = \\ = P(t)B_1(t)P(t), \\ A'(t)Q(t,\theta) + P(t)G(t,\theta) + R(t,0,\theta) + \left(\frac{\partial}{\partial t} - \frac{\partial}{\partial\theta}\right)Q(t,\theta) = \\ = P(t)B_1(t)Q(t,\theta), \\ G'(t,\theta)Q(t,\rho) + Q'(t,\theta)G(t,\rho) + \left(\frac{\partial}{\partial t} - \frac{\partial}{\partial\theta} - \frac{\partial}{\partial\rho}\right)R(t,\theta,\rho) = \\ = Q'(t,\theta)B_1(t)Q(t,\rho), \\ B_1 := BN_1^{-1}B'. \end{array}\right\} \tag{2.5}$$

The boundary conditions have the form

$$\left.\begin{array}{c} P(t) = N_0, \qquad Q(T,\theta) = 0, \qquad R(T,\theta,\rho) = 0, \\ A_1'(t)P(t) = Q'(t,-h), \\ A_1'(t)Q(t,\theta) + Q'(t,\theta)A_1(t) - R(t,-h,\theta) - R'(t,\theta,-h) = 0, \\ -h \le \theta,\rho \le 0, \qquad 0 \le t \le T. \end{array}\right\} \tag{2.6}$$

Taking of the lefthand side of (2.3) the gradient with respect to u, we conclude that the minimum of this expression (2.3) is at the point $u(t,\psi)$:

$$u(t,\psi) = -N_1^{-1}(t)B'(t)\left[P(t)\psi(0) + \int_{-h}^{0} Q(t,\theta)\psi(\theta)\,d\theta\right]. \tag{2.7}$$

So the optimal control is constructed as a closed-loop control.

Note that under these assumptions the boundary value problem (2.5)–(2.6) has a unique solution P, Q, R in the class of piecewise continuously differentiable functions [284]. It has also been proved that it is possible to find a Bellman functional of the form (2.4).

By (2.7), the optimal process can be obtained as the solution of the equation

$$\dot{x}(t) = [A(t) - B(t)N_1^{-1}(t)B'(t)P(t)]x(t) + A_1(t)x(t-h) + \\ + \int_{-h}^{0} [G(t,\theta) - B(t)N_1^{-1}(t)B'(t)Q(t,\theta)]x(t+\theta)\,d\theta \\ (t_0 \le t \le T)$$

with initial condition (1.2).

2.2. Exact solution. Here we will give the exact solution of the problem (2.5)–(2.6) for the case $G = 0$, $N_2 = 0$. We introduce matrices Z, B_2, P_1 by

$$\begin{gathered} \dot{Z}(t) = A(t)Z(t), \qquad Z(0) = I, \\ \left.\begin{aligned} \dot{B}_2(t) = -B_2(t+h)Z^{-1}(t+h)A_1(t+h)Z(t), \\ B_2(T) = I, \qquad B_2(t) = 0 \quad (t > T), \end{aligned}\right\} \qquad (2.8) \\ \dot{P}_1(t) = P_1(t)B_2'(t)Z^{-1}(t)B_1(t)[Z'(t)]^{-1}B_2(t)P_1(t), \\ P_1(T) = Z'(T)N_0Z(T), \end{gathered}$$

where $t_0 \le t \le T$ and I is the identity matrix. Then

$$\left.\begin{aligned} P(t) &= [Z'(t)]^{-1}B_2'(t)P_1(t)B_2(t)Z^{-1}(t), \\ Q(t,\theta) &= -[Z'(t)]^{-1}B_2'(t)P_1(t)\dot{B}_2(t+\theta)Z^{-1}(t+\theta), \\ R(t,\theta,\rho) &= [Z'(t+\theta)]^{-1}\dot{B}_2'(t+\theta)P_1(t)\dot{B}_2(t+\rho)Z^{-1}(t+\rho). \end{aligned}\right\} \qquad (2.9)$$

The equation (2.8) for B_2 can be explicitly solved (e.g., by the method of steps), and the equation for P_1 can be reduced to a linear equation by a change of variables $D(t) = P_1^{-1}(t)$, as $P_1(t)$ is a nondegenerate matrix. For this it is necessary and sufficient that $\det N_0 \ne 0$ and

$$\det\left[Z^{-1}(T)N_0^{-1}[Z'(T)]^{-1} + \int_t^T B_2'(\tau)Z^{-1}(\tau)B_1(\tau)[Z'(\tau)]^{-1}B_2(\tau)\,d\tau\right] \ne 0 \\ (t_0 \le t \le T).$$

In [16] a procedure for approximately solving (2.5)–(2.6) with arbitrary matrices G and N_2 has been proposed.

2.3. Systems with delays in the control. Linear quadratic equations with delays in the control can be studied similarly to (2.1)–(2.2). For this reason we restrict ourselves to the final formulas related to the equation

$$\left.\begin{gathered} \dot{x}(t) = A_1(t)x(t-h) + B(t)u(t) + K(t)u(t-h_1), \\ t_0 \le t \le T, \qquad x\colon [t_0 - h, T] \to \mathbf{R}^n, \qquad u\colon [t_0 - h_1, T] \to \mathbf{R}^m, \\ h \ge 0, \qquad h_1 \ge 0. \end{gathered}\right\} \qquad (2.10)$$

Here, A_1, B, K are piecewise continuous matrices. The solution x of (2.10) is defined by the initial condition $u(t_0+\theta_1) = \phi_1(\theta_1)$, $-h_1 \le \theta_1 \le 0$, where ϕ_1 is a given piecewise continuous function. The performance index has the form (2.2). At time t the optimal control depends on t and the previous values $x(t+\theta) = \psi(\theta)$ $(-h \le \theta \le 0)$ and $u(t+\theta) = \psi_1(\theta_1)$ $(-h_1 \le \theta_1 \le 0)$. The Bellman functional for the problem (2.10), (2.2) is $V_0(t,\psi) + V_1(t,\psi,\psi_1)$, with V_0 defined by (2.4) and V_1 equal to

$$\begin{aligned} V_1(t,\psi,\psi_1) = \psi'(0)\int_{-h_1}^{0} P_1(t,\theta_1)\psi_1(\theta_1)\,d\theta_1 + \\ \int_{-h_1}^{0} \psi_1'(\theta_1)P_1'(t,\theta_1)\psi(0)\,d\theta_1 + \\ \int_{-h}^{0} d\theta \int_{-h_1}^{0} \psi'(\theta)P_2(t,\theta,\theta_1)\psi_1(\theta_1)\,d\theta_1 + \\ \int_{-h}^{0} d\theta \int_{-h_1}^{0} \psi_1'(\theta_1)P_2'(t,\theta,\theta_1)\psi(\theta)\,d\theta_1 + \\ \int_{-h_1}^{0}\int_{-h_1}^{0} \psi_1'(\theta_1)P_3(t,\theta_1,\theta_2)\psi_1(\theta_2)\,d\theta_1\,d\theta_2. \end{aligned} \tag{2.11}$$

The optimal control of the problem (2.10), (2.2) is given by

$$\begin{aligned} u_0(t) = -N_1^{-1}(t)B'(t)\Big[(P(t) + P_1'(t,0))x(t) + \\ + \int_{-h_1}^{0} (P_1(t,\theta_1) + P_3(t,0,\theta_1))u_0(t+\theta_1)\,d\theta_1 + \\ + \int_{-h}^{0} (Q(t,\theta) + P_2'(t,\theta,0))x(t+\theta)\,d\theta\Big]. \end{aligned}$$

But (2.10) implies

$$x(t) = \int_{t_0}^{t} [A_1(\tau)x(\tau-h) + B(\tau)u_0(\tau) + K(\tau)u_0(\tau-h_1)]\,d\tau, \qquad t_0 \le t \le T,$$

where the initial conditions for x and u_0 must be taken into consideration. The last two equations can be regarded as Volterra equations in x and u_0. By solving these equations we obtain an expression for the optimal solution and control.

The matrices P, Q, R satisfy the equations

$$\left.\begin{aligned} &\dot P(t) + Q(t,0) + Q'(t,0) + N_2(t) = \\ &= [P(t)B(t) + P_1(t,0)]N_1^{-1}(t)[B'(t)P(t) + P_1'(t,0)], \\ &\left(\tfrac{\partial}{\partial t} - \tfrac{\partial}{\partial \theta}\right) Q(t,\theta) + R(t,0,\theta) = \\ &= [P(t)B(t) + P_1(t,0)]N_1^{-1}(t)[B'(t)Q(t,\theta) + P_2'(t,\theta,0)], \\ &\left(\tfrac{\partial}{\partial t} - \tfrac{\partial}{\partial \theta} - \tfrac{\partial}{\partial \theta_1}\right) R(t,\theta,\theta_1) = \\ &= [Q'(t,\theta)B(t) + P_2(t,\theta,0)]N_1^{-1}(t)[B'(t)Q(t,\theta_1) + P_2'(t,\theta_1,0)], \\ &0 \le t \le T. \end{aligned}\right\} \tag{2.12}$$

The equations for P_i take the form

$$\left.\begin{array}{c}\left(\frac{\partial}{\partial t}-\frac{\partial}{\partial\theta_1}\right)P_1(t,\theta_1)+P_2(t,0,\theta_1)=\\ =[P(t)B(t)+P_1(t,0)]N_1^{-1}(t)[B'(t)P_1(t,\theta_1)+P_3(t,0,\theta_1)],\\ \left(\frac{\partial}{\partial t}-\frac{\partial}{\partial\theta}-\frac{\partial}{\partial\theta_1}\right)P_2(t,\theta,\theta_1)=\\ =[Q'(t,\theta)B(t)+P_2(t,\theta,0)]N_1^{-1}(t)[P_1(t,\theta_1)+P_3(t,0,\theta_1)],\\ \left(\frac{\partial}{\partial t}-\frac{\partial}{\partial\theta_1}-\frac{\partial}{\partial\theta_2}\right)P_3(t,\theta_1,\theta_2)=\\ =[P_1'(t,\theta_1)B(t)+P_3(t,\theta_2,0)]N_1^{-1}(t)[B'(t)P_1(t,\theta_2)+P_3'(t,\theta_2,0)],\\ -h_1\le\theta_1,\theta_2\le 0.\end{array}\right\}\tag{2.13}$$

And the boundary conditions have the form

$$\left.\begin{array}{c}\text{If } -h\le\theta,\overline{\theta}\le 0,\ -h_1\le\theta_1,\theta_2\le 0,\text{ then}\\ P(T)=N_0,\qquad Q(T,\theta)=R(T,\theta,\overline{\theta})=0,\\ P_1(T,\theta_1)=P_2(T,\theta,\theta_1)=0,\qquad P_3(T,\theta_1,\theta_2)=0.\end{array}\right\}\tag{2.14}$$

For $t<T$, we have

$$\left.\begin{array}{c}P(t)K(t)=P_1(t,-h_1),\qquad K'(t)Q(t,\theta)=P_2'(t,\theta,-h_1),\\ A_1'(t)P(t)=Q_2'(t,-h),\\ A_1'(t)Q(t,\theta)+Q'(t,\theta)A_1(t)=R(t,\theta,-h)+R'(t,\theta,-h),\\ A_1'(t)P_1(t,\theta_1)=P_2(t,-h,\theta_1),\qquad P_3(t,\theta_1,\theta_2)=P_3'(t,\theta_2,\theta_1),\\ K'(t)P_1(t,\theta_1)+P_1'(t,\theta_1)K(t)=P_3(t,\theta_1,-h)+P_3'(t,\theta_1,-h).\end{array}\right\}\tag{2.15}$$

Under our assumptions the existence and uniqueness of the solution of the problem (2.12)–(2.15) has been proved in [284], in which is also given a method for approximately determining it.

2.4. Effects of delays in regulators. Usually, optimal control synthesis for arbitrary actual systems must include some delays in the control because of the finiteness in the time necessary to perform the measurements, process the results of these, and generate a control action. Sometimes an arbitrarily small delay in the feedback regulators radically changes the qualitative behavior of the system.

E.g., consider the dimensionless Euler–Bernoulli equation of transversal oscillation of a beam:

$$\frac{\partial^2\omega(t,x)}{\partial t^2}+\frac{\partial^4\omega(t,x)}{\partial x^4}=0,\qquad 0\le x\le 1,\qquad t>0,$$

with boundary conditions

$$\omega(t,0)=\frac{\partial\omega(t,0)}{\partial x}=\frac{\partial^3\omega(t,1)}{\partial x^3}=0$$

and feedback

$$\frac{\partial^2\omega(t,1)}{\partial x^2}=-\frac{\partial^2\omega(t-h,1)}{\partial t\partial x}.$$

Here, $\omega(t,x)$ is the displacement of the beam at point x and time t. It is known that for $h=0$ the Euler–Bernoulli beam is uniformly exponentially stabilized by the given feedback. However (see [142, 143]), the system becomes unstable if an arbitrarily small delay $h>0$ is introduced into the feedback, and arbitrarily large divergence rates are generated. E.g., for a delay h of order 10^{-182} there is a nontrivial solution which diverges at the rate of approximately $\exp(2.23\cdot 10^{181}t)$. Clearly, this means that despite its superficial innocence this equation cannot adequately describe the actual process.

The same phenomenon takes place for the one-dimensional wave equation

$$\frac{\partial^2\omega(t,x)}{\partial t^2}=\frac{\partial^2\omega(t,x)}{\partial x^2},\qquad 0\le x\le 1,\qquad t>0,$$

with boundary condition

$$\omega(t,0)=0$$

and velocity feedback

$$\frac{\partial\omega(t,1)}{\partial x}=-k\frac{\partial\omega(t-h,1)}{\partial t},$$

where $k>0$ and $h\ge 0$ are given and fixed. For $h=0$ this system is exponentially stable, but for arbitrarily small $h>0$ it becomes unstable [142]. The similar problem of boundary feedback stabilization of a one-dimensional viscoelastic beam has been considered in [241], in which a comparison of solutions of the elastic and viscoelastic cases has been give.

On the other hand we should not believe that delay has only a harmful effect. On the contrary, sometimes the performance of a system can be improved by a judicious use of a delay. This has already been shown by the simplest example in subs. 2.6.3.

In [354, 424] it has been shown that a regulator with delay can eliminate overcontrol and damp the oscillation. The possibility to replace a PD regulator by a regulator with delay has been proved in [471]. The use of regulators with discrete or distributed delay to stabilize a system and to improve the performance index has been demonstrated in [300, 444, 525].

2.5. Neutral type equations. Here we will consider certain formulas for the optimal control of a neutral type system

$$\left.\begin{aligned}\dot x(t)=A_1(t)x(t-h)+A_2(t)\dot x(t-h_1)+B(t)u(t),\\ t_0\le t\le T,\qquad h\ge 0,\qquad h_1>0,\\ x\colon[t_0-\max\{h,h_1\},T]\to\mathbf{R}^n,\qquad u\colon[t_0,T]\to\mathbf{R}^m,\end{aligned}\right\}\tag{2.16}$$

with cost functional (2.2) and initial condition (1.2). We assume that the initial function $\phi\in C^1[-h,0]$, the matrix function $A_2\in C[t_0,T]$; the remaining parameters of the problem (2.16), (2.2) being as in (2.1)–(2.2). The Bellman functional for the problem (2.16), (2.2) is equal to $V_0(t,\psi)+V_1(t,\psi,\dot\psi)$. Here, V_0 is defined by (2.4) and V_1 by (2.11), with ψ_1 replaced by $\dot\psi$. The equations and boundary conditions for P,Q,R have the form (2.5)–(2.6). The matrices P_i satisfy the equations ($B_1=$

$BN_1^{-1}B'$):

$$\left(\frac{\partial}{\partial t}-\frac{\partial}{\partial\theta_1}\right)P_1(t,\theta)+P_2(t,0,\theta_1)+P(t,\theta_1,0)=$$
$$=-P(t)B_1(t)P_1(t,\theta),$$
$$\left(\frac{\partial}{\partial t}-\frac{\partial}{\partial\theta}-\frac{\partial}{\partial\theta_1}\right)P_2(t,\theta,\theta_1)=-Q'(t,\theta)B_1(t)P_1(t,\theta_1),$$
$$\left(\frac{\partial}{\partial t}-\frac{\partial}{\partial\theta_1}-\frac{\partial}{\partial\theta_2}\right)P_3(t,\theta_1,\theta_2)=-P_1'(t,\theta_1)B_1(t)P_1(t,\theta_2).$$

The boundary conditions for $t=T$ are given by (2.14), and for $t<T$ by

$$P(t)A_2(t)=P_1(t,-h_1),\qquad P_1(t,0)=0,$$
$$A_2'(t)Q(t,\theta)=P_2'(t,\theta,-h_1),\qquad A_2'(t)P_1(t,\theta)=P_2'(t,-h,\theta),$$
$$P_2(t,-h,\theta_1)=P_2(t,\theta,-h_1)=0,$$
$$-h\le\theta\le 0,\qquad -h_1\le\theta_1\le 0.$$

3. Optimal control of bilinear hereditary systems

In this section we propose a method for constructing optimal controls for bilinear hereditary systems.

3.1. Optimality conditions. Consider the bilinear equation

$$\left.\begin{array}{l}\dot x(t)=A_1(t)x(t-h)+[A(t)x(t)+B(t)]u(t,x_t),\\ h\ge 0,\quad x\colon[t_0-h,T]\to\mathbf{R}^n,\quad u\colon[t_0,T]\times C[-h,0]\to\mathbf{R}^m,\quad t_0\le t\le T\end{array}\right\}\tag{3.1}$$

with performance index

$$J(u)+\int_0^T f(t,x_t)\,dt\to\min_u.\tag{3.2}$$

Here, $J(u)$ is given by (2.2), the matrix functions A, A_1, B are piecewise continuous, and the nonnegative continuous functional f will be chosen in the sequel. The initial condition has the form (1.2). By (1.4), (1.5), the Bellman functional $V(t,\psi)$ for (3.1)–(3.2) satisfies

$$\left.\begin{array}{l}\inf_{u\in\mathbf{R}}[L_uV(t,\psi)+\psi'(0)N_2(t)\psi(0)+N_1(t)u^2+f(t,\psi)]=0,\\ V(T,\psi)=\psi'(0)N_0\psi(0).\end{array}\right\}\tag{3.3}$$

Let us look for the solution of (3.3) in the form (2.4). We compute L_uV, substitute it in (3.3), replace $\dot x$ according to (3.1), and finally determine the control u_0 for which the lefthand side of (3.3) is minimal. We obtain

$$u_0(t,\psi)=-N_1^{-1}(t)[A(t)\psi(0)+B(t)]'\times$$
$$\times\left[P(t)\psi(0)+\int_{-h}^0 Q(t,\theta)\psi(\theta)\,d\theta\right].\tag{3.4}$$

Now we define $f(t,\psi)$ as follows:

$$f(t,\psi) = N_1^{-1}(t)\Big[(A(t)\psi(0)+B(t))'\times$$
$$\times\Big(P(t)\psi(0)+\int_{-h}^{0} Q(t,\theta)\psi(\theta)\,d\theta\Big)\Big]^2. \tag{3.5}$$

Then the nonlinear terms in (3.3) containing P, Q, R will vanish. Substituting (2.4), (3.4), (3.5) into (3.3) and equating to zero the quadratic forms in $\psi(0)$ and ψ, we obtain a system of linear partial differential equations for the matrices $P(t)$, $Q(t,\theta)$, $P(t,\theta,\rho)$. These equations, defining the optimal cost value and the optimal control synthesis, are:

$$\left.\begin{aligned} &\dot P(t)+Q(t,0)+Q'(t,0)+N_2(t)=0,\\ &\Big(\tfrac{\partial}{\partial t}-\tfrac{\partial}{\partial\theta}\Big)Q(t,\theta)+R(t,\theta,0)=0,\\ &\Big(\tfrac{\partial}{\partial t}-\tfrac{\partial}{\partial\theta}-\tfrac{\partial}{\partial\rho}\Big)R(t,\theta,\rho)=0,\\ &0\le t\le T,\qquad -h\le\theta,\rho\le 0.\end{aligned}\right\} \tag{3.6}$$

By equating to zero the quadratic form in $\psi(-h)$ we obtain the boundary conditions (2.6).

Under the conditions imposed there exists a unique solution of the boundary value problem (3.6), (2.6), as well as a unique solution of the problem (3.1), (1.2), (3.4).

3.2. Construction of the optimal control synthesis. We represent the solution of the boundary value problem (3.6), (2.6) as a sum of two terms: the first for $N_2=0$ and arbitrary N_0, and the second for $N_0=0$ and arbitrary N_2.

By (2.9), the first solution is

$$\begin{gathered} P(t)=B_2'(t)N_0B_2(t),\\ Q(t,\theta)=-B_2'(t)N_0\dot B_2(t+\theta),\\ R(t,\theta,\rho)=\dot B_2'(t+\theta)N_0\dot B_2(t+\rho),\\ 0\le t\le T,\qquad -h\le\theta,\rho\le 0.\end{gathered}$$

Here, $B_2\in\mathbf{R}^{n\times n}$ satisfies the equations

$$\begin{gathered} \dot B_2(t)=-B_2(t+h)A_1(t+h),\qquad 0\le t\le T,\\ B_2(s)=0,\qquad s>T,\qquad B_2(T)=I.\end{gathered}$$

At the discontinuity point $T-h$ of the derivative $\dot B_2(t)$ we take it to be lefthand continuous; I is the identity matrix.

The second solution is given by recurrence formulas. Let $t_i=T-ih$, where $i=0,1,\dots$. For $t_1\le t\le T$ we find

$$P(t)=\int_t^T N_2(r)\,dr,\qquad Q(t,\theta)=\int_{t+\theta+h}^T N_2(r)\,dr\,A_2(t+\theta+h),$$
$$R(t,\theta,\rho)=A_2'(t+\theta+h)\int_{t+h+\max\{\theta,\rho\}}^T N_2(r)\,dr\,A_2(t+\rho+h).$$

Here, the matrix A_2 is defined for $t_{i+1} \le t \le t_i$ and $-h \le \theta \le 0$ by

$$A_2(t+\theta+h) = \begin{cases} A_1(t+\theta+h), & -h \le \theta \le -t+t_{i+1}, \\ 0, & -t+t_{i+1} < \theta \le 0. \end{cases}$$

We also put

$$Q_1(t_i, \theta+t-t_i) = \begin{cases} Q(t_i, \theta+t-t_i), & -t+t_{i+1} \le \theta \le 0, \\ 0, & -h \le \theta < -t+t_{i+1}, \end{cases}$$

$$R_1(t_i, \theta+t-t_i, \rho) = \begin{cases} R(t_i, \theta+t-t_i, \rho), & -t+t_{i+1} \le \theta \le 0, \\ 0, & -h \le \theta < -t+t_{i+1}. \end{cases}$$

The matrix R_2 is defined for $t_{i+1} \le t \le t_i$, $-t+t_{i+1} \le \theta, \rho \le 0$ by

$$R_2(t_i, \theta+t-t_i, \rho+t-t_i) = R_1(t_i, \theta+t-t_i, \rho+t-t_i).$$

If one of the arguments θ, ρ does not belong to the interval $[-t+t_{i+1}, 0]$, then the function $R_2(t_i, \theta+t-t_i, \rho+t-t_i)$ is put equal to zero. The solution of the boundary value problem (3.6), (2.6) for $t_{i+1} \le t \le t_i$, $N_0 = 0$, is

$$P(t) = \int_t^{t_i} N_2(r)\,dr + P(t_i) + \\ + \int_{t-t_i}^0 [Q(t_i, r) + Q'(t_i, r)]\,dr + \int_{t-t_i}^0 \int_{t-t_i}^0 R(t_i, r, \xi)\,dr\,d\xi.$$

The matrix Q satisfies the relations

$$Q(t, \theta) = \Bigg[\int_{t+\theta+h}^{t_i} N_2(r)\,dr + P(t_i) + \\ + \int_{t-t_i}^0 Q(t_i, r)\,dr + \int_{t-t_{i+1}+\theta}^0 Q'(t_i, r)\,dr + \\ + \int_{t-t_i}^0 d\xi \int_{t-t_{i+1}+\theta}^0 R(t_i, r, \xi)\,dr\Bigg] A_2(t+\theta+h) + \\ + Q_1(t_i, t+\theta-t_i) + \int_{t-t_i}^0 R_1(t_i, t+\theta-t_i, r)\,dr, \\ t_{i+1} \le t \le t_i, \qquad -h \le \theta \le 0.$$

Finally, for R we have

$$R(t,\theta,\rho) = A_2'(t+\theta+h)\left[\int_{t+h+\max\{\theta,\rho\}}^{t_i} N_2(r)\,dr + P(t_i) + \right.$$
$$+\int_{t-t_{i+1}+\rho}^{0} Q'(t_i,r)\,dr + \int_{t-t_{i+1}+\theta}^{0} Q(t_i,r)\,dr +$$
$$\left. + \int_{t-t_{i+1}+\theta}^{0} d\xi \int_{t-t_{i+1}+\rho}^{0} R(t_i,r,\xi)\,dr\right] A_2(t+\rho+h) +$$
$$+A_2'(t+\theta+h)\left[Q_1'(t_i,t+\rho-t_i) + \int_{t-t_{i+1}+\theta}^{0} R_1(t_i,r,t+\rho-t_i)\,dr\right] +$$
$$+\left[Q_1(t_i,t+\theta-t_i) + \int_{t-t_{i+1}+\rho}^{0} R_1(t_i,t+\theta-t_i,r)\,dr\right] A_2(t+\rho+h) +$$
$$+R_2(t_i,t+\theta-t_i,t+\rho-t_i),$$
$$t_{i+1}\le t\le t_i, \qquad -h\le\theta,\rho\le 0.$$

REMARK 3.1. It is not difficult to surmise from the above said that similar expansions for the solution are valid for more general bilinear equations [16, 283]. □

3.3. Model of optimal feedback control for microbial growth. The results of this section can be applied to certain practical problems. E.g., consider the process of microbial growth described by (1.5.4). The initial conditions for this equation are

$$x_1(t_0) = x_{10}, \qquad x_1(t_0+\theta) = 0, \quad -h\le\theta<0, \qquad x_2(t_0) = 0.$$

The problem is to achieve a desired value of resulting product $\overline{x}_2$ at time T, and to minimize the spenditure of nourishing substances. We take the cost functional to be

$$C_1[x_2(T)-\overline{x}_2]^2 + C_2x_1(T) +$$
$$+\int_{t_0}^{T} [(x_2(t)-\overline{x}_2)^2 + C_3u^2(t) + f(t,x_1(t),x_2(t))]\,dt \to \min_u, \tag{3.7}$$

where C_i are nonnegative constants.

The method described in this section was applied to th problem (1.5.4), (3.7), and a numerical solution was obtained for the following parameter values:

$$t_0 = 0, \quad T=3, \quad h=1, \quad b=3.5, \quad \overline{x}_2 = 1.5, \quad \alpha = 52,$$
$$\gamma(t)\equiv 0.5, \quad C_i = 1, \quad x_{10} = 4.$$

Fig. 6.3.1 shows the optimal trajectories $x_1(t)$ and $x_2(t)$. It can be seen that $x_2(T)$ is close to $\overline{x}_2$, and $x_1(T)$ is close to zero.

4. Control problems with phase constraint formula

4.1. General optimality conditions. We will consider the problem (1.1)–(1.3) with the additional constraints

$$x(t)\in X(t), \qquad t_0\le t\le T, \tag{4.1}$$

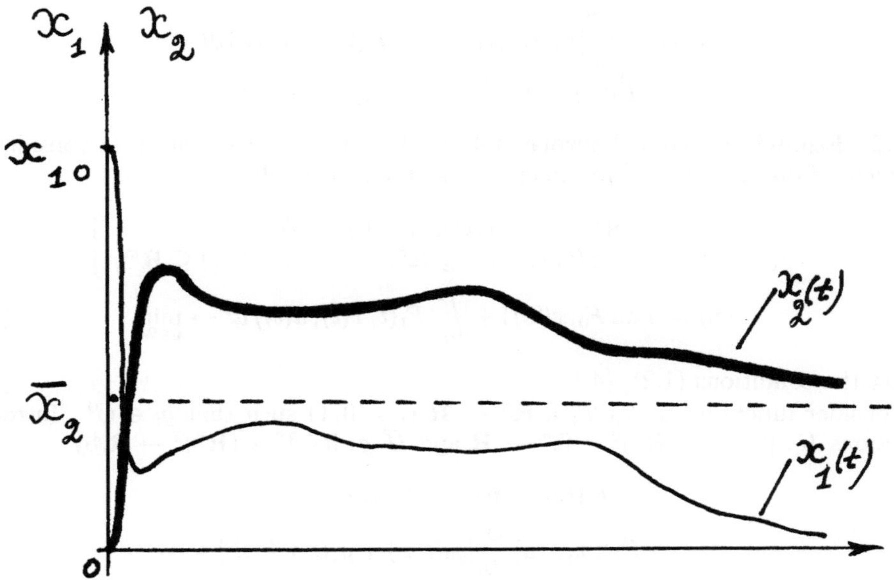

FIGURE 6.3.1. Optimal trajectories for microbial growth

where $X(t) \subseteq \mathbf{R}^n$ is a given set. In this connection the definition of the set W of admissible controls must include the requirement (4.1). In this section we let $V\colon [t_0,T] \times C[-h,0] \to \mathbf{R}$ be a functional such that for any $u \in W$, $\tau \in [t_0,T]$, $t \in [\tau,T]$ we have

$$V(t,x_t(u)) - V(\tau,x_\tau(u)) = \int_\tau^t L_u V(s,x_s(u))\,ds, \tag{4.2}$$

where $x_t(u)(\theta) = x(t+\theta;t_0;\phi;u)$, $-h \le \theta \le 0$.

THEOREM 4.1. *Suppose there are a functional $(t,\psi) \mapsto V(t,\psi)$ satisfying (4.2) and an admissible control u_0 such that*

$$\int_{t_0}^T [F_1(t,x_t(u),u) - L_u V(t,x_t(u))]\,dt \ge 0, \qquad \forall u \in W,$$

$$F_1(t,x_t(u_0),u_0) - L_{u_0} V(t,x_t(u_0)) = 0,$$

$$\alpha := F_0(x_T(u_0)) + V(T,x_T(u_0)) = \inf_{\psi \in X(T+\theta)} [F_0(\psi) + V(T,\psi)].$$

Then u_0 is the optimal control, and

$$J(u_0) = \alpha - V(t_0,\phi).$$

The assertion of the theorem follows from the formula

$$J(u) = \int_{t_0}^{T} [F_1(t, x_t(u), u) - L_u V(t, x_t(u))]\, dt + \\ + F_0(x_T(u)) + V(T, x_T(u)) - V(t_0, \phi).$$

4.2. Equations with discrete delays. For simplicity we will first consider th problem of optimal control for equations with a single delay:

$$\left.\begin{array}{ll} & \dot{x}(t) = f(t, x(t), x(t-h), u(t)), \\ t_0 \le t \le T, & x(t) \in X(t) \subseteq \mathbf{R}^n, \qquad u(t) \in U(t) \subseteq \mathbf{R}^m, \end{array}\right\} \tag{4.3}$$

$$J(u) := \min F_0(x(T)) + \int_{t_0}^{T} F_1(t, x(t), u(t))\, dt \to \min_{u \in W}, \tag{4.4}$$

under the conditions (1.2), (4.1).

Consider functions $g_i\colon [t_0, T] \times \mathbf{R}^n \to \mathbf{R}$ $(i = 0, 1)$ such that $g_i \in C^1$. Introduc functions $H\colon [t_0, T] \times (\mathbf{R}^n)^2 \times \mathbf{R}^m \to \mathbf{R}$ and $\overline{H}, g\colon [t_0, T] \times (\mathbf{R}^n)^2 \to \mathbf{R}$ by

$$\begin{aligned} H(t, x, y, u) &= -F_1(t, x, u) + \\ &+ f'(t, x, y, u)\frac{\partial}{\partial x}[g_0(t, x) + g_1(t + h, x)] \\ (g_1(t + h, x) &:= 0 \qquad \text{if } t + h > T), \\ \overline{H}(t, x, y) &= \sup_{u \in U(t)} H(t, x, y, u), \\ g(t, x, y) &= g_0(t, x) + g_1(t, y). \end{aligned}$$

Note that

$$\frac{d}{dt} g(t, x(t), x(t-h)) = \frac{\partial g}{\partial t} + \frac{\partial g}{\partial x(t)}\dot{x}(t) + \frac{\partial g}{\partial x(t-h)}\dot{x}(t-h).$$

Integration of both parts of this equality from t_0 to T, we obtain by (4.3), (4.4)

$$J(u) = \int_{t_0}^{T} [\overline{H}(t) - H(t)]\, dt - \int_{t_0}^{T} [g_t(t) + \overline{H}(t)]\, dt + C + F_0(x(T)) + \\ + g(T, x(T), x(T-h)) - g(t_0, \phi(t_0), \phi(t_0 - h)). \tag{4.5}$$

Here, $H(t) := H(t, x(t), x(t-h), u(t))$, $\overline{H}(t) := \overline{H}(t, x(t), x(t-h))$,

$$C := \int_{-h}^{0} \dot{\phi}'(\theta) \frac{\partial g_1(\theta + h, \phi(\theta))}{\partial \phi(\theta)}\, d\theta,$$

$$g_t(t) = \frac{\partial g(t, x(t), x(t-h))}{\partial t}.$$

Let u_0 be the optimal control of the problem (4.3)–(4.4), (1.2), (4.1), and let u b an admissible control. The inequality $J(u) - J(u_0) \ge 0$ and the representation (4.5 imply the following theorem.

THEOREM 4.2. *Suppose there exist functions g_i and a process $t \mapsto (u(t), x(t))$, $u \in W$, such that*

$$\begin{aligned}
\overline{H}(t, x(t), x(t-h)) &= H(t, x(t), x(t-h), u(t)),\\
C_1 := F_0(x(T)) + g(T, x(T), x(T-h)) &=\\
&= \min_{\substack{x \in X(t)\\ y \in X(T-h)}} [F_0(x) + g(T, x, y)],\\
0 = \frac{\partial g(t, x(t), x(t-h))}{\partial t} + \overline{H}(t, x(t), x(t-h)) &=\\
&= \max_{\substack{x \in X(t)\\ y \in X(T-h)}} \left[\frac{\partial g(t, x, y)}{\partial t} + \overline{H}(t, x, y)\right],\\
t_0 \le t \le T.
\end{aligned}$$

Then the process $t \mapsto (u(t), x(t))$ is optimal, and

$$J(u) = C + C_1 - g(t_0, \phi(t_0), \phi(t_0 - h)).$$

REMARK 4.1. The same optimality criterion can be stated for equation with several discrete delays h_i, $0 < h_1 < \cdots < h_k = h$:

$$\dot{x}(t) = f(t, x(t), x(t-h_1), \ldots, x(t-h_k), u(t)), \qquad t_0 \le t \le T.$$

If we are also given a cost functional (4.4), constraints (4.1), and an initial condition (1.2), the functions $H, \overline{H}, g$ must be introduced as follows:

$$\begin{aligned}
H(t, x, y_1, \ldots, y_k, u) &= -F_1(t, x, u) +\\
&+ f'(t, x, y_1, \ldots, y_k, u) \frac{\partial}{\partial x} \sum_{i=0}^{k} g_i(t + h_i, x)\\
&(h_0 := 0),\\
\overline{H}(t, x, y_1, \ldots, y_k) &= \sup_{u \in U(t)} H(t, x, y_1, \ldots, y_k, u),\\
g(t, x, y_1, \ldots, y_k) &= g_0(t, x) + \sum_{i=1}^{k} g_i(t, y_i). \qquad \square
\end{aligned}$$

THEOREM 4.3. *Suppose there exist functions $g_i \colon [t_0, T] \times \mathbf{R}^n \to \mathbf{R}$ $(i = 0, \ldots, k)$*

and a process $t \mapsto (u(t), x(t))$, $u \in W$, *such that*

$$\begin{aligned}
&\overline{H}(t, x(t), x(t-h_1), \dots, x(t-h_k)) = \\
&= H(t, x(t), x(t-h_1), \dots, x(t-h_k), u(t)), \\
C_k &:= F_0(x(t)) + g(T, x(T), x(T-h_1), \dots, x(T-h_k) = \\
&= \min_{\substack{x \in X(T) \\ \dots \\ y_k \in X(T-h_k)}} [F_0(x) + g(T, x, \dots, y_k)], \\
0 &= \frac{\partial}{\partial t} g(t, x(t), x(t-h_1), \dots, x(t-h_k)) + \\
&+ \overline{H}(t, x(t), x(t-h_1), \dots, x(t-h_k)) = \\
&= \max_{\substack{x \in X(t) \\ \dots \\ y_k \in X(t-h_k)}} \left[\frac{\partial}{\partial t} g(t, x, \dots, y_k) + \overline{H}(t, x, \dots, y_k)\right], \\
&t_0 \le t \le T.
\end{aligned}$$

Then the process $t \mapsto (u(t), x(t))$ *is optimal, and*

$$\begin{aligned}
J(u) = C_k &- g(t_0, \phi(t_0), \phi(t_0 - h_1), \dots, \phi(t_0 - h_k)) + \\
&+ \sum_{i=1}^{k} \int_{-h_k}^{0} \dot{\phi}'(\theta) \frac{\partial g_i(\theta + h_i, \phi(\theta))}{\partial \phi(\theta)} \, d\theta.
\end{aligned}$$

5. Necessary optimality conditions

In this section we will give some necessary optimality conditions in the form of the Pontrjagin principle, also called the maximum principle. A rigorous proof of this principle for hereditary systems is quite long, and can be found in [405]. According to this principle, the determination of the optimal control is reduced to the solution of a boundary value problem for a set of differential equations with both retarded and advanced arguments, together with a maximization condition of the Hamiltonian.

5.1. Systems with state delays. Consider the optimal control problem

$$\dot{x}(t) = f(t, x(t), x(t-h(t)), u(t)), \qquad t_0 \le t \le T, \tag{5.1}$$

$$\left.\begin{aligned} x(t) \in \mathbf{R}^n, &\qquad u(t) \in U(t) \subseteq \mathbf{R}^m, \\ x(t_0 + \theta) = \phi(\theta), &\qquad \theta \le 0, \end{aligned}\right\} \tag{5.2}$$

$$J(u) := \int_{t_0}^{T} F_1(t, x(t), u(t)) \, dt + F_0(x(T)) \to \min_u . \tag{5.3}$$

Here, $h(t)$ is a positive continuously differentiable function, and $\dot{h}(t) < 1$.On the functions f, F_0, u, F_1, ϕ we impose the same restrictions as in §1. Let H be the scalar function

$$\left.\begin{aligned} H(t, x, y, u, \psi) = -F_1(t, x, y, u) + \psi' f(t, x, y, u), \\ t_0 \le t \le T. \end{aligned}\right\} \tag{5.4}$$

For $t > T$ we put $H(t, x, y, u, \psi) \equiv 0$ for all x, y, u, ψ.

THEOREM 5.1. *If a function $t \mapsto (u_0(t), x_0(t))$ is an optimal control and optimal trajectory, then there is a function $\psi\colon [t_0, T] \to \mathbf{R}^n$ such that*

$$\begin{aligned} &H(t, x_0(t), x_0(t-h(t)), u_0(t), \psi(t)) = \\ &= \max_{u \in U(t)} H(t, x_0(t), x_0(t-h(t)), u, \psi(t)), \end{aligned} \tag{5.5}$$

$$\begin{aligned} &\dot\psi(t) = -H_x(t, x_0(t), x_0(t-h(t)), u_0(t), \psi(t)) + \\ &-H_y(r(t), x_0(r(t)), x_0(t), u_0(r(t)), \psi(r(t)))\dot r(t), \\ &t_0 \le t \le T, \end{aligned} \tag{5.6}$$

$$\psi(T) = -F_{0x}(x_0(T)), \qquad \psi(s) := 0, \qquad s > T. \tag{5.7}$$

Here, $H_x(t,x,y,u,\psi) = \partial H(t,x,y,u,\psi)/\partial x$, the functions H_y, F_{0x} are similarly defined, and the function $t \mapsto r(t)$ is defined by the equation $t = r(t) - h(r(t))$.

If in the problem (5.1)–(5.3) the delay $h(t)$ is constant, then $r(t) = t + h$, $\dot r(t) = 1$, and (5.6) takes the form

$$\begin{aligned} &\dot\psi(t) = -H_x(t, x_0(t), x_0(t-h), u_0(t), \psi(t)) + \\ &-H_y(t+h, x_0(t+h), x_0(t), u_0(t+h), \psi(t+h)). \end{aligned}$$

Hence the solution of the optimal control problem (5.1)–(5.3) is reduced to the solution of the boundary value problem (5.1), (5.6), under the conditions (5.2), (5.4), (5.5), (5.7).

We can similarly state necessary optimality conditions for systems with several delays h_i:

$$\left.\begin{aligned} &\dot x(t) = f(t, x(t), x(t-h_1), \ldots, x(t-h_k), u(t)), \\ &t_0 \le t \le T, \qquad x(t) \in \mathbf{R}^n, \qquad u(t) \in U(t) \subset \mathbf{R}^m, \\ &0 < h_1 < \cdots < h_k = h. \end{aligned}\right\} \tag{5.8}$$

Here, h_i are certain constants, and the Carathéodory function $(t, x, y_1, \ldots, y_k, u) \mapsto f(t, x, y_1, \ldots, y_k, u)$ is differentiable with respect to x, y_i, $i = 1, \ldots, k$. We will consider the problem (5.2)–(5.3) for the system (5.8). Put

$$\begin{aligned} &H(t, x, y_1, \ldots, y_k, u, \psi) = \\ &= -F_1(t, x, u) + \psi' f(t, x, y_1, \ldots, y_k). \end{aligned}$$

THEOREM 5.2. *Suppose there is an optimal control $t \mapsto u_0(t)$ and corresponding optimal trajectory $x_0(t)$ for th problem* (5.8), (5.2)–(5.3). *Then the conditions* (5.7) *and the equations*

$$\begin{gathered}
H(t, x_0(t), x_0(t-h_1), \dots, x_0(t-h_k), u_0(t), \psi(t)) = \\
\max_{u \in U(t)} H(t, x_0(t), x_0(t-h_1), \dots, x_0(t-h_k), u, \psi(t)), \\
\dot{\psi}(t) = -H_x(t, x_0(t), x_0(t-h_1), \dots, x_0(t-h_k), u_0(t), \psi(t)) + \\
-\sum_{i=1}^{k} H_{y_i}(t+h_i, x_0(t+h_i), x_0(t+h_i-h_1), \dots, x_0(t+h_i-h_k), u_0(t+h_i), \psi(t+h_i)), \\
t_0 \le t \le T,
\end{gathered}$$

hold for the function ψ.

5.2. Systems with delays in the control. Suppose the unknown vector function $t \mapsto x(t) \in \mathbf{R}^n$ is the solution of

$$\dot{x}(t) = f(t, x(t), u(t), u(t-h)), \qquad t_0 \le t \le T, \tag{5.9}$$

with initial conditions

$$x(t_0) = x_0 (\in \mathbf{R}^n), \qquad u(t_0 + \theta) = \phi_1(\theta), \qquad -h \le \theta \le 0. \tag{5.10}$$

Suppose the cost functional is given by (5.3), there is a constraint $u(t) \in U(t) \subseteq \mathbf{R}^m$, and T is a fixed time moment. Suppose also that the continuous functions $(t, x, u, v) \mapsto f(t, x, u, v), F_1(t, x, u, v)$, $x \mapsto F_0(x)$ are continuously differentiable with respect to x. We take the class of admissible controls to consist of the piecewise continuous functions u on an interval $[t_0 - h, T]$ satisfying (5.10) and $u(t) \in U(t)$.

THEOREM 5.3. *If there is an optimal process* $t \mapsto (u_0(t), x_0(t))$ *for the problem* (5.9)–(5.10), (5.3), *then there is a function* $\psi\colon [t_0, T+h] \to \mathbf{R}^n$ *satisfying* (5.7) *and*

$$\begin{gathered}
\dot{\psi}(t) = -H_x(t, x_0(t), u_0(t), v_0(t), \psi(t)), \\
\Delta_{u_0(t)} H(t, x_0(t), u_0(t), v_0(t), \psi(t)) + \\
+\Delta_{v_0(t+h)} H(t+h, x_0(t+h), u_0(t+h), v_0(t+h), \psi(t+h)) \le 0, \\
t_0 \le t \le T.
\end{gathered}$$

Here,

$$\begin{gathered}
H(t, x, u, v, \psi) := -F_1(t, x, u, v) + \psi' f(t, x, u, v), \\
v_0(t) = u_0(t-h), \\
\Delta_{u_0(t)} H(t, x_0(t), u_0(t), v_0(t), \psi(t)) := \\
H(t, x_0(t), u, v_0(t), \psi(t)) - H(t, x_0(t), u_0(t), v_0(t), \psi(t)),
\end{gathered}$$

and $\Delta_{v_0(t+h)}$ is similarly defined.

5.3. Systems with distributed delays. Suppose we are given a controlled system

$$\left.\begin{array}{c}\dot{x}(t) = f(t, x(t), u(t)) + \int_{t_0}^{t} K(t, s, x(s), u(s))\, ds, \\ t_0 \le t \le T,\end{array}\right\} \tag{5.11}$$

with cost functional (5.3) and conditions

$$x(t) \in \mathbf{R}^n, \qquad x(t_0) = x_0, \qquad u(t) \in U(t) \subseteq \mathbf{R}^m. \tag{5.12}$$

Put

$$H(t, x, u, \psi) = -F_1(t, x, u) + \psi' f(t, x, u) + \int_t^T \psi'(s) K(s, t, x, u)\, ds.$$

We assume that the continuous functions $(t, x, u) \mapsto f(t, x, u), F_1(t, x, u)$, $(t, s, x, u) \mapsto K(t, s, x, u)$, $x \mapsto F_0(x)$ are continuously differentiable in x.

THEOREM 5.4. *Suppose there is an optimal process $t \mapsto (u_0(t), x_0(t))$ for the problem(5.11)–(5.12), (5.3). Then there is a function ψ: $[t_0, T] \to \mathbf{R}^n$ such that*

$$\begin{gathered}\dot{\psi}(t) = -H_x(t, x_0(t), u_0(t), \psi(t)), \\ \psi(T) = -F_{0x}(x_0(T)), \\ H(t, x_0(t), u_0(t), \psi(t)) = \max_{u \in U(t)} H(t, x_0(t), u, \psi(t)).\end{gathered}$$

Now we will derive optimality conditions for a problem which is linear in x:

$$\left.\begin{array}{l}\dot{x}(t) = A(t)x(t) + \beta(t, u(t)) + \int_{t_0}^{t} [K(t, s)x(s) + \alpha(t, s, u(s))]\, ds, \\ t_0 \le t \le T, \qquad x(t) \in \mathbf{R}^n, \qquad x(t_0) = x_0, \qquad u(t) \in U(t) \subseteq \mathbf{R}^m,\end{array}\right\} \tag{5.13}$$

$$J(u) := \int_{t_0}^{T} [a_0'(t)x(t) + \alpha_0(t, u(t))]\, dt + b'x(T) \to \min_u. \tag{5.14}$$

Here, $\alpha_0(t, u) \in \mathbf{R}$, $a_0(t), \beta(t, u), \alpha(t, s, u) \in \mathbf{R}^n$, $A(t), K(t, s) \in \mathbf{R}^{n \times n}$ are continuous functions, and the vector $b \in \mathbf{R}^n$. We define the function ψ: $[t_0, T] \to \mathbf{R}^n$ by

$$\begin{gathered}\dot{\psi}(t) = a_0(t) - A'(t)\psi(t) - \int_t^T K(s, t)\psi(s)\, ds, \\ \psi(T) = -b.\end{gathered}$$

If an optimal control u_0 exists for the problem (5.13)–(5.14), then

$$\begin{aligned}&-\alpha_0(t, u_0(t)) + \psi'(t)\beta(t, u_0(t)) + \int_t^T \psi'(s)\alpha(s, t, u_0(t))\, ds = \\ &= \max_{u \in U(t)} \left\{-\alpha_0(t, u) + \psi'(t)\beta(t, u) + \int_t^T \psi'(s)\alpha(s, t, u)\, ds\right\}.\end{aligned}$$

5.4. Linear systems with discrete and distributed delays. Suppose we are given a vector $b \in \mathbf{R}^n$ and two continuous functions $t \mapsto a_0 \in \mathbf{R}^n$, $(t,u) \mapsto \alpha_0(t,u) \in \mathbf{R}$. The problem is to find the minimum value of the functional (5.14), where $t \mapsto x(t)$ denotes the solution of the linear equation

$$\dot{x}(t) = \sum_{i=1}^{N} A_i(t)x(t-h_i) + \\ + \int_{-h}^{0} A(t,\theta)x(t+\theta)\,d\theta + B(t)u(t), \tag{5.15}$$

under the conditions (5.2). Here, A_i, A, B are continuous matrix functions, and the constants $h_i \geq 0$, $h \geq 0$.

We define the function $\psi\colon [t_0,\infty) \to \mathbf{R}^n$ as the solution of the problem

$$\dot{\psi}(t) = a_0(t) - \sum_{i=1}^{N} A_i'(t+h_i)\psi(t+h_i) - \int_t^{t+h} A'(\theta, t-\theta)\psi(\theta)\,d\theta,$$
$$t_0 \leq t \leq T, \qquad \psi(T) = -b, \qquad \psi(s) = 0, \qquad s > T.$$

Because of the identity

$$\psi'(T)x(T) - \psi'(t_0)x(t_0) = \int_{t_0}^{T} d[\psi'(t)x(t)],$$

we find

$$J(u) + \int_{t_0}^{T} \Bigg\{ a_0'(t)x(t) + \alpha_0(t,u(t)) - \dot{\psi}'(t)x(t) + \\ -\psi'(t)\left[Bu(t) + \sum_{i=1}^{N} A_i(t)x(t-h_i) + \int_{-h}^{0} A(t,\theta)x(t+\theta)\,d\theta\right]\Bigg\}\,dt - \psi'(t_0)x(t_0).$$

Set $\delta(s) = 0$ for $s < 0$ and $\delta(s) = 1$ for $s > 0$ ($\delta(s) + \delta(-s) = 1$), and let C be equal to

$$C := -\int_{t_0}^{T} \left[\psi'(t)\int_{-h}^{0} \delta(-\theta - t + t_0)A(t,\theta)\phi(t+\theta-t_0)\,d\theta\right]dt.$$

So, $J(u)$ can be written as

$$J(u) = \int_{t_0}^{T} \Bigg[a_0'(t)x(t) + \alpha_0(t,u(t)) - \psi'(t)Bu(t) + \\ -\dot{\psi}'(t)x(t) - x'(t)\int_t^{t+h} A'(s,t-s)\psi(s)\,ds\Bigg]\,dt + \\ -\sum_{i=1}^{N}\int_{t_0}^{T} \psi'(t)A_i(t)x(t-h_i)\,dt - \psi'(t_0)x(t_0) + C = \\ = \int_{t_0}^{T} [\alpha_0(t,u(t)) - \psi'(t)Bu(t)]\,dt - \psi'(t_0)x(t_0) + \\ +C - \sum_{i=1}^{N}\int_{-h_i}^{0} \psi'(t_0+\theta+h_i)A_i(t_0+\theta+h_i)\phi(\theta)\delta(T-h_i-\theta)\,d\theta.$$

This implies the following theorem.

THEOREM 5.5. *A control $t \mapsto u_0(t)$ in the problem* (5.14), (5.15) *is optimal if and only if*

$$-\alpha_0(t, u_0(t)) + \psi'(t)Bu_0(t) = \max_{u \in U(t)} \{-\alpha_0(t, u + \psi'(t)Bu\},$$
$$t_0 \leq t \leq T.$$

5.5. Neutral type systems. It is well known that in general the maximum principle for NDEs is not valid without additional assumptions. Here we will consider necessary optimality conditions for the neutral type system

$$\left. \begin{array}{c} \dot{x}(t) = f(t, x(t), x(t-h), \dot{x}(t-h), u(t)), \\ t_0 \leq t \leq T, \end{array} \right\} \tag{5.16}$$

with initial condition (5.2). We assume that the functions $(t, x, y, z, u) \mapsto f(t, x, y, z, u)$ and $\theta \mapsto \phi(\theta)$ are continuous, and continuously differentiable in x, y, z, u, respectively, θ. The cost functional has the form

$$F(x(T)) \to \min_{u(t) \in U}, \qquad U \subseteq \mathbf{R}^m, \tag{5.17}$$

where U is a given convex set. We define the function ψ: $[t_0, T + h] \to \mathbf{R}^n$ by

$$\begin{aligned} \psi(t) = \psi(T) + \int_t^T [f_1'(s, x(s), x(s-h), \dot{x}(s-h), u(s))\psi(s) + \\ + f_2'(s+h, x(s+h), x(s), \dot{x}(s), u(s+h))\psi(s+h)]\, ds + \\ + f_3'(t+h, x(t+h), x(t), \dot{x}(t), u(t+h))\psi(t+h), \end{aligned} \tag{5.18}$$

where

$$\left. \begin{array}{c} f_i(t, x_1, x_2, x_3, u) := \frac{\partial f(t, x_1, x_2, x_3, u)}{\partial x_i}, \\ \psi(T) = -\frac{\partial F(x(T))}{\partial x}, \qquad \psi(t) \equiv 0, \qquad t > T. \end{array} \right\} \tag{5.19}$$

THEOREM 5.6. *If a process $t \mapsto (u_0(t), x_0(t))$ is optimal for the problem* (5.16)–(5.17), (5.2), *then there is a function ψ: $[t_0, T + h] \to \mathbf{R}^n$ such that* (5.18)–(5.19) *hold, as well as*

$$\psi'(t) \left[\frac{\partial}{\partial u} f(t, x_0(t), x_0(t-h), \dot{x}_0(t-h), u_0(t)) \right] u_0(t) =$$
$$= \max_{u \in U} \left\{ \psi'(t) \left[\frac{\partial}{\partial u} f(t, x_0(t), x_0(t-h), \dot{x}_0(t-h), u) \right] u \right\}.$$

REMARK 5.1. The papers [380, 425, 514] are concerned with the control of parabolic systems with delays in either the equations or the boundary conditions. A maximum principle has been formulated for systems with delays in the state variables. In [514] the author has considered a scalar linear time diffusion system as an example, and have obtained an analytic solution using the maximum principle. □

CHAPTER 7

Optimal control of stochastic delay systems

In practical applications the behavior of many dynamical systems depends not only on their previous history, but also on unknown disturbances.

This phenomenon can be brought about by the presence of delays and stochastic factors in the equations under consideration. In view of the intrinsic difficulties in solving such problems, progress in this is slow, especially so when stochastic processes are involved. In this chapter we will give some methods and results of the optimal control of stochastic delay systems.

1. Dynamic programming method for controlled stochastic hereditary processes

In this section we will derive the Bellman equation for controlled SRDEs. In the sequel we will use some notation mentioned in §6.1.

1.1. Problem statement. Suppose we are dealing with a controlled stochastic system of Itô type

$$\left.\begin{aligned} dx(t) = f(t,x_t,u)\,dt + \sigma(t,x_t,u)\,d\xi(t), \qquad & t_0 \le t \le T, \\ x_t(\theta) = x(t+\theta), \quad -h \le \theta \le 0, \qquad x(t) \in \mathbf{R}^n, \qquad & u(t) \in U \subseteq \mathbf{R}^m. \end{aligned}\right\} \tag{1.1}$$

Here, $t \mapsto \xi(t) \in \mathbf{R}^l$ is the standard Wiener process, and the continuous functions f, σ are defined on $[t_0,T] \times C[-h,0] \times U$. The control u is chosen from the set U, at any time t. In this situation, a feedback control is a function u: $[t_0,T] \times C[-h,0] \to U$. The initial condition for (1.1) has the form

$$x_{t_0} = \phi \in C[-h,0]. \tag{1.2}$$

The criterion $J(u)$ to be minimized is

$$J(u) = \mathcal{E}\left[F_0(x_T) + \int_{t_0}^{T} F_1(t,x_t,u(t,x_t))\,dt\right] \to \min, \tag{1.3}$$

where $\mathcal{E}$ is expectation, and F_0, F_1 are continuous scalar functions. The observations of the sate $x(t)$ are assumed to be complete. This means that for any time moment $t \in [t_0,T]$ the realization of $x(t)$ can be exactly measured. A Carathéodory feedback control $(t,\psi) \mapsto u(t,\psi)$ is admissible if for any $t_1 \in [t_0,T)$, $\psi \in C[-h,0]$ there is in

$[t_0, T]$ a solution of (1.1) with initial condition $x_{t_1} = \psi$, while the value of the cos functional (1.3) (with t_1 instead of t_0) is finite under this control. Let W be the clas of admissible controls. The problem is to find an admissible control minimizing (1.3 among all controls.

The Bellman functional $V\colon [t_0, T] \times C[-h, 0] \to \mathbf{R}$ is defined as follows:

$$V(t, \psi) = \inf_{u \in W} \left\{ \mathcal{E} \left[F_0(x_T(t, \psi; u) + \int_t^T F_1(s, x_s(t, \psi; u), u(s, x_s))\, ds \right] \right\}.$$

Here, $x_s(t, \psi; u)(\theta) = x(s + \theta; t, \psi; u)$, $-h \le \theta \le 0$, and $x(\cdot; t_1, \psi; u)$ is the solution o (1.1) with control $u \in W$ and initial condition $x(t_1 + \cdot; t_1, \psi; u) = \psi$. The operato $L_u V(t, \psi)$ is defined by the relation

$$L_u V(t, \psi) =$$
$$= \limsup_{\Delta t \to 0^+} \frac{1}{\Delta t} \mathcal{E}[V(t + \Delta t, x_{t+\Delta t}(t, \psi; u)) - V(t, \psi)]. \tag{1.4}$$

Our optimality conditions are given in the following theorem, whose proof is com pletely analogous to the proof of thm. 6.1.1.

THEOREM 1.1. *Suppose there are a functional* $(t, \psi) \mapsto V_0(t, \psi)$ *and an admissibl control* $(t, \psi) \mapsto u_0(t, \psi)$ *satisfying*

$$\inf_{u \in U} [L_u V_0(t, \psi) + F_1(t, \psi, u)] =$$
$$= L_{u_0(t,\psi)} V_0(t, \psi) + F_1(t, \psi, u_0(t, \psi)) = 0, \tag{1.5}$$
$$V_0(T, \psi) = F_0(\psi). \tag{1.6}$$

Then $(t, \psi) \mapsto u_0(t, \psi)$ *is the optimal control, and* V_0 *is the Bellman function, for th problem* (1.1)–(1.3).

Formally, the control u_0 defined by (1.5) depends on V_0, so we will sometimes denot it by $u_0(t, \psi; V_0)$ to emphasize this dependence.

2. The linear quadratic problem

2.1. Bellman functional and optimal control. We encounter substantial diffi culties when trying to use (1.5)–(1.6) to solve the optimal control problem (1.1)–(1.3) Up till now, mainly linear and quasilinear systems with quadratic criterion have beer treated in this manner.

Here we will consider the Itô type linear stochastic equation

$$\left.\begin{aligned} dx(t) = \Big[&A(t)x(t) + A_1(t)x(t-h) + \textstyle\int_{-h}^0 G(t, \theta)x(t+\theta)\, d\theta + \\ &+ B(t)u(t, x_t)\Big]\, dt + \sigma(t)\, d\xi(t), \\ t_0 \le t \le T, &\qquad x(t) \in \mathbf{R}^n, \qquad u(t, x_t) \in \mathbf{R}^m. \end{aligned}\right\} \tag{2.1}$$

Here, T is arbitrary but fixed, $\sigma(t)$ is a matrix with continuous entries, $t \mapsto \xi(t)$ is ε Wiener process, and the remaining terms are the same as in (6.2.1). The expressior to be minimized is the expectation of (6.2.2), i.e.

$$J(u) := \mathcal{E}\left[x'(T)N_0x(T) + \int_{t_0}^{T}(u'(t,x_t)N_1(t)u(t,x_t) + x'(t)N_2(t)x(t))\,dt\right] \to \to \min_u, \tag{2.2}$$

where the symmetric matrices $N_0 \geq 0$, $N_1(t) > 0$, $N_2(t) \geq 0$, and N_1, N_2 are continuous.

The Bellman equation (1.5) for the problem (2.1)–(2.2) has the same form (6.2.3) as in the deterministic case:

$$\left.\begin{aligned} \inf_u[L_uV_0(t,\psi) + u'N_1(t)u + \psi'(0)N_2(t)\psi(0)] = 0, \\ V_0(T,\psi) = \psi'(0)N_0\psi(0), \end{aligned}\right\} \tag{2.3}$$

where the operator L_u is defined by (1.4).

We try to find a solution $(t,\psi) \mapsto V(t,\psi)$ of (2.3) in the form

$$V(t,\psi) = V_0(t,\psi) + g(t). \tag{2.4}$$

Here, V_0 is given by (6.2.4) and g is a scalar function. Since we have the representation (2.4) for V, the functional L_uV can be calculated by Itô's formula (5.1.5). As before, a direct calculation shows that the optimal control $(t,\psi) \mapsto u(t,\psi)$ is defined by (6.2.7):

$$u(t,\psi) = -N_1^{-1}(t)B'(t)\left[P(t)\psi(0) + \int_{-h}^{0} Q(t,\theta)\psi(\theta)\,d\theta\right].$$

Substitute (2.4) and (6.2.7) into (2.3). Equating to zero the quadratic form in $\psi(\theta)$, $-h \leq \theta \leq 0$, we conclude that P, Q, R satisfy (6.2.5) with boundary conditions (6.2.6). The function g is equal to

$$g(t) = \int_t^T \mathrm{Tr}[\sigma'(s)P(s)\sigma(s)]\,ds. \tag{2.5}$$

Sufficient conditions for the existence of the matrices P, Q, R, as well as certain exact formulas for them, are given in §6.2.

2.2. Approximate solution. Here we will describe the method of successive approximation for solving (2.3). Let u_0 be a control, defined on $[t_0, T] \times C[-h, 0]$:

$$u_0(t,\psi) = -N_1^{-1}(t)B'(t)\left[P_0(t)\psi(0) + \int_{-h}^{0} Q_0(t,\theta)\psi(\theta)\,d\theta\right],$$

where P_0, Q_0 are arbitrary continuous matrix functions, and P_0 is symmetric and nonnegative definite. We find the solution $(t,\psi) \mapsto V_{k+1}(t,\psi)$ $(k = 0, 1, \dots)$ of the problem

$$\begin{aligned} L_{u_k}V_{k+1}(t,\psi) + u_k'N_1(t)u_k + \psi'(0)N_2(t)\psi(0) = 0, \\ V_{k+1}(T,\psi) = \psi'(0)N_0\psi(0), \end{aligned}$$

assuming that the functional u_k is known. Then we will determine a functional u_{k+1} minimizing the expression

$$\min_u\{L_uV_{k+1}(t,\psi) + u'N_1(t)u + \psi'(0)N_2(t)\psi(0)\}.$$

By this procedure we construct sequences of controls $(t, P) \mapsto u_k(t, \psi)$ and functional $(t, \psi) \mapsto V_k(t, \psi)$ $(k = 0, 1, \dots)$. Moreover, for all $k \geq 1$,

$$u_k(t, \psi) = -N_1^{-1}(t)B'(t)\left[P_k(t)\psi(0) + \int_{-h}^{0} Q_k(t, \theta)\psi(\theta)\,d\theta\right],$$

$$\begin{aligned} V_k(t, \psi) = \psi'(0)P_k(t)\psi(0) + \psi'(0)\int_{-h}^{0} Q_k(t, \theta)\psi(\theta)\,d\theta + \\ + g_k(t) + \int_{-h}^{0} \psi'(\theta)Q_k'(t, \theta)\psi(0)\,d\theta + \\ + \int_{-h}^{0}\int_{-h}^{0} \psi'(\theta)R_k(t, \theta, \theta_1)\psi(\theta_1)\,d\theta\,d\theta_1. \end{aligned}$$

The matrices P_k, Q_k, R_k are the solutions of the linear equations (6.2.5)–(6.2.6), wher the nonlinear terms are linearized by the rule

$$\begin{aligned} PB_1P &\to P_kB_1P_{k-1} + P_{k-1}B_1P_k - P_{k-1}B_1P_{k-1}, \\ PB_1Q &\to P_kB_1Q_{k-1} + P_{k-1}B_1Q_k - P_{k-1}B_1Q_{k-1}, \\ Q'B_1Q &\to Q_k'B_1Q_{k-1} + Q_{k-1}'B_1Q_k - Q_{k-1}'B_1Q_{k-1}. \end{aligned}$$

Using the linear equations for P_k, Q_k, R_k it has been proved [284] that $V_k(t, \psi) \geq V_{k+1}(t, \psi) \geq 0$, that the sequences $\{u_k\}$, $\{V_k\}$ tend to u, V as $k \to \infty$, where $(t, \psi) \mapsto u(t, \psi), V(t, \psi)$ are the optimal control and Bellman functional for the problem (1.1)–(1.3), and that

$$\begin{aligned} \max_{t, \theta, \theta_1}\{\|P_k(t) - P(t)\| + \|Q_k(t, \theta) - Q(t, \theta)\| + \\ + \|R_k(t, \theta, \theta_1) - R(t, \theta, \theta_1)\|\} = \mathcal{O}\left(\frac{1}{k!}\right), \\ t_0 \leq t \leq T, \qquad -h \leq \theta, \theta_1 \leq 0. \end{aligned}$$

Here, $\|\cdot\|$ is the euclidean matrix norm. Analytic formulas have been derived for th matrices P_k, Q_k, R_k, and each $g_k(t)$ is given by (2.5) with $P(t)$ replaced by $P_k(t)$.

2.3. Some generalizations. *2.3.1.* The above mentioned results have been generalized in [16] to linear systems with several discrete delays and under disturbance:

$$\sigma(t)\,d\xi(t) + \int_{y \in \mathbf{R}^n} f(t, y)q(dt, dy),$$

where q is a Poisson measure with independent increments.

2.3.2. Another generalization is that to systems

$$\begin{aligned} dx(t) = \Big[A(t)x(t) + A_1(t)x(t-h) + \\ + \int_{-h}^{0} G(t, \theta)x(t+\theta)\,d\theta + B(t)u(t)\Big]\,dt + \\ + \sigma(t)\,d\xi(t) + \sum_{i=1}^{l} \sigma_i(t)x(t)\,d\xi_i(t), \qquad t_0 \leq t \leq T, \end{aligned} \tag{2.6}$$

where $t \mapsto \xi(t), \xi_i(t)$ are mutually independent scalar Wiener processes. The solution of the problem (2.6), (1.3) is given by (2.4), (6.2.7). The functions P, Q, R, g satisfy the same equations (6.2.5)–(6.2.6), (2.5), but on the lefthand side of the first equation of (6.2.5) we must add

$$\sum_{i=1}^{l} \sigma_i'(t) P(t) \sigma_i(t)$$

(see [16]).

3. Approximate optimal control for systems with small parameters

3.1. Formal algorithm. Here we will consider an optimal control problem of the type (1.1)–(1.3) depending on a small parameter:

$$\begin{aligned} dx(t) &= f(t, x_t, u(t, x_t); \epsilon)\, dt + \\ &+ \sigma(t, x_t, u(t, x_t); \epsilon)\, d\xi(t), \qquad t_0 \le t \le T, \end{aligned} \tag{3.1}$$

$$J(u; \epsilon) := \mathcal{E}\left[F_0(x_T; \epsilon) + \int_{t_0}^{T} F_1(t, x_t, u(t, x_t); \epsilon)\, dt\right] \to \inf_{u(t,x_t)\in U}. \tag{3.2}$$

Then the optimal control u_0, Bellman functional V, and operator L_u also depend on the parameter ϵ. We assume that the dependence of u_0 on t, ψ, V, ϵ given by the Bellman equation for (3.1)–(3.2) is known.

We represent $u_0, L_u, V, F_0.F_1$ by series in ϵ:

$$\left.\begin{aligned} u_0(t, \psi; V, \epsilon) &= u^0(t, \psi) + \epsilon u^1(t, \psi) + \dots, \\ V(t, \psi; \epsilon) &= V^0(t, \psi) + \epsilon V^1(t, \psi) + \dots, \\ L_{u_0}(\epsilon) &= L^0 + \epsilon L^1 + \dots, \\ F_0(\psi; \epsilon) &= F_0^0(\psi) + \epsilon F_0^1(\psi) + \dots, \\ F_1(t, \psi, u_0; \epsilon) &= F_1^0(t, \psi) + \epsilon F_1^1(t, \psi) + \dots. \end{aligned}\right\} \tag{3.3}$$

The coefficients of all the expansions on the righthand side of (3.3) are independent of ϵ.

Substituting the expansions (3.3) into (1.5)–(1.6) and equating to zero the coefficients at the successive powers of ϵ, we obtain

$$\left.\begin{aligned} \textstyle\sum_{j=0}^{i} L^{i-j} V^j(t, \psi) + F_1^i(t, \psi) &= 0, \\ V^i(T, \psi) &= F_0^i(\psi), \end{aligned}\right\} i = 0, 1, \dots. \tag{3.4}$$

By solving (3.4) for $i = 0, \dots, k$, we obtain the functionals V^i.

Consider the sequence of controls $(t, \psi) \mapsto u_0(t, \psi; Q_k; \epsilon)$, where $Q_k = V^0 + \dots + \epsilon^k V^k$. We will denote the controls obtained in this way by

$$v_k(t, \psi; \epsilon) = u_0(t, \psi; Q_k; \epsilon).$$

Let $(t, \psi) \mapsto \delta_k(t, \psi; \epsilon)$ be the functional

$$\begin{aligned} \delta_k(t, \psi; \epsilon) = \sum_{i=0}^{k} \epsilon^i &\left[\sum_{j=0}^{i} L^{i-j} V^j(t, \psi) + F_1^i(t, \psi)\right] + \\ &- L_{v_k} Q_k(t, \psi; \epsilon) - F_1(t, \psi, v_k; \epsilon). \end{aligned} \tag{3.5}$$

Equations (3.4)–(3.5) imply that the boundary value problem

$$\inf_{u\in U}[L_u V(t,\psi;\epsilon) + F_1(t,\psi,u;\epsilon)] + \delta_k(t,\psi;\epsilon) = 0,$$
$$V(T,\psi;\epsilon) = Q_k(T,\psi;\epsilon),$$

has the solution $Q_k(t,\psi;\epsilon)$, since

$$\left.\begin{aligned} L_{v_k}Q_k(t,\psi;\epsilon) + F_1(t,\psi,v_k;\epsilon) + \delta_k(t,\psi;\epsilon) = 0,\\ Q_k(T,\psi;\epsilon) = F_0^0(\psi) + \cdots + \epsilon^k F_0^k(\psi). \end{aligned}\right\} \tag{3.6}$$

Comparing (3.6) with (1.5)–(1.6), we conclude that v_k and Q_k are the optimal control and Bellman functional for the problem (3.1) with minimization criterion $J_k(u;\epsilon)$:

$$\left.\begin{aligned} J_k(u;\epsilon) = J(u;\epsilon) + r_k,\\ r_k = \mathcal{E}\left[-F_0(x_T;\epsilon) + \textstyle\sum_{j=0}^k \epsilon^j F_0^j(x_T) + \int_{t_0}^T \delta_k(s,x_s;\epsilon)\,ds\right]. \end{aligned}\right\} \tag{3.7}$$

If we could show that $r_k = \mathcal{O}(\epsilon^{k+1})$, then v_k could be shown to be the kth approximation to the optimal control. (The latter means that

$$0 \le J(v_k;\epsilon) - J(u_0;\epsilon) = \mathcal{O}(\epsilon^{k+1}).)$$

This estimate has been proved in [16] for quasilinear systems with quadratic cost functional.

3.2. Quasilinear systems with quadratic cost. We will consider in some detail the zeroth approximation of the optimal control for a quasilinear system of the type (1.1):

$$dx(t) = [\epsilon f(t,x_t) + B(t)u(t,x_t)]\,dt + \sigma(t)\,d\xi(t), \tag{3.8}$$
$$t_0 \le t \le T, \qquad x(t)\in\mathbf{R}^n, \qquad u(t,x_t)\in\mathbf{R}^m.$$

The initial condition for (3.8) is (1.2), and the cost functional $J(u)$ is given by (2.2). We assume f to be a continuous functional, and

$$|f(t,\psi)| \le a_0 + \int_{-h}^0 |\psi(\theta)|^2\,dr(\theta), \tag{3.9}$$

where r is a nondecreasing bounded function on $[-h,0]$. The matrices B and σ are taken measurable and bounded.

In the expansions (3.3) we find

$$\left.\begin{aligned} L^0 = \tfrac{\partial}{\partial t} + \tfrac{1}{2}\operatorname{Tr}\left(\sigma\sigma'\tfrac{\partial^2}{\partial x^2}\right),\\ V^0(t,\psi) = \psi'(0)P(t)\psi(0) + \textstyle\int_t^T \operatorname{Tr}[\sigma(s)\sigma'(s)P(s)]\,ds,\\ u^0(t,\psi) = -N_1^{-1}B'(t)P(t)\psi(0). \end{aligned}\right\} \tag{3.10}$$

Here, the matrix $P(t)$ satisfies the conditions

$$\dot P(t) + N_2(t) = P(t)B_1(t)P(t), \qquad P(t) = N_0, \qquad B_1 := BN_1^{-1}B'.$$

Taking into account (3.7), we see that the control $v_0 = u^0(t,\psi)$ is optimal for equation (3.8) with cost functional

$$\left.\begin{array}{c} J_0(u;\epsilon) = J(u) + \mathcal{E}\int_{t_0}^T \delta_0(s,x_s;\epsilon)\,ds \to \min_u, \\ \delta_0(t,\psi;\epsilon) := -2\epsilon f'(t,\psi)P(t)\psi(0). \end{array}\right\} \tag{3.11}$$

The Bellman functional for (3.8), (3.11) is V^0. By (3.9)–(3.10),

$$|\delta_0(t,\psi;\epsilon)| \le \epsilon C(1+\|\psi\|^2). \tag{3.12}$$

In the sequel C will denote various constants not depending on t,ψ,ϵ.

Let u be the optimal control in the problem (3.8), (2.2). Then

$$J(u) \le J(u^0) = J_0(u^0;\epsilon) - \mathcal{E}\int_{t_0}^t \delta_0(s,x_s^{u^0};\epsilon)\,ds, \tag{3.13}$$

where x_s^u is the solution of the problem (3.8), (1.2) with control u. However, the second moment $\mathcal{M}|x_s^{u^0}|^2$ is bounded, because of (3.8)–(3.10). This, (3.12), and (3.13) imply that

$$-J(u) + J_0(u^0;\epsilon) \le \epsilon C(1+\|\phi\|^2). \tag{3.14}$$

On the other hand,

$$J_0(u^0;\epsilon) \le J_0(u;\epsilon) = J(u) + E\int_{t_0}^t \delta_0(s,x_s^u;\epsilon)\,ds. \tag{3.15}$$

But $J(u) \le J(0)$. This inequality and the positive definiteness of $N_1(t)$ imply that

$$\mathcal{E}\int_{t_0}^T |u(s,x_s)|^2\,ds < \infty.$$

Consequently we may conclude, as above, that the second moment $\mathcal{M}|x_s^u|^2$ is bounded. So, by (3.15),

$$J(u) - J_0(u^0;\epsilon) \le \epsilon C(1+\|\phi\|^2). \tag{3.16}$$

Using (3.14), (3.16) we find

$$\begin{aligned} 0 &\le J(u^0) - J(u) \le \\ &\le |J(u) - J_0(u^0;\epsilon)| + |J(u^0) - J_0(u^0;\epsilon)| \le \\ &\le \epsilon C(1+\|\phi\|^2). \end{aligned}$$

Hence u^0 is the zeroth approximation of the optimal control.

4. Another approach to the problem of optimal synthesis control

One of the main difficulties in applying the dynamic programming equation for hereditary systems is that for a general functional $(t,\psi) \mapsto V(t,\psi)$ the form of the operator L_u is unknown. However, if we suitably restrict the set of admissible functionals V, then L_u can be given explicitly. For such classes of hereditary systems we can sometimes obtain analytic formulas for the successive approximations to the optimal control. This approach has been developed in [16], and we will give it here.

4.1. Admissible functionals. Let us once again consider the optimal control problem (1.1)–(1.3). Let $PC[-h,0]$ be the space of piecewise continuous functions $\phi\colon [-h,0] \to \mathbf{R}^n$. For a functional $V\colon [t_0,T]\times PC[-h,0] \to \mathbf{R}$ we define the function $((t,x)\mapsto V_\psi(t,x)) : [t_0,T]\times \mathbf{R}^n \to \mathbf{R}$ as $V(t,\psi^x)$, where $\psi^x(\theta) := \psi(\theta)$ $(-h \le \theta < 0)$, $\psi^x(0) := x$. We assume that

$$\begin{aligned}
&L_u V(t,\psi) = L_u V_\psi(t,x) = \\
&\frac{\partial}{\partial t} V_\psi(t,x) + f'(t,\phi,u)\nabla V_\psi(t,x) + \\
&+\frac{1}{2}\operatorname{Tr}[\sigma'(t,\psi,u)\nabla^2 V_\psi(t,x)\sigma(t,\psi,u)].
\end{aligned}\tag{4.1}$$

Here, L_u is defined by (1.4), and $\nabla V_\psi(t,x) := \partial V_\psi(t,x)/\partial x$. We will denote the class of such functional by D; it is sometimes used in control problems. Of course, all previous assertions in this chapter can be reformulated, with minor modifications, in terms of the functionals $V \in D$ ([16]).

E.g., the modification of thm. 1.1 is

THEOREM 4.1. *Suppose there are a functional $V_0 \in D$ satisfying* (1.6) *and an admissible control u_0 such that*

$$\begin{aligned}
&\inf_{u\in U}\{L_u V_0(t,\psi) + F_1(t,\psi,u)\} = \\
&= \frac{\partial}{\partial t} V_{0\psi}(t,x) + f'(t,\psi,u_0)\nabla V_{0\psi}(t,x) + \\
&+\frac{1}{2}\operatorname{Tr}[\sigma'(t,\psi,u_0)\nabla^2 V_{0\psi}(t,x)\sigma(t,\psi,u_0)] = 0.
\end{aligned}\tag{4.2}$$

Then u_0 is the optimal control and V_0 the Bellman function of the problem (1.1)–(1.3).

For the linear quadratic problem (2.1)–(2.2) the solution of (4.2), (1.6) is given by (2.4). Note that the functional V defined by (2.4) belongs to D.

4.2. Quasilinear quadratic problems. We will apply the above described approach to the problem (3.8), (2.2). Then

$$L_u V_\psi(t,x) = L^0 V_\psi(t,x) + [\epsilon f(t,\psi) + B(t)u]'\nabla V_\psi(t,x),$$

where L^0 is defined by (3.10). The solution of (3.8), (2.2) is, by (4.1), related to the investigation of the equation

$$\left.\begin{aligned}
&L^0 V_\psi(t,x) + \epsilon f'(t,\psi)\nabla V_\psi(t,x) + x' N_2(t)x = \\
&= \tfrac{1}{4}\nabla V_\psi'(t,x)B_1(t)\nabla V_\psi(t,x), \qquad t_0 \le t \le T, \\
&V(T,\psi) = x' N_0 x, \qquad x = \psi(0), \qquad B_1 = BN_1^{-1}B'.
\end{aligned}\right\}\tag{4.3}$$

Under certain restrictions, the optimal control u_0 can be written as

$$u_0(t,\psi) = -\frac{1}{2}N_1^{-1}(t)B'(t)\nabla V_\psi(t,x).\tag{4.4}$$

Using the general scheme (3.3)–(3.5) we find that an approximate solution of (4.3) may be looked for as the cut-off expansions

$$\left.\begin{aligned} V(t,\psi;\epsilon) &= V^0(t,\psi) + \epsilon V^1(t,\psi) + \dots, \\ u_0(t,\psi;\epsilon) &= u^0(t,\psi) + \epsilon u^1(t,\psi) + \dots. \end{aligned}\right\} \tag{4.5}$$

Here, V^0 and u^0 are defined by (3.10), (4.4).

By (4.3)–(4.3), for $i \geq 1$,

$$\left.\begin{aligned} &L^0 V_\psi(t,x) + f'(t,\psi)\nabla V_\psi^{i-1}(t,x) = \\ &= \tfrac{1}{4}\textstyle\sum_{j=0}^{i-1} \nabla V_\psi^{j\prime}(t,x) B_1(t) \nabla V_\psi^{i-j}(t,x), \\ &V_\psi^i(T,x) = 0, \qquad x = \psi(0). \end{aligned}\right\} \tag{4.6}$$

Solving (4.6) for $i = 1, \dots, k$, we find the approximate optimal control v_k:

$$v_k(t,\psi) = -\frac{1}{2} N_1^{-1}(t) B'(t) \nabla[V_\psi^0(t,x) + \dots + \epsilon^k V_\psi^k(t,x)]. \tag{4.7}$$

Equation (4.6) implies that the representation

$$V^i(t,\psi) = \mathcal{E}\int_t^T S_i(t,y_s)\,ds \tag{4.8}$$

holds. Here,

$$\begin{gathered} S_i(t,P) = f'(t,\psi)\nabla V_\psi^{i-1}(t,x) + \\ -\frac{1}{4}\sum_{j=0}^{i-1} \nabla V_\psi^{j\prime}(t,x) B_1(t) \nabla V_\psi^{i-j}(t,x), \\ x = \psi(0). \end{gathered}$$

The process $s \mapsto y(s)$ satisfies the linear stochastic equation

$$\left.\begin{aligned} dy(s) = -B_1(s)P(s)y(s)\,ds + \sigma(s)\,d\xi(s), \qquad t \leq s \leq T, \\ y_t = \psi \in H[-h,0]. \end{aligned}\right\} \tag{4.9}$$

If $\sigma\sigma'$ is nondegenerate, then $s \mapsto y(s)$ is the Gaussian process with expectation $m(s)$ and covariance matrix $D(s)$, where

$$\left.\begin{aligned} m(s) &= z(s,t)x, \qquad x = \psi(0), \\ D(s) &= \textstyle\int_t^s z(s,\tau)\sigma(\tau)\sigma'(\tau)z'(s,\tau)\,d\tau. \end{aligned}\right\} \tag{4.10}$$

Here, the Cauchy matrix $z(s,t)$ is defined by

$$\frac{\partial z(s,t)}{\partial s} = -B_1(s)P(s)z(s,t), \qquad z(t,t) = I,$$

where I is the identity matrix.

In certain cases the righthand side of (4.8) may be reduced by quadratures. E.g., suppose $f(t,x_t) = f(t,x(t-h))$. Then

$$V^1(t,\psi) = 2\int_t^{t+h} ds \int_{\mathbf{R}^n} f'(s,\psi(d-t-h))P(s)yp(t,\psi(0),s,y)\,dy +$$

$$+2\int_{t+h}^{T} ds \int_{\mathbf{R}^n} dy \int_{\mathbf{R}^n} f'(s,y^1)P(s)yp(t,\psi(0),s-h,y^1)p(s-h,y^1,s,y)\,dy^1,$$

$$0 \le t \le T-h,$$

$$V^1(t,\psi) = 2\int_t^{T} ds \int_{\mathbf{R}^n} f'(s,\psi(s-t-h))P(s)yp(t,\psi(0),s,y)\,dy,$$

$$T-h \le t \le T.$$

Here, the transition function p of the process (4.9) is

$$p(t,x,s,y) = [(2\pi)^n\Delta(s)]^{-1/2}\exp\left[-\frac{1}{2}(y-m(s))'D^{-1}(s)(x-m(s))\right],$$

where $m(s), D(s)$ are given by (4.10), and $\Delta(s) = \det D(s)$.

We will finally consider an auxiliary problem for which the control (4.7) is optimal. Equation (3.6) implies that the functionals $(t,\psi) \mapsto Q_k(t,\psi;\epsilon) = V^0 + \cdots + \epsilon^k V^k$ satisfy

$$L^0 Q_{k\psi}(t,x) + \epsilon f'(t,\psi)\nabla Q_{k\psi}(t,x) + x'N_2(t)x + \delta_k(t,\psi;\epsilon) =$$

$$= \frac{1}{4}\nabla Q'_{k\psi}(t,x)B_1(t)\nabla Q_{k\psi}(t,x),$$

$$Q_k(T,\psi) = x'N_0x, \qquad x = \psi(0),$$

where

$$\delta_k(t,\psi;\epsilon) =$$

$$= \left[\frac{1}{4}\sum_{j=1}^{k}\sum_{i=0}^{j-1}\epsilon^i\nabla V_\psi^{j\prime}(t,x)B_1(t)\nabla V_\psi^{k+1+i-j}(t,x) - f'(t,\psi)\nabla V_\psi^k(t,x)\right]\epsilon^{k+1}.$$

Suppose we are give the system (3.8). The cost functional to be minimized is taken to be $u \mapsto J_k(u;\epsilon)$, which is according to (3.7) defined by

$$J_k(u;\epsilon) = J(u) + \mathcal{E}\int_{t_0}^{T}\delta_k(t,x_t;\epsilon)\,dt \to \inf_u. \tag{4.11}$$

We assume that

$$J\nabla V_\psi^i(t,x)|^2 \le C(1+\|\psi\|^2), \qquad i = 2,\ldots,k.$$

Then the control v_k in (4.7) and the functional Q_k are the optimal control and Bellman functional for the problem (3.8), (4.11); moreover, $0 \le J(v_k) - J(u) = \mathcal{O}(\epsilon^{k+1})$. The details of the proof are the same as for the zeroth approximation, and can be found in [16].

CHAPTER 8

State estimates of stochastic systems with delay

In this chapter we will consider filtering problems for Gaussian unobservable processes and linear observations with delays. We will investigate the dependence of optimal estimates on the delays. We use a least-squares method to estimate the coordinates of the system at each time moment, based on input data and corresponding measurements with noise.

1. Filtering of Gaussian processes

1.1. Problem statement. We are given a basic probability space $\{\Omega, \sigma, P\}$, a family of σ-algebras $\{f_t\}$, $t \geq t_0$, and an f_t-adapted stochastic process $t \mapsto (x(t), y(t))$, where $t \mapsto x(t) \in \mathbf{R}^n$ is the Gaussian unobservable component and $t \mapsto y(t) \in \mathbf{R}^l$ the observable component, with bounded second moments. The filtering problem is to find the optimal (in the mean square sense) estimate $m_0(T)$ for a value $x(T)$ under observations $y(t)$, $t_0 \leq t \leq T$. It is well known that this optimal estimate is the conditional expectation

$$m_0(T) = \mathcal{E}(x(T)|f_T^y), \tag{1.1}$$

where f_t^y is the minimal σ-algebra generated by the observations $y(t)$, $t_0 \leq t \leq T$. We assume that y satisfies the equation

$$\left.\begin{aligned} dy(t) &= A(t)x(t-h)\,dt + d\xi(t), \qquad t_0 \leq t \leq T, \\ y(t_0) &= 0, \qquad x(s) = 0, \quad t_0 - h \leq s \leq t_0. \end{aligned}\right\} \tag{1.2}$$

Here, $t \mapsto \xi(t) \in \mathbf{R}^l$ is a Gaussian stochastic process with independent increments:

$$\left.\begin{aligned} &\mathcal{E}\xi(t) = 0, \qquad \mathcal{E}x(t) = 0, \qquad \mathcal{E}(x(t)x'(s)) = R(t,s), \\ &\mathcal{E}(x(t)\,d\xi'(s)) = Q(t,s)\,ds, \qquad \mathcal{E}(d\xi(s)\,d\xi'(s)) = N(s)\,ds. \end{aligned}\right\} \tag{1.3}$$

The matrices A, R, Q, N are bounded and piecewise continuous, $Q(t,s) \equiv 0$ for $t < s$, and $N(s)$ is uniformly positive definite with respect to s.

The delay h in the observation equation (1.2) is due to the finiteness of the time necessary to perform the observation process and process its results.

1.2. Integral representation for the optimal estimate

LEMMA 1.1. *There is a definite matrix function* $u\colon [t_0, T] \to \mathbf{R}^{n\times l}$ *such that under the conditions of subs.* 1.1 *we find for the function* (1.1)

$$m_0(T) = \int_{t_0}^{T} u_0(t)\, dy(t), \qquad \int_{t_0}^{T} \|u_0(t)\|^2\, dt < \infty. \tag{1.4}$$

PROOF. Without loss of generality we may assume that $t_0 = 0$. Let

$$t_k^{(i)} = 2^{-i}kT, \qquad k = 0, 1, \dots, 2^i, \qquad i = 1, 2, \dots.$$

Let $f_{t,i}^y$ be the minimal σ-algebra generated by the quantities $y(t_0^{(i)})$, $y(t_1^{(i)}) - y(t_0^{(i)}), \dots,$ $y(t_{2^i}^{(i)}) - y(t_{2^i-1}^{(i)})$. Then, by [331]

$$\lim_{i\to\infty} \mathcal{E}|\mathcal{E}(x(T)|f_{t,i}^y) - \mathcal{E}(x(T)|f_t^y)|^2 = 0. \tag{1.5}$$

Moreover, for any integer $i = 1, 2, \dots$ we find

$$\mathcal{E}(x(t)|f_{t,i}^y) = \int_0^T u_i(t)\, dy(t), \tag{1.6}$$

where $\{u_i\}$ is a sequence of determined (non-stochastic) piecewise constant functions. We will prove that $\{u_i\}$ is a fundamental sequence in $L_2[0, T]$. Put

$$\left.\begin{array}{c} x_0(t) = \int_0^t \alpha(t,s)\, d\xi(s), \qquad \alpha(t,s) := Q(t,s)N^{-1}(s), \\ \beta_i(\tau) = u_i(\tau) + \int_{\tau+h}^T u_i(t)A(t)\alpha(t-h,\tau)\, dt, \\ \Delta_{ij}u(t) = u_i(t) - u_j(t), \qquad \Delta_{ij}\beta(t) = \beta_i(t) - \beta_j(t). \end{array}\right\} \tag{1.7}$$

Then

$$\begin{aligned} &\mathcal{E}([x(t) - x_0(t)]\, d\xi'(s)) = \\ &= \mathcal{E}(x(t)\, d\xi'(s)) - \int_0^t \alpha(t,\tau)\mathcal{E}(d\xi(\tau)\, d\xi'(s)) = \\ &= Q(t,s)\, ds - Q(t,s)N^{-1}(s)N(s)\, ds = 0. \end{aligned}$$

Consequently,

$$\begin{aligned} &\mathcal{E}|\mathcal{E}(x(T)|f_{t,i}^y) - \mathcal{E}(x(T)|f_{t,j}^y)|^2 = \\ &= \left|\int_0^T \Delta_{ij}u(t)A(t)[x(t-h) - x_0(t-h)]\, dt\right|^2 + \\ &\quad + \operatorname{Tr}\left[\int_0^T \Delta_{ij}\beta(t)N(t)\Delta_{ij}\beta'(t)\, dt\right]. \end{aligned} \tag{1.8}$$

Equation(1.5) implies that the lefthand side of (1.8) tends to zero as $i, j \to \infty$. However, the righthand side of (1.8) consists of two nonnegative terms, and so both these terms tend to zero as $i, j \to \infty$. Since the matrix $N(t)$ is positive definite,

$$\lim_{i,j\to\infty} \int_0^T |\Delta_{ij}\beta(t)|^2\, dt = 0.$$

Hence the sequence $\{\beta_i\}_{i=1}^{\infty}$ is fundamental in $L_2[0, T]$.

But the u_i are solutions of the linear Volterra equation (1.7), hence the values of a linear functional on β_i. So the sequence $\{u_i\}$ is also fundamental in $L_2[0,T]$. This implies the existence of a function $u_0 \in L_2[0,T]$ such that

$$\lim_{i\to\infty}\int_0^T |u_i(t)-u_0(t)|^2\,dt = 0. \tag{1.9}$$

Now (1.4) follows from (1.5)–(1.6), (1.8)–(1.9), using the same arguments as in [331]. □

1.3. The fundamental filtering equation. Here we will investigate manners to determine the kernel u_0 of the optimal estimate (1.4).

THEOREM 1.1. *Under the assumptions of subs.* 1.1 *the kernel* u_0 *is the unique solution of the integral equation*

$$u_0(s)N(s) + \int_{t_0}^T u_0(t)Z(t,s)\,dt = P(s), \tag{1.10}$$

where

$$\left.\begin{array}{l} P(s) = R(T,s-h)A'(s) + Q(T,s),\\ Z(t,s) = A(t)R(t-h,s-h)A'(s)+\\ +A(t)Q(t-h,s) + Q'(s-h,t)A'(s). \end{array}\right\} \tag{1.11}$$

PROOF. Let $m_0(T)$ be the mean-square optimal estimate of $x(T)$, and let $m(T)$ be an arbitrary f_T^y-adapted random variable for which $\mathcal{E}|m(T)|^2<\infty$, Then

$$\mathcal{E}([x(T)-m_0(T)]m'(T)) = 0. \tag{1.12}$$

We assume that $m(T)$ has the form (1.4) with kernel $u(s)$. Then (1.12) implies

$$\int_{t_0}^T \left[\mathcal{E}(x(T)\,dy'(s)) - \int_{t_0}^T u_0(t)\mathcal{E}(dy(t)\,dy'(s))\right] u'(s) = 0.$$

Taking into account that here $u(s)$ is arbitrary, we obtain an equation for u_0:

$$\mathcal{E}(x(T)\,dy'(s)) = \int_{t_0}^T u_0(t)\mathcal{E}(dy(t)\,dy'(s)). \tag{1.13}$$

Using (1.2)–(1.3) we find

$$\mathcal{E}(x(T)\,dy'(s)) = \mathcal{E}(x(T)x'(s-h))A'(s)\,ds + \mathcal{E}(x(T)\,d\xi'(s)) =$$
$$[R(T,s-h)A'(s) + Q(T,s)]\,ds = P(s)\,ds,$$
$$\int_{t_0}^{T} u_0(t)\mathcal{E}(dy(t)\,dy'(s)) =$$
$$\int_{t_0}^{T} u_0(t)A(t)\mathcal{E}(x(t-h)\,dy'(s))\,dt + \int_{t_0}^{T} u_0(t)\mathcal{E}(d\xi(t)\,dy'(s)) =$$
$$\int_{t_0}^{T} u_0(t)A(t)[R(t-h,s-h)A'(s) + Q(t-h,s)]\,ds\,dt +$$
$$+ \int_{t_0}^{T} u_0(t)Q'(s-h,t)A'(s)\,ds\,dt + u_0(s)N(s)\,ds =$$
$$\int_{t_0}^{T} u_0(t)[A(t)R(t-h,s-h)A'(s) + A(t)Q(t-h,s) + Q'(s-h,t)A'(s)]\,dt\,ds +$$
$$+u_0(s)N(s)\,ds =$$
$$\left[\int_{t_0}^{T} u_0(t)Z(t,s)\,dt + u_0(s)N(s)\right]\,ds.$$

This implies (1.10)–(1.11), as well as the existence of a solution of (1.10).

Now we prove uniqueness. Let $s \mapsto \Delta u(s)$ be the difference between two solutions of (1.10). Then by (1.13) (which is equivalent to (1.10))

$$\int_{t_0}^{T} \Delta u(t)\mathcal{E}(dy(t)\,dy'(s)) = 0.$$

Multiplying both sides of this equation by $\Delta u'(s)$ and integrating with respect to $s \in [t_0, T]$, we find

$$S := \int_{t_0}^{T}\int_{t_0}^{T} \Delta u(t)\mathcal{E}(dy(t)\,dy'(s))\Delta u'(s) = 0.$$

Hence $\operatorname{Tr} S = 0$, i.e.

$$\mathcal{E}\left|\int_{t_0}^{T} \Delta u(t)\,dy(t)\right|^2 = 0.$$

We put

$$x_0(t) = \int_{t_0}^{t} \alpha(t,s)\,d\xi(s), \qquad \alpha(t,s) = Q(t,s)N^{-1}(s),$$
$$\beta(s) = \Delta u(s) + \int_{s+h}^{T} \Delta u(t)A(t)\alpha(t-h,s)\,dt. \tag{1.14}$$

But

$$\mathcal{E}([x(t) - x_0(t)]\,d\xi'(s)) = Q(t,s)\,ds - \alpha(t,s)N(s)\,ds = 0.$$

Consequently,

$$\mathcal{E}\left|\int_{t_0}^{T} \Delta u(t)\, dy(t)\right|^2 =$$

$$= \mathcal{E}\left|\int_{t_0}^{T} \Delta u(t)[A(t)(x(t-h) - x_0(t-h))\, dt + A(t)x_0(t-h)\, dt + d\xi(t)]\right|^2 =$$

$$= \mathcal{E}\left|\int_{t_0+h}^{T} \Delta u(t)A(t)[x(t-h) - x_0(t-h)]\, dt + \right.$$

$$\left. + \int_{t_0}^{T} \Delta u(t)A(t) \int_{t_0}^{t-h} \alpha(t-h,s)\, d\xi(s)\, dt + \int_{t_0}^{T} \Delta u(s)\, d\xi(s)\right|^2 =$$

$$= \mathcal{E}\left|\int_{t_0+h}^{T} \Delta u(t)A(t)[x(t-h) - x_0(t-h)]\, dt + \int_{t_0}^{T} \beta(s)\, d\xi(s)\right|^2 =$$

$$= \mathcal{E}\left|\int_{t_0}^{T} \Delta u(t)A(t)[x(t-h) - x_0(t-h)]\, dt\right|^2 +$$

$$+ \operatorname{Tr} \int_{t_0}^{T} \beta(s)N(s)\beta'(s)\, ds = 0.$$

Since $N(s)$ is positive definite, we obtain $\beta(s) = 0$ almost everywhere on $[t_0, T]$. The relations (1.14) show that the equation $\beta(s) = 0$ is a linear homogeneous Volterra equation in Δu. Therefore $\Delta u(s) = 0$ almost everywhere. □

1.4. Dual optimal control problem. We will show that the solution of the filtering problem can be obtained as the solution of an auxiliary optimal control problem.

Consider the controlled process $t \mapsto m_u(t)$ governed by the relation

$$m_u(t) = \int_{t_0}^{t} u(t,s)\, dy(s), \tag{1.15}$$

with cost functional to be minimized:

$$J(u) = \mathcal{E}\left[z_u'(T)N_0 z_u(T) + \int_{t_0}^{T} z_u'(s)N_1(s)z_u(s)\, ds\right], \tag{1.16}$$

$$z_u(s) := x(s) - m_u(s).$$

Here N_0 and the values of the piecewise continuous function N_1 are symmetric nonnegative definite matrices, x is a stochastic process, described by (1.2), and u is a control used to minimize (1.16). The set of admissible controls consists of the measurable functions u such that

$$\sup_{t_0 \le t \le T} \int_{t_0}^{t} |u(t,s)|^2\, ds < \infty, \qquad u\colon [t_0,T] \times [t_0,T] \to \mathbf{R}^{n\times m}.$$

Let u_0 be an optimal control for (1.15)–(1.16), and $u_\epsilon = u_0 + \epsilon u$, $\epsilon > 0$, an admissible one. It is clear that the optimal control u_0 must satisfy the condition

$$J_0(u_0) := \limsup_{\epsilon\to 0+} \frac{1}{\epsilon}[J(u_\epsilon) - J(u_0)] \ge 0.$$

Computing $J_0(u_0)$ we find

$$J_0(u_0) = -2\,\mathrm{Tr}\left(N_0 \int_{t_0}^{T} \left[\mathcal{E}(x(T)\,dy'(\tau)) - \int_{t_0}^{T} u_0(T,s)\mathcal{E}(dy(s)\,dy'(\tau))\right] u(T,\tau)\right) + \\ -2\,\mathrm{Tr} \int_{t_0}^{T} \left(N_1 \int_{t_0}^{t} \left[\mathcal{E}(x(t)\,dy'(\tau)) - \int_{t_0}^{t} u_0(t,s)\mathcal{E}(dy(s)\,dy'(\tau))\right] u(t,\tau)\right) dt. \tag{1.17}$$

For (1.17) to be nonnegative for any admissible control u, the expression in square brackets must vanish. Hence there follows, similarly to (1.10), an equation for u_0:

$$u_0(t,s)N(s) + \int_{t_0}^{t} u_0(t,\tau)Z(\tau,s)\,d\tau = P(t,s),$$

where

$$P(t,s) := R(t,s-h)A'(s) + Q(t,s),$$

and the matrix $Z(t,s)$ is defined by (1.11).

If the optimal control u_0 for the problem (1.15)–(1.16) has been found, then the optimal estimate is defined by (1.4) with $u_0(t) = u_0(T,t)$. In particular, if $N_0 = I$ and $N_1 = 0$, then the error estimate is

$$J(u_0) = \mathrm{Tr}\left[R(T,T) - \int_{t_0}^{T} Q(T,s)N^{-1}(s)Q'(T,s)\,ds\right].$$

1.5. Particular cases. In certain cases the method of optimal estimation described above can be simplified, and the solution can be obtained in analytical form. We will consider some of these cases.

1.5.1. Suppose $h \geq T - t_0$. Then

$$u_0(s) = Q(T,s)N^{-1}(s),$$

$$J(u_0) = \mathrm{Tr}\left[R(T,T) - \int_{t_0}^{T} Q(T,s)N^{-1}(s)Q'(T,s)\,ds\right].$$

1.5.2. Suppose $h \leq T - t_0$. We assume that $x(t) = \chi(t)x_0$, where χ is a bounded piecewise continuous stochastic matrix function on $[t_0,T]$, $\chi(t_0) = I$, the gaussian variable x_0 is independent of $\xi(t)$, and $\mathcal{E}x_0 = 0$, $\mathcal{E}(x_0x_0') = D_0$, with I the identity matrix. Then $Q(t,s) = 0$ and $R(t,s) = \chi(t)D_0\chi'(s)$. In this case $u_0(s)$ for $t_0 \leq s \leq T$ satisfies the equation

$$u_0(s) = \left[\chi(T) - \int_{t_0+h}^{T} u_0(t)A(t)\chi(t-h)\,dt\right] D_0\chi'(s-h)A'(s)N^{-1}(s). \tag{1.18}$$

We try to find a solution of this equation in the form

$$u_0(s) = \chi(T)F\chi'(s-h)A'(s)N^{-1}(s), \tag{1.19}$$

where the matrix F is to be determined. Substituting (1.19) into (1.18) we obtain

$$\left.\begin{aligned} F = (D_0^{-1} + G)^{-1}, \qquad G &:= \textstyle\int_{t_0}^{T-h} \chi'(t)A_1(t+h)\chi(t)\,dt, \\ A_1 &:= A'N^{-1}A. \end{aligned}\right\} \tag{1.20}$$

Therefore, under assumptions stated earlier, the optimal estimate is defined by (1.4), (1.19)–(1.20). By (1.19)–(1.20) the error of this estimate is

$$
\begin{aligned}
J(u_0) = \operatorname{Tr}\Big[&\chi(T)D_0\chi'(T)+\\
&+\int_{t_0+h}^{T}\chi(T)F\chi'(s-h)A'(s)N^{-1}(s)A(s)\chi(s-h)F\chi'(T)\,ds+\\
&-2\int_{t_0+h}^{T}\chi(T)D_0\chi'(s-h)A'(s)N^{-1}(s)A(s)\chi(s-h)F\chi'(T)\,ds+\\
&+\int_{t_0+h}^{T}\int_{t_0+h}^{T}\chi(T)F\chi'(t-h)A'(t)N^{-1}(t)A(t)\chi(t-h)\times\\
&\times D_0\chi'(s-h)A'(s)N^{-1}(s)A(s)\chi(s-h)F\chi'(T)\,ds\,dt\Big]=\\
&=\operatorname{Tr}[\chi(T)(D_0+FGF-2D_0GF+FGD_0GF)\chi'(T)].
\end{aligned}
$$

But $F = D_0(I - GF) = F' = D_0 - FGD_0$, so

$$
\begin{aligned}
&D_0 + FGF - 2D_0GF + FGD_0GF =\\
&= D_0 + (F - D_0)GF - D_0GF + FGD_0GF =\\
&= D_0 - FGD_0GF - D_0GF + FGD_0GF = F.
\end{aligned}
$$

Hence the error of the optimal estimate is

$$
J(u_0) = \operatorname{Tr}[\chi(T)F\chi'(T)]. \tag{1.21}
$$

1.6. Dependence of the error of the optimal estimate on the delay. Here we will study the interesting problem of the dependence of the error of the optimal estimate on the value of the delay h obtained in observations. Since $x(s) = 0$ for $s < t_0$, and $x(t)$ and $\xi(t)$ are independent, the observations (1.2) on the interval $[t_0, t_0 + h]$ do not contain information on the process x, and the general observation time for x is equal to $T - t_0 - h$, which decreases as h grows. Hence it is natural to assume that, as a function of h, the error of the estimate is nondecreasing as $h \in [0, T - t_0]$ increases, and is constant for $h \geq T - t_0$.

In particular, if A_1 is a constant matrix, then this assumption holds, since (1.21) means that

$$
\frac{d}{dh}J(u_0) = \operatorname{Tr}\left[\chi(T)\left(\frac{dF}{dh}\right)\chi'(T)\right].
$$

But

$$
\frac{dF}{dh} = -F\left(\frac{d}{dh}F^{-1}\right)F = -F\frac{dG}{dh} = F\chi'(T-h)A_1\chi(T-h)F.
$$

Consequently,

$$
\frac{d}{dh}J(u_0) = \operatorname{Tr}[\chi(T)F\chi'(T-h)A_1\chi(T-h)F\chi'(T)] \geq 0.
$$

Thus, in this case the function $J(u_0)$, regarded as a function of h, does not decrease.

However, there are situations in which increase of h leads to decrease of $J(u_0)$. As an example we will consider a scalar process. Suppose the observable process $t \mapsto x(t)$ and the observations $y(t)$ are scalar, and such that

$$\left.\begin{aligned} &x(t) = \chi(t)x_0, \qquad \chi(t_0) = 1, \qquad x(s) = 0, \quad s \in [t_0 - h, t_0), \\ &dy(t) = A(t)x(t-h)\,dt + d\xi(t), \quad t_0 \le t \le T, \qquad y(t_0) = 0. \end{aligned}\right\} \tag{1.22}$$

Here, x_0 is a gaussian variable with expectation zero and variance σ_0^2, independent of the gaussian process $t \mapsto . \xi(t)$, which has independent increments, $\mathcal{E}\xi(t) = 0$, $\mathcal{E}(d\xi(t))^2 = \sigma^2(t)\,dt$, $\sigma(t) > 0$. In this case the solution of the filtering problem is

$$m(T) = \int_{t_0+h}^{T} u_0(s)\,dy(s), \qquad u_0(s) = \chi(T)F(h)\xi(s-h)A(s)\sigma^{-2}(s),$$

$$F(h) = [\sigma_0^{-2} + G(h)]^{-1}, \qquad G(h) = \int_{t_0}^{T-h} \chi^2(t)A_1(t+h)\,dt,$$

$$A_1(t) = A^2(t)\sigma^{-2}(t).$$

The error of the estimate equals

$$J(h) = \chi^2(T)F(h).$$

So, minimization of $J(h)$ with respect to h is equivalent to minimization of the function $h \mapsto F(h)$, or maximization of the function $G(h)$, $h \in [0, T - h_0]$. Note that

$$\max G(h) = G(0), \qquad 0 \le h \le T - t_0,$$

if $A_1(t) = \text{const}$ or if $x(t) \equiv x_0$, i.e. $\chi(t) \equiv 1$. The interior extremal points of G satisfy the equation

$$\begin{aligned} \frac{d}{dt}G(h) &= \\ &= -\chi^2(T-h)A_1(t) + \int_{t_0}^{T-h} \chi^2(t)\,dA_1(t+h) = \\ &= -\chi^2(T-h)A_1(T) + \chi^2(t)A_1(t+h)\Big|_{t_0}^{T-h} - \int_{t_0}^{T-h} A_1(t+h)\,d\chi^2(t) = \\ &= -A_1(t) - \int_{t_0}^{T-h} A_1(t+h)\,d\chi^2(t) = 0. \end{aligned} \tag{1.23}$$

We will give some examples in which (1.23) has a unique solution $h_0 > 0$, defining an interior maximum of G, i.e. an interior minimum of J.

EXAMPLE 1.1. Let $t_0 = 0$, $\chi(t) = e^{-\alpha t}$, $A(t) = e^{\beta t}$, $\sigma^2(t) = \sigma_0^2 = 1$, $\beta \ge \alpha > 1/(2T)$. Then (1.23) takes the form

$$e^{2\beta h} = 2\alpha \int_0^{T-h} e^{2\beta(t+h) - 2\alpha t}\,dt.$$

By solving this equation we find for h_0:

$$h_0 = T - \frac{1}{2\alpha} g\left(\frac{\beta}{\alpha}\right),$$

$$g(r) := \begin{cases} (r-1)^{-1}\ln r, & r > 1, \\ 1, & r = 1. \end{cases}$$

The corresponding minimum value of the error is

$$J(h_0) = e^{-2\alpha T}[1 + G(h_0)]^{-1},$$

$$G(h) := \frac{e^{2\beta h}}{2(\beta - \alpha)}[e^{2(\beta-\alpha)(T-h)} - 1].$$

Fig. 8.1.1 shows the graph of the function $h \mapsto J(h)$ for $T = 1$ and various values of α, β:

1) $\alpha = 0.6$, $\beta = 1$, $h_0 = 0.361$;
2) $\alpha = 0.75$, $\beta = 1$, $h_0 = 0.425$;
3) $\alpha = \beta = 1$, $h_0 = 0.5$;
4) $\alpha = 0.6$, $\beta = 2$, $h_0 = 0.570$;
5) $\alpha = 0.75$, $\beta = 2$, $h_0 = 0.608$;
6) $\alpha = 1$, $\beta = 2$, $h_0 = 0.653$. □

EXAMPLE 1.2. Consider the filtering problem (1.22) with $t_0 = 0$, $\chi(t) = \sqrt{1 - \alpha(t/T)}$, $A(t) = \sqrt{t}$, $\sigma^2(t) = \sigma_0^2 = 1$, $\alpha < 1$. Then (1.23) takes the form

$$h + \int_0^{T-h} (t+h)\left(-\frac{\alpha}{T}\right) dt = 0.$$

Hence

$$\frac{Th}{\alpha} = \int_0^{T-h} (t+h)\, dt = \frac{T^2 - h^2}{2}.$$

By solving this equation we find

$$h_0 = T\left(\sqrt{1 + \frac{1}{\alpha^2}} - \frac{1}{\alpha}\right),$$

which corresponds to the minimum of the function

$$J(h) = (1-\alpha)[1 + G(h)]^{-1},$$

$$G(h) := \int_0^{T-h} \left(1 - \alpha\frac{t}{T}\right)(t+h)\, dt =$$

$$= (T-h)\left[h + \frac{T-h}{2}\left(1 - \frac{\alpha h}{T}\right) - \frac{\alpha}{3T}(T-h)^2\right].$$

Fig. 8.1.2 gives the graph of the function $h \mapsto J(h)$ for $T = 1$ and

1) $\alpha = 0.1$, $h_0 = 0.0498$;
2) $\alpha = 0.3$, $h_0 = 0.147$;

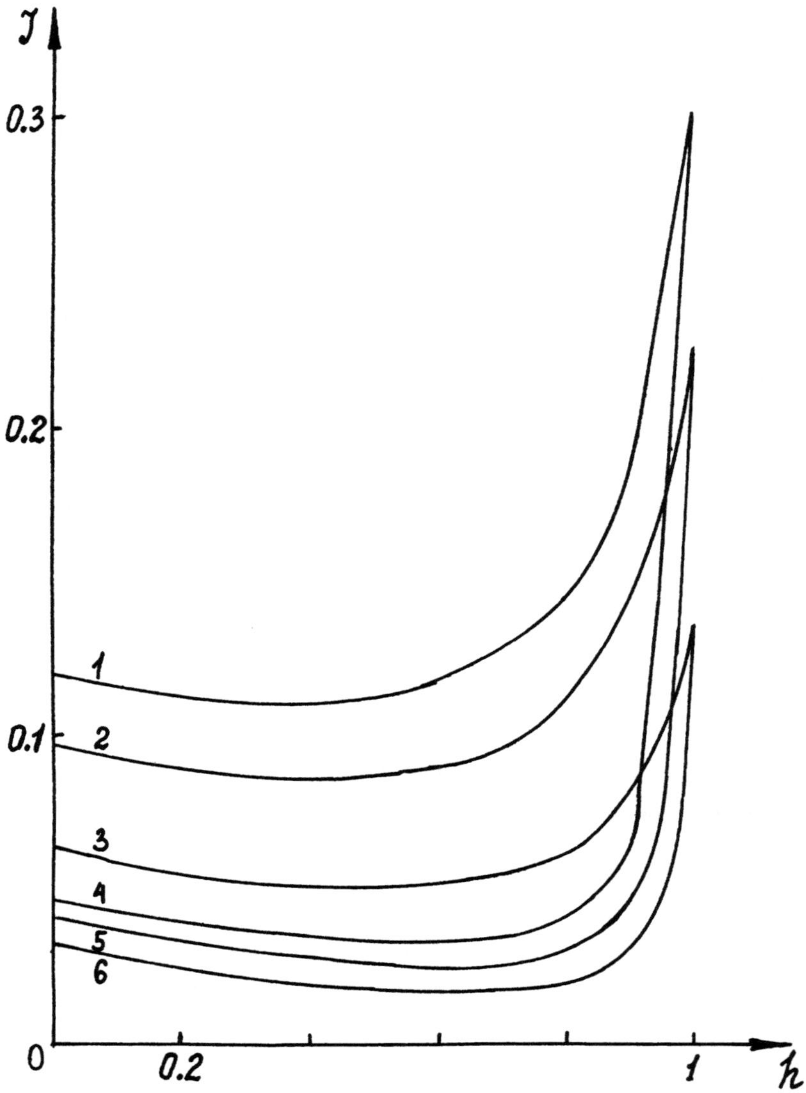

FIGURE 8.1.1. Dependence of the error of the optimal estimate on the delay h obtained by observation

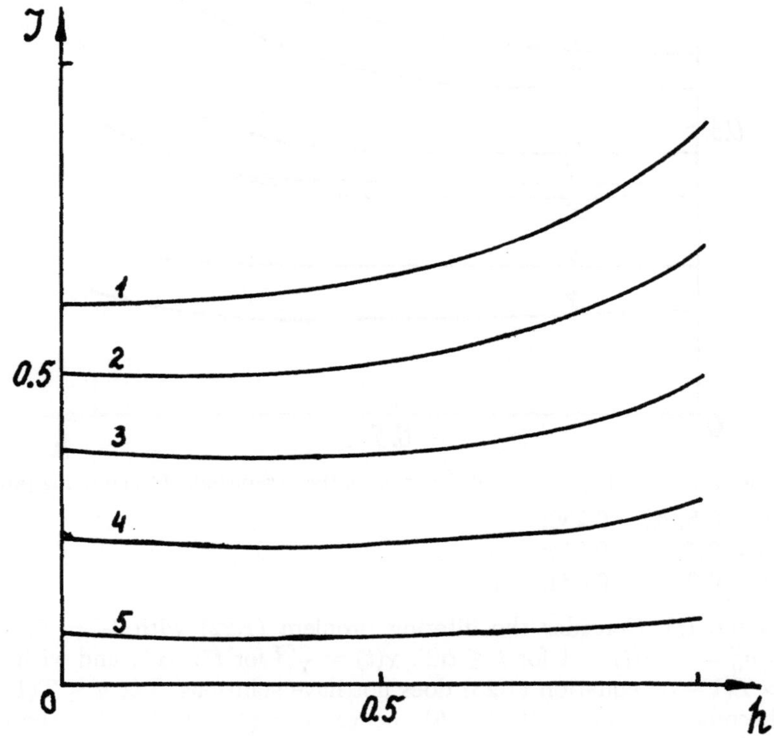

FIGURE 8.1.2. Influence of various parameters on the error

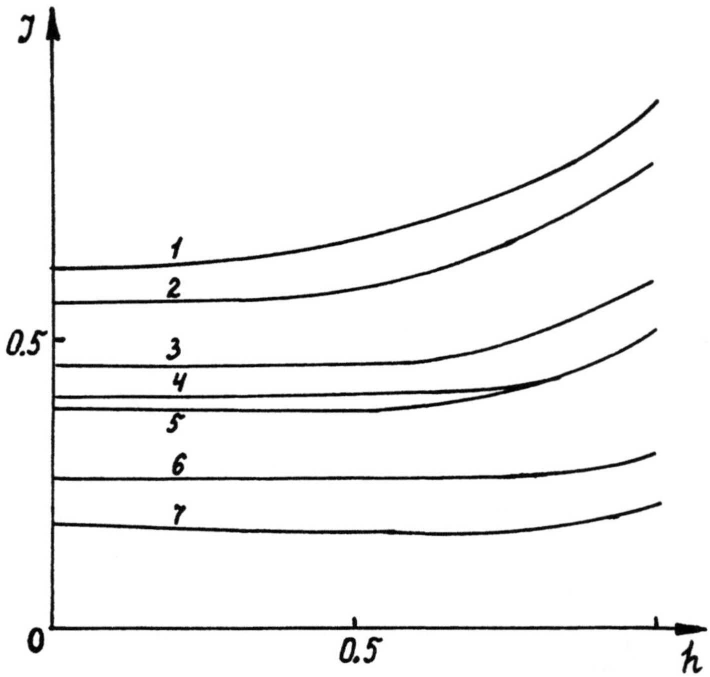

FIGURE 8.1.3. Dependence of the error on the parameters of the filtering problem

3) $\alpha = 0.5$, $h_0 = 0.236$;
4) $\alpha = 0.7$, $h_0 = 0.315$;
5) $\alpha = 0.9$, $h_0 = 0.384$. □

EXAMPLE 1.3. Consider the filtering problem (1.22) with $t_0 = 0$, $A(t) = \sqrt{t}$, $\sigma^2(t) = \sigma_0^2 = 1$, $\chi(t) = 1$ for $t \leq \alpha T$, $\chi(t) = \sqrt{\beta}$ for $t > \alpha T$, and with $\alpha \leq \beta < 1$. For $h > T(1-\alpha)$ equation (1.23) does not have solutions. Let $h \leq T(1-\alpha)$. Then (1.23) becomes $h = (1-\beta)(\alpha T + h)$. hence $h_0 = (1-\beta)\alpha T \beta^{-1}$ is the minimum of $J(h) = \beta[1 + G(h)]^{-1}$, where

$$2G(t) := \begin{cases} \beta T^2 - h^2 + (1-\beta)(\alpha T + h)^2, & h \leq T(1-\alpha), \\ T^2 - h^2, & h > T(1-\alpha). \end{cases}$$

Fig. 8.1.3 gives the graph of $J(h)$ for $T = 1$ and

1) $\alpha = 0$, $\beta = 0.9$, $h_0 = 0$;
2) $\alpha = 0.5$, $\beta = 0.8$, $h_0 = 0.125$;
3) $\alpha = 0.4$, $\beta = 0.6$, $h_0 = 0.267$;
4) $\alpha = 0.2$, $\beta = 0.5$, $h_0 = 0.2$;
5) $\alpha = \beta = 0.5$, $h_0 = 0.5$;
6) $\alpha = 0.1$, $\beta = 0.3$, $h_0 = 0.233$;
7) $\alpha = \beta = 0.2$, $h_0 = 0.8$. □

1.7. Some generalizations. *1.7.1.* Suppose that the observation equations involve many delays, i.e.

$$dy(t) = \sum_{i=1}^{r} A_i(t)x(t-h_i)\,dt + d\xi(t), \qquad y(t_0) = 0, \tag{1.24}$$

where $h_i \geq 0$, the processes $t \mapsto x(t), \xi(t)$ are as in subs. 1.1, and the matrix functions A_i are piecewise continuous.

Then the optimal estimate is given by (1.4), where the matrix function u_0 is the solution of (1.10) with

$$P(s) = \sum_{i=1}^{r} R(T, s-h_i)A_i(s) + Q(T,s),$$

$$Z(t,s) = \sum_{i,j=1}^{r} A_i(t)R(t-h_i, s-h_j)A_j'(s) +$$

$$+\sum_{i=1}^{r} A_i(t)Q(t-h_i, s) + \sum_{j=1}^{r} Q'(s-h_j, t)A_j'(s).$$

1.7.2. One can show that smoothing and interpolation problems can be reduced to filtering problems. As an example we consider a smoothing problem. The process $t \mapsto (x(t), y(t))$ is supposed to be as in subs. 1.1. The smoothing problem consists of constructing a mean-square optimal estimate $x(T_1)$, $T_1 > T$, under the observation (1.2) on the interval $t_0 \leq t \leq T$. This problem can be reduced to the filtering problem. In fact, putting $\overline{A}(t) = A(t)$ for $t_0 \leq t \leq T$, $\overline{A}(t) = 0$ for $t > T$, and $\overline{Q}(t,s) = Q(t,s)$ for $t_0 \leq s \leq t \leq T$, $\overline{Q}(t,s) = 0$ otherwise. Consider the auxiliary filtering problem of estimating $x(T_1)$ under the observations (1.2) on $[t_0, T_1]$, with $A(t)$ replaced by $\overline{A}(t)$ and $\mathcal{E}(x(t)\,d\xi(s)) = \overline{Q}(t,s)\,ds$. For $t \geq s \geq T$ the processes x and ξ are uncorrellated and gaussian, and are independent. Hence the solution of the auxiliary filtering problem is at the same time the solution of the initial smoothing problem. Therefore the solution of the smoothing problem is given by

$$m_0(T_1) = \int_{t_0}^{T} u_0(t)\,dy(t).$$

Here u_0 is the solution of (1.10) with $T_1, \overline{A}, \overline{Q}$ instead of T, A, Q.

2. Filtering of solutions of Itô equations with delay

In this section we will consider the form of the filtering equations for the particular case when the unobservable process $t \mapsto x(t)$ is described by a linear stochastic equation. We confine ourselves to stating the problem and the resulting filtering equations. The reader is referred to [16, 112, 281.9, 287] for a comprehensive coverage of the subject.

2.1. Problem statement. The unobservable process $t \mapsto x(t)$ is such that

$$\left.\begin{array}{ll} dx(t) = \sum_{i=1}^{k} g_i(t)x(t-\tau_i)\,dt + \sigma_1(t)\,d\xi_1(t), & t_0 \le t \le T, \\ 0 \le \tau_1 < \cdots < \tau_k, \quad x(s) = 0, \quad s < t_0, & x(t_0) = x_0. \end{array}\right\} \tag{2.1}$$

The observation process $t \mapsto y(t)$ satisfies (1.24), i.e.

$$dy(t) = \sum_{i=1}^{r} A_i(t)x(t-h_i)\,dt + d\xi(t), \qquad y(t_0) = 0.$$

The processes ξ_1, ξ_2, x are mutually independent, ξ_1 is the standard Wiener process, x_0 is a gaussian vector, $\mathcal{E}x_0 = 0$, $\mathcal{E}(x_0x_0') = D_0$, the matrix functions g_i and σ_1 are piecewise continuous. The problem is to find the optimal mean-square estimate $m_0(T)$ for $x(T)$ under the observations $y(t)$, $t_0 \le t \le T$.

2.2. Dual control problem. The optimal estimate of the vector $x(T)$ under the observations $y(t)$, $t_0 \le t \le T$, is defined by (1.4), where the matrix function u_0 is the optimal control of the deterministic system

$$\left.\begin{array}{c} \dot\alpha(t) = -\sum_{i=1}^{k} g_i'(t+\tau_i)\alpha(t+\tau_i) + \sum_{j=1}^{r} A_j'(t+h_j)u(t+h_j), \\ \alpha(T) = I, \qquad \alpha(s) = 0, \quad s > T. \end{array}\right\} \tag{2.2}$$

Here, $\alpha(t) \in \mathbf{R}^{n\times n}$ is a square matrix function. The cost functional is

$$J(u) = \operatorname{Tr}\Big[\alpha'(t_0)D_0\alpha(t_0) + \int_{t_0}^{T} (\alpha'(t)N_1(t)\alpha(t) + u'(t)N(t)u(t))\,dt\Big] \to \min_u, \qquad N_1 := \sigma_1\sigma_1'. \tag{2.3}$$

Linear quadratic problems of the type (2.2)–(2.3) were investigated in chapter 6, where it has been proved that the solution of (2.2)–(2.3) can be reduced to the study of a boundary value problem for nonlinear differential equations. Here we will illustrate the application of the results in chapter 6 for $k = r = 1$, $h_1 = h < T - t_0$. In this case the optimal control $u_0(t)$ is

$$u_0(t) = \left\{\begin{array}{l} 0, \\ \qquad t_0 \le t \le t_0 + h, \\ N^{-1}(t)A_1(t)\left[P(T-h)\alpha(t-h) + \int_{-\tau_1}^{0} Q_1(T-t,\tau)\alpha(t-\tau-h)\,d\tau\right], \\ \qquad t_0 + h < t \le T. \end{array}\right\} \tag{2.4}$$

The covariance matrix $D(T)$ is equal to

$$D(T) = \mathcal{E}([x(T) - m(T)][x(T) - m(T)]') = P(T). \tag{2.5}$$

The matrices $P(t), Q_1(t,\tau)$ and $R_1(t,\tau,\rho)$ are the solutions of the equations

$$\left.\begin{aligned}&\dot P(t)+P(t)A_0(t+h)P(t)-Q_1(t,0)-Q_1'(t,0)-N_1(t)=0,\\&\left(\tfrac{\partial}{\partial t}+\tfrac{\partial}{\partial\tau}\right)Q_1(t,\tau)+P(t)A_0(t+h)Q_1(t,\tau)-R_1(t,0,\tau)=0,\\&\left(\tfrac{\partial}{\partial t}+\tfrac{\partial}{\partial\tau}+\tfrac{\partial}{\partial\rho}\right)R_1(t,\tau,\rho)+Q_1'(t,\tau)A_0(t+\tau)Q_1(t,\rho)=0,\end{aligned}\right\}\tag{2.6}$$

$$t_0\le t\le T,\quad -\tau_1\le\tau,\rho\le 0,\quad A_0:=A_1'N^{-1}A_1,\quad A_0(s)\equiv 0,\quad s>T.$$

The boundary conditions for (2.6) are

$$\left.\begin{aligned}P(t_0)=D_0,\qquad Q_1(t_0,\tau)=R_1(t_0,\tau,\rho)=0,\\Q_1'(t,-\tau_1)=g(t+\tau_1)P(t),\\R_1(t,-\tau_1,\tau)+R_1'(t,\tau,-\tau_1)=2A(t+\tau_1)Q_1(t,T_1),\end{aligned}\right\}\tag{2.7}$$

$$t_0\le t\le T,\qquad -\tau_1\le\tau,\rho\le 0.$$

Hence the construction of the optimal estimate $m(T)$ and covariance matrix $D(T)$ consists of the following steps:

1) solve the boundary value problem (2.6)–(2.7), and determine the matrices $P(t)$, $Q_1(t,\tau)$;
2) find the solution α of (2.2), with control $t\mapsto u_0(t)$ given by (2.4);
3) substitute the matrices P, Q_1, α in (2.4), and determine the kernel $u_0(t)$ of the optimal estimate (1.4) and error (2.5).

In certain cases this algorithm can be simplified. We will consider some such cases.

2.2.1. If $k=r=1$, $\sigma_1=0$, then by Cauchy's formula (subs. 2.5.3) the solution oft he problem (2.1) can be represented as $x(t)=\chi(t)x_0$. In this case the solution of the filtering problem has been obtained in subs. 1.5. Moreover, for $\sigma_1=0$ the exact analytical solution of (2.6)–(2.7) is given by (6.2.9).

2.2.2. Suppose $k=r=1$, $\tau_1=0$, and (2.1) holds for all $t\ge t_0-h$. Then we obtain for the covariance matrix $D(t)$ the equation

$$\begin{gathered}\dot D-DA'-AD=\\=-Dz_1D+Dz_1\overline{\sigma}_1+\overline{\sigma}_1z_1D+\sigma_1\sigma_1-\overline{\sigma}_1z_1\overline{\sigma}_1,\\D(t_0)=D_0.\end{gathered}$$

Here, the matrix functions $z_1,\overline{\sigma}_1$ are given by

$$z_1(t)=z'(t-h,t)A_1'(t)N_1^{-1}(t)A(t)z(t-h,t),$$

$$\overline{\sigma}_1(t)=\int_{t-h}^{t}z(t,s)N_1(s)z'(t,s)\,ds,$$

where the matrix function z is defined by

$$\frac{\partial z(t,s)}{\partial t}=g_1(t)z(t,s),\qquad z(s,s)=I.$$

For the optimal estimate $m(t)$ we have the equation

$$dm(t) - g_1(t)m(t)\,dt =$$
$$\left[D(t) - \int_{t-h}^{t} \sigma_1(s)\,ds\right] z'(t, t-h)A_1'(t)N^{-1}(t) \times$$
$$\times [dy(t) - A_1(t)z(t-h, t)m(t)\,dt],$$
$$m(t_0) = m_0.$$

REMARK 2.1. The results in this section have been obtained by reducing the initial filtering problem to the dual optimal control problem. A similar procedure has been applied in [16] for solving a filtering problem under both continuous and discrete observations. The same filtering algorithms can also be obtained by the maximum likelihood method. The latter was used in [16] to evaluate the parameters of equation (1.5.2), which describes the Volterra population model in a stochastic environment. □

Bibliography

[1] M.I. Ablowits, I.F. Ladik, 1) Nonlinear differential equations, *J. Math. Phys.* **16: 3** (1975), pp. 598–603. 2) Nonlinear differential-difference equations and Fourier analysis, *J. Math. Phys.* **17: 6** (1978), pp. 1011–1018.

[2] D.L. Abrahamson, E.F. Infante, A Liapunov functional for linear Volterra integrodifferential equations, *Quaterly of Appl. Math.* **11: 1** (1983), pp. 35–44.

[3] I. Ackermann, H. Hu, C.V. Hollot, R. Tempo, On extreme point results for time-delay systems, *Proc. Amer. Control Conf. San Diego* (1990), pp. 1499–1501.

[4] G.E. Adomian, Solutions of the Marchuk model of infections disease and immune response, *Math. Model.* **7: 5/8** (1986), pp. 803–807.

[5] G.E. Adomian, R. Rach, 1) Nonlinear stochastic differential delay equations, *J. Math. Anal. & Appl.* **91:1** (1983), pp. 94–101. 2) A nonlinear differential delay equation, *J. Math. Anal. & Appl.* **91: 2** (1983), pp. 301–304.

[6] Adaptive control binary systems with delay, *Itogi Nauk. i Tekhn. Ser. Tekhn. Kibernetik.* **21** (1987). (In Russian.)

[7] M. Agarval, C. Canudas, On-line estimation of time delay and continuous-time process parameters, *Int. J. Control* **46: 1** (1987), pp. 295–311.

[8] P. Agathoklis, S. Foda, Stability and the matrix Lyapunov equations for delay systems, *Int. J. Control* **49: 2** (1989), pp. 417–432.

[9] Z. Agur, Population dynamics in harshly varying environments, *Int. Centre Theoret. Physics, Italy* , World Scientific, 1986, pp. 440–454.

[10] W.G. Aiello, H.I. Freedman, A time-delay model of single-species growth with stage structure, *Math. Biosciences* **101** (1990), pp. 139–153.

[11] S. Aizicovici, On semilinear Volterra integrodifferential equations, *Israel J. Math.* **36: 3/4** (1980), pp. 273–284.

[12] G. Alain, R. Colly, Mathematical study of a muscle construction model in which the fibre is a continuum of elements, *Adv. Appl. Math.* **9: 1** (1988), pp. 104–126.

[13] V. Alexiades, Almost periodic solutions of an integrodifferential system with infinite delay, *Nonlinear Anal., Th., Methods & Applic.* **5: 4** (1981), pp. 401–410.

[14] T. Amemiya, Arazawak H. Tokumaru, Delay independent stabilization and decay rate assignability of linear systems with limited measurable state variables, *Int. J. Control* **47: 1** (1988), pp. 65–73.

[15] I.M. Ananjevski, V.B. Kolmanovskii, Stabilization of some nonlinear hereditary mechanical systems, *Nonlinear Anal., Th., Methods & Applic.* **15: 2** (1990), pp. 101–114.

[16] E.A. Andreeva, V.B. Kolmanovskii, L.E. Shaikhet, *Control of Systems with Delay*, Nauka, Moscow, 1992. (In Russian.)

[17] D. Andrencci, A. Fasano, M. Primicerio, On a mathematical model for the crystallization of polymers, *Proc. 4th Europ. Conf. Math. in Industry* , Kluwer, 1991, pp. 3–16.

[18] A.A. Andronov, A.G. Mayer, The simplest linear systems with delay, *Autom. Remote Control* **7: 2/3** (1946), pp. 95–106. (In Russian.)

[19] T.S. Angell, A. Kirsch, On the necessary conditions for optimal control of retarded systems, *Appl. Math. Optim.* **22: 2** (1990), pp. 117–145.

[20] J.S. Ansary, Stability of differential-difference equations representing heat exchanger and certain other systems, *Int. J. Control* **17: 1** (1973), pp. 193–198.

[21] S.S. Antman, The paradoxical asymtotic status of massless springs, *SIAM J. Appl. Math.* **48: 6** (1988), pp. 1319–1334.

[22] F. Argentesi, R. de Bernardi, C. Ricci, A simple parameter estimation technique for single species population models, *Mem. Ist. Ital. Idrobiol. Dott. M. Marchi* **37, Suppl.** (1979), pp. 201–212.

[23] O. Arino, M. Kimmel, 1) Asymtotic analysis of a functional-integral equation related to cell population kinetics, *Trends, Theory & Pract. Nonlinear Anal.* , Amsterdam, 1985, pp. 27–32. 2) Stability analysis of model of cell production systems, *Math. Modell.* **7: 9/12** (1986), pp. 1269–1300. 3) Asymptotic behavior of a nonlinear functional-integral equation of cell kinetics with unregular division, *J. Math. Biol.* **27: 3** (1989), pp. 341–354.

[24] Z. Artstein, G. Zarakostas, Convergence in the dealy population equation, *SIAM J. Appl. Math.* **38: 2** (1980), pp. 261-272.

[25] N.H. Arutunian, V.B. Kolmanovskii, *Creep Theory of Nonhomogeneous Bodies*, Nauka, Moscow, 1983. (In Russian.)

[26] V.G. Babskii, A.D. Myshkis, Mathematical models in biology connected with regard of delays, Appendix to: J.D. Murray, *Lectures on Non-linear Differential Equations. Models in Biology*, Mir, Moscow, 1983, pp. 383–394. (In Russian.)

[27] A. Bagchi, Approximation of Ito integral arising in stochastic time-delayed systems, *SIAM J. Control Optim.* **22: 6** (1984), pp. 878–888.

[28] A.A. Bakaev, V.P. Melnichuk, Stability of the quasidynamical forecasting model, *Appl. Math. Methods in Economic Investigation & Planning* , Kiev (1979), pp. 3–17. (In Russian.)

[29] K. Balachandran, Nonlinear perturbations of quasilinear delay control systems, *Kibernetika* **25: 1** (1989).

[30] Yu.D. Balanchus, D.I. Shvitra, Problem of the interaction of thyroid gland harmons with hypophis, *Math. Model. in Biol. & Medicine, Vilnius* **1** (1985),

pp. 141–156. (In Russian.)

[31] C. Banfi, Su una mova impostazione per l'analisi dei sistemi ereditari, *Ann. Univ. Ferrara, sez. 7* **23** (1977), pp. 29–38.

[32] H.T. Banks, Delay systems in biological models: approximation techniques, *Nonlinear Systems & Appl.* , Acad. Press, New York, 1971, pp. 21-28.

[33] H.T. Banks, R.H. Fatiano, Y. Wang, D.J. Inman, H.J. Cudney, Spatial versus time hysteresis in damping mechanisms, *Proc. 27th IEEE Conf. Decision & Control Dec. 7–9, Austin, Texas* **2** (1988), pp. 1674–1677.

[34] H.T. Banks, J.M. Mahaffy, Global asymptotic stability of certain models for protein synthesis and repression, *Quaterly Appl. Math.* **36: 3** (1978), pp. 209–221.

[35] J. Bansal, K.S. Chang, Optimal control of dispersion type distributed-parameter systems with time-delay in interacting two-point boundary conditions, *Int. J. Control* **16: 3** (1972), pp. 481–500.

[36] J.J. Baptista, Venturi A. da Silva, Su un'equazione integro-differenziale collegata alla propagazione del calore in materiali con memoria, *Rend. Sem. Mat. Univ. e Politecn. Torino* **40: 3** (1982), pp. 81–104.

[37] M. Bardi, Exchange of stability along a branch of periodic solutions of a single species model, *Lect. Notes in Meth.* **1017** (1983), pp. 73–82.

[38] B.R. Barmish, Z. Shi, Robust stability of perturbed systems with time dealy, *Automatica* **25: 3** (1989), pp. 371–381.

[39] H.A.M. Basher, R. Mukundan, Model reference control of uncertain systems with delay in plant state and control, *Int. J. Systems Sci.* **18: 9** (1987), pp. 1609–1626.

[40] L. Bass, A.I. Bracken, K. Holmaker, B.R. Jefferies, Integro-differential equations for the selforganization of liver zones by competitive exclusion of cell types, *J. Austal. Math. Soc.* **29: 2** (1987), pp. 156–194.

[41] J. Baumeister, Boundary control of an integrodifferential equation, *J. Math. Anal. & Appl.* **93: 2** (1983), pp. 550–570.

[42] L.S. Becker, T.A. Burton, T. Krisztin, Floquet theory for a Volterra equation, *J. London Math. Soc.* **2: 37** (1988), pp. 141–147.

[43] J.R. Beddington, R.M. May, Time delays are not necessarily destabilizing, *Math. Biosc.* **27: 1/2** (1975), pp. 108–117.

[44] I. Belair, M.C. Mackey, Consumer memory and price fluctuations in commodity markets: an integrodifferential model, *J. Dynamics & Diff. Eq.* **1: 3** (1989), pp. 299–325.

[45] J.W. Belernes, Ignition for a gaseous thermal reaction, *Equadiff. 5th Proc. Czech. Conf. Diff. Eq. & Appl., Bratislava, Aug. 24–28, 1981* (1982), pp. 34–36.

[46] D. Bell, S. Mohammed, On the solutions of stochastic ordinary differential equations with small elays, *Stochastics* **28** (1989), pp. 293–299.

[47] R. Bellman, K.L. Cooke, *Differential-Difference Equation*, Acad. Press, New York–London, 1963.

[48] P.A. Belmisov, A.D. Drozdov, V.B. Kolmanovskii, *Stability of Viscoelastic Systems*, Saratov Univ. Press, Saratov, 1991. (In Russian.)

[49] L.N. Belykh, Recent results in mathematical modelling of infections diseases, *Acta Applic. Math.* **14: 1/2** (1989), pp. 125–133.

[50] L.N. Belykh, G.I. Marchuk, Chronic forms of disease and their treatment according to mathematical immune respons models, *Lect. Notes Control & Inform. Sci.* **18** (1979), pp. 79–86.

[51] E.A. Bencer, L.P. Neuwirth, Traffic flow: Laplace transform, *Amer. Math. Monthly* **80: 4** (1973), pp. 417–423.

[52] J. Bentsman, J. Fakhfakh, R.B. Lehman, Vibrational stabilization of linear time delay systems and their robustness with respect to delay size, *Systems, Control Letters* **12: 3** (1989), pp. 267–272.

[53] E.L. Beretta, G.I. Bischi, F. Solimano, 1) Oscillations of a biotic species feeding on an abiotic resource, *Biomath. & Relat. Comput. Probl., Dordrecht* (1988), pp. 121–131. 2) Stability in chemostat equations with delayed nutrient recycling, *J. Math. Biol.* **28: 1** (1990), pp. 99–111.

[54] E. Beretta, F. Solimano, A generalization of Volterra models with continuous time delay in population dynamics: boundedness and global asymptotic stability, *SIAM J. Appl. Math.* **48: 3** (1988), pp. 607–626.

[55] E. Beretta, F. Solimano, Y. Takeuchi, Global stability and periodic orbits for two-patch predator prey diffusion-delay models, *Math. Biosc.* **85: 2** (1987), pp. 153–183.

[56] E. Beretta, Y. Takeuchi, Homotopy techniques in stability problems of delay-diffusion Volterra models, *Biomath. & Relat. Comput. Probl., Dordrecht* (1988), pp. 245–254.

[57] M.A. Berger, V.J. Mizel, Volterra equations with Ito integrals, *J. Int. Eq.* **2** (1980), pp. 187–245.

[58] M.L. Bernardi, C.A. Candido, On a differential problem coming from cell biology, *Ann. Mat. Pura ed Appl.* **141** (1985), pp. 159–175.

[59] M.L. Bernardi, A.C. Capevo, P. Periti, A mathematical model for the involution of cell populations under the action of mutagenic agents, *Math. Biosc.* **71: 1** (1984), pp. 19–39.

[60] Y. Biran, B. Innis, Optimal control of bilinear systems: time-varying effects of cancer drugs, *Automatica* **5** (1979).

[61] G. Birkhoff, Numerical solution of the reactor kinetics equations, *Num. Sol. Nonlinear Diff. Eq.* , Wiley, New York, etc., (1966), pp. 3–20.

[62] M.L. Blank, Finite-dimensional attractors in one turbulence model, *Diff. Eq.* **24: 11** (1988), pp. 1854–1862. (In Russian.)

[63] B.I. Blanter, *Probabilistic Models in Physiology*, Nauka, Moscow, 1977. (In Russian.)

[64] A.M. Blohin, V.R. Isimerman, Some differential-difference models investigation for acoustics equations systems, *Some Appl. of Function. Anal. to Probl. in Math. Phys.* (1988), pp. 21–32. (In Russian.)

[65] F. Bloom, 1) Continuous data dependence for abstract Volterra integro-differential equations in Hilbert space with applications to viscoelasticity, *Ann. Scuola Norm. Sup. Pisa. Cl. Sci. Ser. 14* **4: 1** (1977), pp. 179–207. 2) *Ill Posed Problems for Integro-Differential Equations in Mechanics and Electro-Magnetics*, Acad. Press, New York, 1981.

[66] S.P. Blythl, R.M. Anderson, Distributed incubation and infection periods in models of transmission dynamics of the human immunodeficiency virus (HIV), *IMA J. Math. Appl. Med. & Biol.* **5: 1** (1988), pp. 1–19.

[67] G.A. Bocharov, Mathematical model of the antiviral immune response of acellular type, *Math. Modell. Immunal & Med. Proc. IFIP TC 7th Work Conf. Moscow, July 5–11, 1982* (1983), pp. 85–95.

[68] F.G. Boese, Delay independent stability of a special sequence of neutral difference-differential equations with one delay, *J. Diff. Eq.* **90** (1991), pp. 397–407.

[69] V.M. Bogdan, Existence of solutions to differential equations of relativistic mechanics involving Lorentzian time delays, *J. Math. Anal. & Appl.* **118: 2** (1986), pp. 561–573.

[70] V.L. Bogomolov, Control of power of hydrostation, *Autom. & Remote Control* **4/5** (1941), pp. 103–129. (In Russian.)

[71] G.N. Bojadziev, R.W. Lardner, Asymptotic solutions of partial differential equations with damping and delay, *Quat. Appl. Math.* **33: 3** (1975), pp. 205–214.

[72] W.R. Boland, W.M. Powers, The solution of a problem in the theory of epidemics, *J. Math. Anal. & Appl.* **67: 2** (1979), pp. 565–585.

[73] R.D. Braddock, P. van den Driessche, 1) A population model with time delay, *Math. Sc.* **5: 1** (1980), pp. 55–66. 2) On a two lag differential delay equation, *J. Austal. Math. Soc.* **B24: 3** (1983), pp. 292–317.

[74] F. Braner, 1) Some applications of the theory of ordinary differential equations to population growth problems, *Ann. Acad. Brasil. Sc.* **48: 3** (1976), pp. 369–385. 2) Characteristic return times for harvested population models with time lag, *Math. Biosc.* **45: 3/4** (1979), pp. 295–311.

[75] F. Braner, Ma Zhien, Stability of stage-structured population models, *J. Math. Anal. & Appl.* **126: 2** (1987), pp. 301–315.

[76] F. Braner, D.A. Sancher, On some models for population growth with harvesting, *Int. Conf. Diff. Eq., 1974* , New York, (1975), pp. 53–64.

[77] R.K. Brayton, Small-signal stability criterion for electrical networks containing lossless transmission lines, *IBM J. Res. & Developm.* **12: 8** (1968), pp. 431–440.

[78] N. Britten, Travelling wave front solutions of a differential-difference equation arising in the modelling of myelinated nerve axon, *Lect. Notes in Math.* **1151** (1985), pp. 77–89.

[79] W.E. Brumley, On the asymptotic behaviour of solutions of differential difference equations of neutral type, *J. Diff. Eq.* **7** (1970), pp. 175–188.

[80] H. Brunner, 1) The numerical treatment of Volterra integro-differential equations with unbounded delay, *J. Comp. & Appl. Math.* **28** (1989), pp. 5–23. 2) On the numerical solution of nonlinear Volterra-Fredholm integral equations by collocation methods, *SIAM J. Numer. Anal.* **27: 4** (1990), pp. 987–1000.

[81] S.A. Brykalov, 1) Nonlinear boundary problems and steady-states existence for heating control systems, *Dokl. Akad. Nauk. SSSR* **307: 1** (1989), pp. 11–14. (In Russian.) 2) Existence of one-dimensional steady-state temperature distributions like given ones in systems with persistent reverse connection and control, *Inst. Mat. Mekh. Ural. Dep. Akad. Nauk. SSSR* (1989). Deposited at VINITI 2.08.89, nr. 5171-B89. (In Russian.)

[82] E.N. Bulange, The investigation of dynamical development model of forest population under vermin influence, *Vestnik Moskov. Gos. Univ., Vychisl. Mat. Kibernet.* **4** (1984), pp. 34–39. (In Russian.)

[83] V.Sh. Burd, L.A. Zafranskaya, Analysis of one immune response model, *Investigation on Stability & Vibration Theory* , Yaroslavl (1979), pp. 63–71. (In Russian.)

[84] D.N. Burghes, Population dynamics. An introduction to differential equations, *Int. J. Math. Educ. Sci. & Techn.* **6: 3** (1975), pp. 265–276.

[85] J.A. Burns, P.D. Hirsch, A difference equation approach to parameter estimation for differential delay equations, *Appl. Math. & Comp.* **7: 4** (1980), pp. 281–311.

[86] J.A. Burns, K. Ito, G. Propst, On nonconvergence of adjoint semigroups for control systems with delays, *SIAM J. Control & Optim.* **26: 6** (1988).

[87] T.A. Burton, *Volterra Intergal and Difference Equations*, Acad. Press, New York, 1983.

[88] T. Burton, L. Hatvani, Stability theorems for nonautonomous functional differential equations by Liapunov functionals, *Tohoku Math. J.* **41** (1989), pp. 65–104.

[89] T. Burton, V. Hudson, Repellers in systems with infinite delay, *J. Math. Anal. & Appl.* **137: 1** (1989), pp. 240–263.

[90] S. Busenberg, Bifurcation phenomena in biomathematics, *Lect. Notes in Math.* **1057** (1984), pp. 1–56.

[91] S. Busenberg, K.L. Cooke, 1) Periodic solutions of a periodic nonlinear delay differential equation, *SIAM J. Appl. Math.* **35: 4** (1978), pp. 704–721. 2) The effect of integral conditions in certain equations modelling epidemics and population growth, *J. Math. Biol.* **10: 1** (1980), pp. 13–32. 3) *Differential Equations and Applications in Ecology, Epidemics, and Population problems*, Acad. Press, New York, 1981.

[92] S.N. Busenberg, M. Martelli (eds.), *Delay Differential Equations and Dynamical Systems*, Springer, Berlin, 1991.

[93] S.N. Busenberg, C.G. Trevis, On the use of reducible-functional differential equations in biological models, *J. Math. Anal. & Appl.* **89: 1** (1982), pp. 46–66.

[94] S.N. Busenberg, M.A. Iannelli, Class of nonlinear diffusion problems in age-dependent population dynamics, *Nonlinear Anal.: Theory, Methods & Appl.* **7: 5** (1983), pp. 501–529.

[95] M.A. Buykis, K.P. Shadurskis, Ecological societies behaviour near the statistical equilibrium, *Random Search Problems (Riga)* **11** (1988), pp. 120–130. (In Russian.)

[96] D. Bylarsky, Lag matrix in Leontief model formd by means of random functions theory, *Ekon. Mat. Obz.* **13: 4** (1977), pp. 405–419.

[97] B. Cahlon, J. Herband, A confidence headway for safe car following, *J. Math. Anal. & Appl.* **69: 2** (1979), pp. 511–530.

[98] Y. Cao, The discrete Lyapunov function for scalar differential delay equations, *J. Diff. Eq.* **87** (1990), pp. 365–390.

[99] V. Capasso, Asymptotic stability for an integro-differential reaction-diffusion system, *J. Math. Anal. Appl.* **103** (1984), pp. 575–588.

[100] D.A. Carlson, An elementary proof of the maximum principle for optimal control problems governed by a Volterra integral equation, *J. Optim. Theory & Appl.* **54: 1** (1987), pp. 43-61.

[101] G.C. Carter, Time-delay estimation, *IEEE Trans. Acoustics, Speech & Signal Processing* **29** (1981), special issue.

[102] A. Casal, M. Freedman, A Poincare–Lindstedt approach to bifurcation problems for dfiiferential-delay equations, *IEEE Trans. Autom. Control* **25: 5** (1980), pp. 967–973.

[103] A. Casal, R. Martiner-Herrero, M.A. Vences, On a two-body problem of classical relativistic electrodynamics, *Rev. Mat. Hisp.-Amer.* **40: 1/2** (1980), pp. 17–24.

[104] P.A. Casal, A. Somolinos, 1) Estudio analitico y numerico de la equacion del girasol, *Rev. Real. Acad. Sci. Exact. Fis. y Natur., Madrid* **73: 4** (1979), pp. 503–519. 2) Entrainement of a nonlinear second order delay differential equation, *Proc. 9th Int. Conf. Nonlinear Vibrations, August 8-Sept. 6 Kiev* **2** (1981), pp. 167–169. 3) Forced oscilations for the sun-flower equation, entrainment, *Nonlinear Anal. Theory, Methods & Appl.* **6: 4** (1982), pp. 397–414.

[105] J.H. Case, A problem in international trade, *J. Optim. Th. & Appl.* **27: 1** (1979), pp. 7-30.

[106] A.M. Castelfranco, H.W. Stech, Periodic solution in a model of recurrent neutral feedback, *SIAM J. Appl. Math.* **47: 3** (1987), pp. 573–588.

[107] C. Castillo-Chavez, Nonlinear character dependent models with constant time delay in population dynamics, *J. Math. Anal. & Appl.* **128** (1987), pp. 1–29.

[108] C. Castllo-Chavez, K. Cooke, W. Huang, S.A. Levin, On the role of long incubation periods in the dynamics of acquired immunodeficiency syndrome (AIDS). Part 1. Single population models, *J. Math. Biol.* **27: 4** (1989), pp. 373–398.

[109] C. Castillo-Chavez, H.R. Thieme, On the role of variable infectivity in the dynamics of the human immunodeficiency virus epidemic, In: C. Castillo-Chavez (ed.), *Math. and Statistical Approaches to AIDS Epidemology*, Springer, Berlin,

1989, pp. 157–176.

[110] J.M.F. Chamayou, A probabilistic approach to a differential-difference equatio: arising in analytic number theory, *Math. Comp.* **27** (1973), pp. 197–203.

[111] W.L. Chan, Guo Bao Z. Hu, Overtaking optimal control problem of age depen dent population with infinite horizon, *J. Math. Anal. & Appl.* **150: 1** (1990) pp. 41–53.

[112] M.H. Chang, Discrete approximation of nonlinear filtering for stochastic dela equations, *Stoch. Anal. & Appl.* **5: 3** (1987), pp. 267–298.

[113] Rong-Yeu Chang, Maw-Ling Wang, Solution of population balance equatio of breakage model via shifting Legendre functions, *Int. J. Systems Sc.* **15:** (1984), pp. 63–74.

[114] N.G. Chebotarev, N.N. Meimen, Problem of Routh–Hurwitz for polynomial and entire functions, *Tr. Mat. Inst. Steklov.* **26** (1949). (In Russian.)

[115] M. Chen, Differential-difference equations for the radial distribution functio: of hard rods and hard spheres under Percus–Jevick approximation, *J. Math Anal. & Appl.* **64: 3** (1978), pp. 629–650.

[116] E. Cheres, Hypersurface attractivity for systems with delays driven by min-ma: controller, *Int. J. Systems Sc.* **19** (1988), p. 10.

[117] E.L. Chernousko, V.B. Kolmanovskii, 1) *Optimal Control under Stochasti Disturbances*, Nauka, Moscow, 1978. (In Russian.) 2) Computational and ap proximate methods of optimal control, *J. Soviet Math.* **12** (1979), pp. 310 353.

[118] Y. Cherruault, *Mathematical Modelling in Biomedicine; Optimal Control c Biomedical Systems*, Reidel, Dordrecht, 1986.

[119] N.G. Chetaev, *Stability of Motion*, Nauka, Moscow, 1965. (In Russian.)

[120] I.D. Cheuskov, About one system of equations with delay appearing in aircon trol, *Funct. Anal., Theory & Appl.* **54** (1990), pp. 123–130. (In Russian.)

[121] Henjin Chi, J. Bell, B. Hassard, Numerical solution of nonlinear advance delay-differential equations from nerve conduction theory, *J. Math. Biol.* **24 3** (1986), pp. 583–601.

[122] U. Chiasson, A method for computing the interval of delay values for which differential delay system is stable, *IEEE Trans. Auom. Control* **33: 12** (1988) pp. 1176–1178.

[123] Lyn-Horng Chou, Ing-Rong Horng, Bor-Sen Chen, Dynamical feedback com pensator for uncertain time-delay systems containing saturating actuator, *In J. Control* **49: 3** (1989), pp. 961-968.

[124] Shui-Nee Chow, Existence of periodic solutions of autonomous functional di ferential equations, *J. Diff. Eq.* **15: 2** (1974), pp. 350–378.

[125] J. Chretien, Automatic control in space, *Proc. IFAC Symp. Toulouse* (1985)

[126] D.H. Chyung, E.B. Lee, Linear optimal control problems with delay, *SIAI J. Control* **4: 1** (1966).

[127] M.D. Ciletti, Differential games with an information time lag, *Topics in Dif Games* , North-Holand, Amsterdam, 1973, pp. 179–268.

[128] D.S. Cohen, P.S. Hagen, H.C. Simpson, Spatial structures in predator-pre

communities with hereditary effects and diffusion, *Math. Biosc.* **44: 3/4** (1979), pp. 167–177.

[129] D.S. Cohen, S. Rosenblatt, A delay logistic equation with variable growth rate, *SIAM J. Appl. Math.* **42: 3** (1982), pp. 608–624.

[130] B.D. Coleman, M.E. Gurtin, Waves in materials with memory: II. On the growth and decay of one-dimensional accelaration waves; III. Thermodynamic influences on the growth and decay of acceleration waves, *Arch. Rat. Mech. & Anal.* **19: 4** (1965), pp. 239–265; 266–298.

[131] B.D. Coleman, M.E. Gurtin, R.I. Herrera, Waves in materials with memory: I. The velocity of one dimensional shock and acceleration waves, *Arch. Rat. Mech. & Anal.* **19: 1** (1965), pp. 1–19.

[132] B.D. Coleman, V.J. Mizel, Stability of functional differential equations, *Arch. Rat. Mech. & Anal.* **30: 3** (1968), pp. 173–196.

[133] B.D. Coleman, D.R. Owen, On the initial value problem for a class of functional-differential equations, *Arch. Rat. Mech. & Anal.* **55: 4** (1974), pp. 275–299.

[134] K.L. Cooke, Functional-differential equations: some models and perturbation problems, *Diff. Eq. & Dynam. Syst.* , Acad. Press, New York–London, 1967, pp. 167–183.

[135] K.L. Cooke, Z. Grossman, Discrete delay, distributed delay and stability switches, *J. Math. Anal. Appl.* **86: 2** (1982), pp. 592–627.

[136] K.L. Cooke, J.L. Kaplan, M. Sorg, Stability of a functional differential equation for the motion of a radiating charged particle, *Nonlinear Anal. Theory, Methods & Appl.* **10: 5** (1981), pp. 1133–1139.

[137] C. Corduneanu, 1) *Integral Equations and Stability of Feeback Systems*, Acad. Press, New York–London, 1973. 2) Stability problems for some feedback systems with delay, *Modern Trends in Cybernetics and Systems* **2** Springer, Berlin, 1977, pp. 321–328.

[138] F. Corpet, C. Chevalet, M. Gillois, A. Micali, Dynamics and stability of induction of the lactose operon of Escherischia Coli, *Lect. Notes Biomath.* **49** (1983), pp. 123–143.

[139] J.M. Cushing, 1) Forced asymptotically periodic solutions of predator-prey systems with or without hereditary effects, *SIAM J. Appl. Math.* **30: 47** (1976), pp. 665–674. 2) Periodic solutions of Volterra's population equation with hereditary effects, *SIAM J. Appl. Math.* **31: 2** (1976), pp. 251–261. 3) Predator-prey interactions with time delay, *J. Math. Biol.* **3: 3/4** (1976), pp. 369–380. 4) Stable positive periodic solutions of the time-dependet logistic equation under possible hereditary influences, *J. Math. Anal. & Appl.* **60: 3** (1977), pp. 747–754. 5) Time delays in single species growth models, *J. Math. Biol.* **4: 3** (1977), pp. 257–264. 6) Bifurcation of periodic solutions of integrodifferential systems with applications to time delay models in population dynamics, *SIAM J. Appl. Math.* **33: 4** (1977), pp. 650–654. 7) Bifurcation of periodic oscillations due to delays in single species growth models, *J. Math. Biol.* **6: 2** (1978), pp. 145–161.

[140] J.M. Cushing, M. Saleen, A predator-prey model with age structure, *J. Math Biol.* **14: 2** (1982), pp. 231–250.

[141] Lo Sheng Dai, Nonconstant periodic solutions in predator prey systems with continuous time delay, *Math. Biosc.* **53: 1/2** (1981), pp. 149–157.

[142] R. Datko, Not all feedback stabilized hyperbolic systems are robust with respect to small time delays in their feedbacks, *SIAM J. Control Optim.* **26: 3** (1988), pp. 687–713.

[143] R. Datko, J. Lagnese, M.P. Polis, An example of the effect of time delays in boundary feedback stabilization of wave equations, *SIAM J. Control Optim.* **24** (1986), pp. 152–156.

[144] P.L. Davis, On the hyperbolicity of the equations of the linear theory of heat conduction for materials with memory, *SIAM J. Appl. Math.* **30: 1** (1976), pp. 75–80.

[145] E.J. Davison, B.R. Copeland, Gain margin and time lag tolerance constraints applied to the stabilization problem and robust servomechanism problem, *IEEE Trans. Autom. Control* **30: 3** (1985), pp. 229–239.

[146] A.G. Degtyarev, Synthesis of stochastic systems optimal control with delay in control contours, *Optimal Problems of Aircraft Technology, Kazan* (1986), pp. 40–45. (In Russian.)

[147] M.C. Delfour, A. Manitius, Control systems with delays; areas of applications and present status of the linear theory, *Anal. Syst. et Orientat. Nouv. Coll. TRTA Versailles-Rocquencourt 1976* (1976), pp. 345–392.

[148] M. Delpech, M. Maurette, Feasibility of time delay compensation for a space teleoperation task, In: J. Chretien (ed.), *Autom. Control in Space*, Toulouse 1985, pp.279–286.

[149] M.W. Derstine, H.M. Gibbs, F.A. Hopf, L.D. Sanders, Distinguishing chaos from noise in an optically bistable system, *IEEE J. Quantum Electronics* **QE 21: 9** (1985), pp. 1419–1422.

[150] W. Desch, R. Grimmer, W. Schappacher, Wellposedness and wave propagation for a class of integrodifferential equations in Banach space, *J. Diff. Eq.* **74: 2** (1988), pp. 391–411.

[151] G. Diblasio, K. Kunisch, E. Sinestrari, Stability for abstract linear functional differential equations, *Israel J. Math.* **50** (1985), pp. 231–263.

[152] J. Diblik, 1) Stability of the trivial solution of retarded functional differential equations, *J. Diff. Eq.* **26: 2** (1990), pp. 219–223. 2) Application of Vajevskii's method to the differential equations with delay, *Dokl. Akad. Nauk. SSSR* **310: 2** (1990), pp. 285–287. (In Russian.)

[153] B.F. Dibrov, M.A. Livshits, M.V. Volkenstein, The effect of a time lag in the immune reaction, *Lect. Notes Control & Inform. Sc.* **18** (1979), pp. 87–94.

[154] O. Diekmann, The cell size distribution and semigroups of linear operators, *Lect. Notes Biomath.* **68** (1986), pp. 46–77.

[155] O. Diekmann, S.A. van Gils, A variation-of-constants formula for nonlinear Volterra integral equations of convolution type, *Math. Centre Dept. Applied Math. Notes* **206** (1980). 10pp.

[156] O. Diekmann, H.J.A.M. Heymans, H.R. Thieme, *J. Math. Biol.* **19: 2** (1984), pp. 227-248.

[157] I. Diory, N.V. Pertzev, 1) Behaviour of solutions of functional equations systems with delays as $t \to \infty$ possessing mixed monotonous property, *Komm. Mat. Akad. Nauk. SSSR* **86** (1987). Preprint 27pp. (In Russian.) 2) Stability of the equilibrium state of functional-differential equation systems possessing mixed monotonous property, *Komm. Mat. Akad. Nauk. SSSR* **126** (1986). Preprint 23pp. (In Russian.)

[158] J.A. Dixon, A nonlinear weakly singular Volterra integrodifferential equation arising from a reaction-diffusion on study in a small cell, *J. Comb. & Appl. Math.* **18: 3** (1987), pp. 289–305.

[159] A.A. Dmitriev, A.P. Shafiro, On the one functional equation from training theory, *Inst. Autom. & Control Process. Far-East Sc. Centre Akad. Nauk. SSSR, Vladivostok* (1979). 19pp. (In Russian.)

[160] R.K. Dodd, Generalised Backlund transformation for some nonlinear partial differential-difference equations, *J. Phys.* **A11: 1** (1978), pp. 81–92.

[161] U.E. Doldii, A.V. Kim, Lyapunov functionals method for after-effect systems, *Diff. Uravn.* **27: 8** (1991), pp. 1313–1318. (In Russian.)

[162] B. Dorizzi, B. Grammaticos, M. le Berre, I. Pomean, E. Ressayre, A. Tallet, Statistics and dimension of chaos in differential delay systems, *Phys. Rev.* **A35: 1** (1987), pp. 328–339.

[163] P. Dormayer, The stability of special symmetric solutions of $\dot{x}(t) = af(x(t-1))$ with small amplitudes, *Nonlinear Anal. TMA* **14: 8** (1990), pp. 701–715.

[164] R.D. Driver, 1) The 'backwards' problem for a delay differential system arising in classical electrodynamics, *Proc. 5th Conf. Nonlinear Oscillation, Kiev* **2** (1970), pp. 137–143. 2) *Ordinary and Delay Differential Equations*, Springer, New York, etc., 1977. 3) A neutral system with state-dependent delay, *J. Dif. Eq.* **54: 1** (1984), pp. 73–86.

[165] D.K. Durdiev, Inverse problems for three-dimensional wave equation in an environment with memory, *Mat. Anal. i Diskret. Mat., Novosibirsk* (1989), pp. 19–26. (In Russian.)

[166] C.J. Duyn, Regularity properties of the solution of an equation arising in the theory of turbulence, *J. Diff. Eq.* **33: 2** (1979), pp. 226–238.

[167] L.C. Edie, Car-following and steady state theory for noncongested traffic, *Oper. Res.* **9: 1** (1961), pp. 66–76.

[168] R.H. Elderkin, Nonlinear, globally age-dependent population models: some basic theory, *J. Math. Anal. & Appl.* **108: 2** (1985), pp. 546–562.

[169] M. El-Doma, Analysis of nonlinear integro-differential equation arising in age-dependent epidemic models, *Nonlinear Anal., Theory, Methods & Appl.* **11: 8** (1987), pp. 913–937.

[170] M.E. El'jasberg, About stability of metals cutting, *Izv. Akad. Nauk. SSSR, OTN* **9** (1958).

[171] H. El-Owaidy, A.A. Ammar, Stable oscillations in a predator-prey model with time lag, *J. Math. Anal. & Appl.* **139: 1** (1988), pp. 191–199.

[172] I.B. Emmerth, On the existence of solution of a nonlinear differential equatioı with delay, *Arch. Rat. Mech. Anal.* **68: 2** (1978), pp. 155–169.

[173] L.H. Erbe, H.J. Freedman, V.S.H. Rao, Three-species food-chain models witl mutual interference and time delay, *Math. Biosc.* **80: 1** (1986), pp. 57–80.

[174] M. Farkas, Stable oscillations in a predator-prey model with time lags, *J. Math Anal. & Appl.* **102: 1** (1984), pp. 175–188.

[175] A. Farkas, M. Farkas, Stable oscillations in a more realistic predator-prey mode with time lag, *Asympt. Methods of Math. Phys., Kiev* (1988), pp. 250–256 (In Russian.)

[176] C.D. Farmer, Chaotic attractors of an infinite-dimensional dynamical system *Physica* **D 4: 3** (1982), pp. 366-393.

[177] N.B. Fedotov, About strict regime of appearance autovibrations in long lin with tunnel diode, *Investigations in Stability and Vibration Theory, Yaroslaı* (1978), pp. 81–99. (In Russian.)

[178] A. Feldstein, C. Grafton, Experimental mathematics: an application to re tarded ordinary differential equations with infinite lag, *Proc. 23-rd ACM Nat Conf.* (1968), pp. 67–71.

[179] M.L. Fernandes, F. Zanolin, Existence and uniqueness results for periodic solu tions of differential systems with application to the competing species problem *J. Math. Anal. & Appl.* **137: 2** (1989), pp. 335–348.

[180] J.M. Ferreira, On the stability of a distributed network, *SIAM J. Math. Anal* **17: 1** (1986), pp. 38–45.

[181] Y.A. Fiagbedzi, A.E. Pearson, A multistage reduction technique for feedbacl stabilizing distributed time-lag systems, *Automatica* **23: 3** (1987), pp. 311-326.

[182] W. Fickett, A mathematical problem from detonation theory, *Quart. Appl* **46: 3** (1988), pp. 459–471.

[183] A.C. Fowler, 1) Approximate solution of a model of biological immune response incorporating delay, *J. Math. Biol.* **13: 1** (1981), pp. 23–45. 2) Asymptoti analysis of the delayed logistic equation when the delay is large, *IMA J. Appl Math.* **28: 1** (1982), pp. 41–49.

[184] L. Fox, D.F. Mayers, J.R. Ockenden, A.B. Tauler, On a functional differentia equation, *J. Inst. Math. Anal. & Appl.* **8: 3** (1971), pp. 271–307.

[185] A. Freedamn, L.A. Romero, Functional differential equation for the determi nation of the viscosity function in a rheometer, *Arch. Rat. Mech. Anal.* **107 1** (1989), pp. 85–97.

[186] H.I. Freedman, K. Gopalsamy, Nonoccurrence of stabilty in systems witl discrete delays, *Canad. Math. Bull.* **31: 1** (1988), pp. 52–58.

[187] H.I. Freedman, J.W.H. So, P. Waltman, Chemostat competition with tim delays, *IMACS Ann. Comput. & Appl. Math.* **5: 1/4** (1989), pp. 171–173 (12th IMACS World Congr. Sci. Com. Paris, July 1988.)

[188] H.I. Freedman, V. Ito, Hari Sree, The trade-off between mutual interferenc and time lags in predator-prey systems, *Bull. Math. Biol.* **45: 6** (1983), pp 991–1004.

[189] M. Fu, A.W. Olbrot, M.P. Polis, Robust stability for time-delay systems: the edge theorem and graphical tests, *IEEE Trans. Autom. Control* **AC-34** (1989), pp. 813–820.

[190] K. Furuta, M. Yamakita, Y. Sato, Computation of optimal control for linear systems with delay, *Int. J. Control* **48: 2** (1988), pp. 577–589.

[191] R.F. Gabasov, F.M. Kirillova, *Maximum Principle in Optimal Control Theory*, Nauka i Tekhn., Minsk, 1974. (In Russian.)

[192] G. Gabisch, H.W. Lorenz, Business cycle theory, *Lect. Notes in Econ. & Math. Systems* **283** (1987). 229pp.

[193] J.I. Gao, L.M. Narducci, H. Sadiky, M. Squicciarini, J.M. Yuan, Higher-order bifurcations in a bistable system with delay, *Phys. Rev.* **A 30: 2** (1984), pp. 901–904.

[194] L. Gastaldi, F. Tomarelli, A uniqueness result for a nonlinear hyperbolic equation, *Ann. Mat. Pura Appl.* **137** (1984), pp. 175–205.

[195] J.A. Gatica, P. Waltman, 1) A singular functional differential equation arising in an immunological model, *Lect. Notes in Math.* **564** (1976), pp. 114–124. 2) A system of functional differential equations modelling Threshold phenomena, *Applicable Anal.* **28** (1989), pp. 39–50.

[196] V.G. Gerasimov, *Theoretical Foundation of Control of Heat Processes*, Gosenergoizdat, Moscow, 1949. (In Russian.)

[197] M. Ghil, Successive bifurcations and the ICE-AGE problem, *Bifurcation Phenom. Math. Phys. & Rel. Topics. NATO Proc., Cargese 1979* (1980), pp. 57–68.

[198] T.R. Gicev, A.L. Dontchev, Convergence of the solution of convex optimal control problem with delay, *Rend. Math.* **11: 4** (1978), pp. 635–643.

[199] M.I. Gil′, *Operator Function Method in Differential Equations Theory*, Nauka, Moscow, 1990. (In Russian.)

[200] L.P. Gilbert, A.M. Johnson, An application of the Jordan canonical form to the epidemic problem, *J. Appl. Probab.* **17: 2** (1980), pp. 313–323.

[201] L. Glass, M.C. Mackey, *From Clocks to Chaos*, Princeton Univ. Press, 1988. 248pp.

[202] E.A. Godin, V.B. Kolmanovskii, E.Sh. Stengold, *Hypertension Disease as a Result of Altered Stress-Strained Vascular State*, Nauka, Moscow, 1990. (In Russian.)

[203] K. Gopalsamy, *Equations of Mathematical Ecology*, Flinders Univ. Australia, Adelaide, 1991.

[204] K. Gopalsamy, B.D. Agarwal, The logistic equation with a diffusionally coupled delay, *Bull. Math. Biol.* **43: 2** (1981), pp. 125–140.

[205] K. Gopalsamy, R.A. Ahlip, Time delays in n-species competition I. Global stability in constant environments, *Bull. Austal. Math. Soc.* **27: 3** (1983), pp. 427–441.

[206] K. Gopalsamy, M.R.S. Kulenovic, G. Ladas, Time lags in a food-limited population model, *Applicable Anal.* **31** (1981), pp. 225–237.

[207] K. Gopalsamy, B. Rai, Stability analysis of synchronous oscilations in a weakly

coupled integro-differential system, *J. Math. Anal. & Appl.* **136: 2** (1988), pp. 368–382.

[208] K. Gopalsamy, B.G. Zhang, On a neutral delay logistic equation, *Dynamics & Stability of Systems* **2: 4** (1988), pp. 183–195.

[209] H. Gorecki, S. Fuksa, P. Grabomski, A. Korytowski, *Analysis and Synthesis of Time Delay Systems*, Wiley, 1989.

[210] V.D. Gorjachenko, S.L. Zolotarev, V.A. Kolchin, *Qualitative Methods in Nuclear Reactor Dynamics*, Energoatomizdat, Moscow, 1988.

[211] R.C. Graef, M.K. Grammatikopoulos, P.W. Spikes, On the asymptotic behavior of solutions of a second order nonlinear neutral delay differential equation, *J. Math. Anal. & Appl.* **156** (1991), pp. 23–39.

[212] D. Greenhalgh, Some results on optimal control applied to epidemics, *Math. Biosc.* **88: 2** (1988).

[213] P.S. Gromova, A.F. Pelevina, Absolute stability of automatic control systems with delay, *Boundary Problems* (1982), pp. 24–30. Perm. (In Russian.)

[214] P.S. Gromova, A.M. Zverkin, About trigonometric series the sum of which is a continuous unbounded function - solutionof an equation with delay, *Diff. Uravn.* **4: 10** (1968), pp. 1774–1784. (In Russian.)

[215] S. Grossberg, 1) Global ratio limit theorems for some nonlinear functional-differential equations, *Bull. Amer. Math. Soc.* **74: 1** (1986), pp. 95–99; 100–105. 2) A predicion theory for some nonlinear functional-differential equations 1. Learning of lists, *J. Math. Anal. & Appl.* **21: 3** (1968), pp. 643–694. 3) A predicion theory for some nonlinear functional-differential equations 2. Learning of patterns, *J. Math. Anal. & Appl.* **22: 2** (1968), pp. 490–522. 4) On variational systems of some nonlinear difference-differential equations, *J. Diff. Eq.* **6: 3** (1969), pp. 534–577. 5) On the global limits and oscillations of a system of nonlinear differential equations describing a flow on a probabilistic network, *J. Diff. Eq.* **5: 3** (1969), pp. 531–563. 6) Learning and energy-entropy dependence in some nonlinear functional differential systems, *Bull. Amer. Math. Soc.* **75: 6** (1969), pp. 1238–1242.

[216] Z. Grossman, Oscillatory phenomena in a model of infections diseases, *Theor. Popul. Biol.* **18: 2** (1980), pp. 204–243.

[217] Z. Grossman, I. Gumowski, Self-sustained oscillations in the Jacob–Monod mode of generegulation, *Lect. Notes Comp. Sci.* **40** (1976), pp. 145–154.

[218] F. Grupp, On difference-differential equations in the theory of sieves, *J. Number Th.* **24** (1984), pp. 154–173.

[219] I. Guintchev, On the stability of the zero solution of retarded functional-differential systems with an application to nuclear reactor dynamics, *Zag. drgan nielin* **22** (1984), pp. 107–126.

[220] V.V. Guljaev, A.S. Dmitriev, V.E. Kislov, Strange attractors in the circle: selfoscillating systems, *Dokl. Akad. Nauk. SSSR* **282: 1** (1985), pp. 53–66. (In Russian.)

[221] I. Gumowski, 1) Sur le calcul des solutions periodiques de l'equation de Cherwell–Wright, *C.R. Acad. Sci.* **268: 3** (1969), pp. A157–A159. 2) Peri-

odic solutions of a system of differential equations with pure delay arising in a biological problem, *Simul. Syst. Proc. 8th AICA Congr., Delft 1976* (1977), pp. 633–639. Amsterdam. 3) Qualitative properties of some dynamical systems with pure delay, *Lect. Notes Biomath.* **41** (1981), pp. 139–152.

[222] I. Gumowski, P. Sonrias, Analysis of the blood platelet control mechanism, *Sim. Syst.'79, Amsterdam* (1980), pp. 1071–1076.

[223] Guo-Bao-Shu, Guang-Tian Zhu, The linear invariant population evolution equation with time delay, *J. Syst. Sci. & Math. Sci.* **8: 4** (1988), pp. 346–350.

[224] Dajun Guo, V. Lakshmikantham, Positive solutions of nonlinear integral equations arising in infections diseases, *J. Math. Anal. & Appl.* **134: 1** (1988), pp. 1–8.

[225] Naresh N. Gupta, J. Houndeshell, A differential-difference equations model of a dynamic aquatic ecosystem, *Int. J. Syst. Sci.* **7: 5** (1976), pp. 481–492.

[226] H. Guretsky, *Analysis and Synthesis of Control Systems with Delay*, Mashinostoenie, Moscow, 1974. (In Russian.)

[227] L.S. Gurin, About a problem connected with Poisson's flow, *Math. Suppor. Space Exp., Moscow* (1978), pp. 51–62. (In Russian.)

[228] W.S. Gurney, S.P. Blythe, R.M. Nisbet, Nicholson's blowflies revisited, *Nature* **287** (1980), pp. 17–21.

[229] P. Guttorp, R. Kulperger, Statistical inference for some Volterra population processes in a random environment, *Canad. J. Stat.* **12: 4** (1984), pp. 289–302.

[230] M. Gyllenberg, H.J.A.M. Heijmans, An abstract delay-differential equation modelling size dependent cell growth and division, *SIAM J. Math. Anal.* **18: 1** (1987), pp. 74–88.

[231] I. Gyori, 1) Connections between compartmental systems with pipes and integro-differential equations, *Math. Modell.* **7: 9/12** (1986), pp. 1215–1238. 2) Remarks about differential equations of the population growth, *Alkalm. Mat. Lapok* **1: 1/2** (1975), pp. 139–150. (In Hungarian.)

[232] I. Gyori, J. Eller, 1) Compartmental systems with pipes, *Math. Biosc.* **53: 3/4** (1981), pp. 223–247. 2) Analytic properties and identifiability problems of compartmental models with time lags, *Math. & Comp. Meth. Physiol. Satell. Symp. 28th Int. Congr. Physiol. Sci. Budapest, 1980* (1981), pp. 61–74. Oxford.

[233] I. Gyori, G. Ladas, *Oscillation Theory of Delay Differential Equations*, Clarendon Press, Oxford, 1991.

[234] J.R. Haddock, W.E. Hornor, A uniform asymptotic stability theorem for fading memory spaces, *J. Math. Anal. & Appl.* **155** (1991), pp. 55–65.

[235] K.P. Hadeler, 1) On the stability of the stationary state of a population growth equation with time-lag, *J. Math. Biol.* **3: 2** (1976), pp. 197–201. 2) Delay equations in biology, *Lect. Notes in Math.* **730** (1979), pp. 136–156. 3) The hypercycle travelling waves and Wright's equation, *J. Math. Biol.* **24: 5** (1986), pp. 473–477. 4) Pair formation in age-structured population, *Acta Appl. Math.* **14: 1/2** (1989), pp. 91–102.

[236] K.P. Hadeler, J. Tomink, Periodic solutions of difference-differential equations, *Arch. Rat. Mech. & Anal.* **65: 1** (1977), pp. 87–95.

[237] J.B.S. Haldane, A contribution to the theory of price fluctuations, *Rev. Econ. Stud.* **1** (1933), pp. 186–195.

[238] J. Hale, *Theory of Functional Differential Equations*, Springer, New York–Berlin, 1977.

[239] A.J. Hall, G.C. Wake, A functional differential equation arising in modelling of cell growth, *J. Austral. Math. Soc.* **B 30: 4** (1989), pp. 424–435.

[240] E.J. Hannan, P.J. Thomson, Time delay estimation, *J. Time Series Anal.* **9: 1** (1988), pp. 21-33.

[241] K.B. Hannsgen, I. Renardy, R.L. Wheeler, Effectiveness and robustness with respect to time delays of boundary feedback stabilization in one-dimensional viscoelasticity, *SIAM J. Contr. Optim.* **26: 5** (1988), pp. 1200–1234.

[242] F.B. Hanson, Analysis of a singular functional differential equation that arises from stochastic modelling of population growth, *J. Math. Anal. & Appl.* **96: 2** (1983), pp. 405–419.

[243] A. Haraux, *Nonlinear Evolution Equations. Global Behavior of Solutions*, Springer, Berlin, 1981.

[244] J. Harband, The existence of monotonic solutions of a nonlinear car-following equation, *J. Math. Anal. & Appl.* **57: 2** (1977), pp. 257–272.

[245] H.W. Hethcote, M.A. Lewis, P. van der Driessche, An epidemiological model with a delay and a nonlinear incidence rate, *J. Math. Biol.* **27: 1** (1989), pp. 49–64.

[246] H.W. Hethcote, H.W. Stech, P. van der Driessche, Nonlinear oscillations in epidemic models, *SIAM J. Appl. Math.* **40: 1** (1981), pp. 1–9.

[247] Y. Hino, S. Murakami, T. Naito, *Functional Differential Equations with Infinite Delay*, Springer, 1991.

[248] J.T. Hoag, R.D. Driver, A delayed-advanced model for the electrodynamic two-body problem, *Nonlinear Anal., Theory, Methods & Appl.* **15: 2** (1990), pp. 165–184.

[249] K.H. Hoffman, J. Sprekels, Automatic delay-control in a two-phase Stefan problem, In: L. Collatz (ed.), *Differential-Difference Equations*, Birkhäuser, Basel, 1983, pp. 119–136.

[250] W.J. Hrusa, A nonlinear functional differential equation in Banach space with applications to material with fading memory, *Arch. Rat. Mech. & Anal.* **84: 2** (1983), pp. 99–137.

[251] E.Ya. Hruslov, Averaging model of medium with great homogeneity with memory, *Uspekhi Mat. Nauk.* **45: 1** (1990), pp. 197–198. (In Russian.)

[252] Tsutomu Ikeda, Stationary solutions of a spatially aggregating population model, *Proc. Japan Acad.* **A60: 2** (1984), pp. 46–48.

[253] M. Ilolov, Stability and periodic solution in an epidemic model, *Proc. 11th Intern. Conf. Nonlinear Oscillations, Budapest, aug. 17–23, 1987* (1987), pp. 832–835.

[254] E.F. Infante, Stability theory for general dynamical systems and some applications, *Lect. Notes Phys.* **21** (1973), pp. 63–82.

[255] D. Israelsson, A. Johnson, A theory of circumnutations of Helianthus annus, *Physiol. Plant* **20** (1967), pp. 957-976.

[256] K. Ito, M. Nisio, On stationary solutions of a stochastic differential equation, *J. Math. Sci. Kyoto Univ.* **4** (1954), pp. 1–75.

[257] A.F. Ize (ed.), *Functional Differential Equations and Bifurcations*, Springer, New York, 1980.

[258] A.F. Ize, A.A. Freiria, Total stability for neutral functional differential equations, *Proc. Amer. Math. Soc.* **81: 3** (1981), pp. 437–442.

[259] Z. Jankiewicz, One step methods for the numerical solution of Volterra functional differential equations of neutral type, *Applicable Anal.* **12: 1** (1981), pp. 1–11.

[260] S. Kakutani, L. Marcus, On the nonlinear difference-differential equation $y'(t) = (A - By(t-1))y(t)$, *Contr. to the Theory of Nonlinear Oscillations* **4** (1958), pp. 1–18.

[261] D.V. Kalyaev, Periodic solutions in the borders of the simplest models of disease with strong organ's debeat, *Meth. Comp. & Appl. Math., Moscow* (1985), pp. 64–73. (In Russian.)

[262] A.M. Kamachkin, Orbital stability of periodical solutions of automatic regulation systems with delay effect of controlling mechanism, *Appl. & Comp. Practice in Shipbuilding, Leningrad* (1981), pp. 94–98. (In Russian.)

[263] D. Kannan, Evolution operators and equations of an interacted prey-predator population incorporating historical actions, *Nonlinear Anal., Theory, Meth. & Appl.* **2: 1** (1978), pp. 59–68.

[264] F. Kappel, Differenzen-Differentialgleichungen, *Ber. Math. Stat. Seh. Forschungszentrum Gras* **1** (1973), pp. 1–77.

[265] F. Kappel, K. Kunisch, Invariance results for delay and Volterra equations in fractional order Sobolev spaces, *Trans. Amer. Math. Soc.* **304: 1** (1987), pp. 1–51.

[266] P. Kappel, D. Salamon, Spline approximations for retarded systems and Riccati equation, *SIAM J. Control Opt.* **25: 4** (1987), pp. 1082–1114.

[267] F. Kappel, W. Schappacher, 1) Some considerations to the fundamental theory of infinite delay equations, *J. Diff. Eq.* **37** (1980), pp. 141–183. 2) *Evolution Equations and their Applications*, Pitman, Boston–London, 1982.

[268] J.N. Kapur, 1) Predator-prey models with discrete time lags, *Nat. Acad. Sci. Letters* **2: 7** (1979), pp. 273–275. 2) Bifurcation theory in population models with time delays, *Indian J. Pure & Appl. Math.* **11: 11** (1980), pp. 1397–1410. 3) Stability analysis of a difference equation population model with delayed recruitment, *Indian J. Pure & Appl. Math.* **12: 1** (1981), pp. 7–21.

[269] G. Karakostas, 1) On the delay population equation with seasonal variations, *Trans. 2nd Conf. Diff. Eq. & Appl., Russe June 29 -July 4, 1981* **4: 1** (1982), pp. 333-336. 2) The effect of seasonal variations to the delay population equation, *Nonlinear Anal., Theory, Methods & Appl.* **6: 11** (1982), pp. 1143–1154.

3) Asymptotic equilibrium in prey-predator interactions with nonlinear "per capita" growth rates, *Funkc Ekv.* **30: 2/3** (1987), pp. 383–397.

[270] S.A. Kashchenko, About staedy-state regimes of the Hutchinson equation with diffusion, *Akad. Nauk SSSR* **292: 2** (1987), pp. 327–330. (In Russian.)

[271] I. Kashiwagi, I. Flugge-Lotz, Periodic solutions of differential-difference equations with delay control, *Int. J. Control* **6: 1** (1967), pp. 1–25.

[272] J. Kato, Liapunov second method in functional differential equations, *Tohoku Math. J.* **32** (1980), pp. 487–497.

[273] D.A. Kault, L.M. Marsh, Modelling AIDS as a function of other sexuality transmitted diseases, *Math. Biosc.* **103: 1** (1991), pp. 17–31.

[274] W. Kerscher, R. Nagel, Positivity and stability for Cauchy problems with delay, *Lect. Notes in Math.* **1424** (1988), pp. 216–235.

[275] A.V. Kim, Method of Liapunov functionals for systems with aftereffect, *Autom. & Remote Control* **1** (1990), pp. 2–31.

[276] A.V. Kim, *Direct Liapunov Method in Stability Theory of Hereditary Systems*, Ural Univ. Press, Sverdlovsk, 1992.

[277] K.D. Kim, N.K. Bose, Vertex implications of stability for a class of delay-differenctial interval systems, *IEEE Trans. Circuits & Systems* **37: 7** (1990), pp. 969–972.

[278] H. Knolle, The general age-dependent endemic with age-specific contact rate, *Biom. J.* **25: 5** (1983), pp. 469–475.

[279] Yu.S. Kolesov, Some problems of mathematical ecology, *Diff. Eq. & Appl.* **29** (1981), pp. 27–35.

[280] Yu.S. Kolesov, D.I. Shvitra, *Autovibrations in Systems with Delay*, Mokslas, Vilnius, 1979.

[281] V.B. Kolmanovskii, 1) Stationary solutions of equations with delay, *Probl. Inform. Transmission* **3: 1** (1967), pp. 50–57. (In Russian.) 2) Application of the Liapunov method to linear systems with lag, *J. Appl. Math. Mech.* **3** (1967), pp. 976–980. (In Russian.) 3) Ergodic properties of stochastic differential equation solutions, *Th. Probab. Appl.* **14: 1** (1969), pp. 142–148. (In Russian.) 4) On the stability of stochastic systems with delay, *Probl. Inform. Transmission* **5: 4** (1969), pp. 59–67. (In Russian.) 5) On the stability of neutral type differential equations, *Diff. Uravn.* **6: 7** (1970), pp. 1235–1246. (In Russian.) 6) The stability of nonlinear systems with lag, *Math Notes* **7** (1970), pp. 446–450. (In Russian.) 7) Solutions of constant signs of neutral-type equations, *Diff. Uravn.* **7: 6** (1971), pp. 1116–1119. (In Russian.) 8) Representation of solutions of neutral type equations, *Ukr. Mat. Zh.* **24: 2** (1972), pp. 171–178. (In Russian.) 9) Filtration of stochastic processes with delay, *Autom. & Remote Control* **1** (1974), pp. 42–49. (In Russian.) 10) On the stability of some stochastic differential equations with retarded arguments, *Th. Probab. & Math. Stat.* **2** (1974), pp. 110–120. (In Russian.) 11) On the equalities for second moments of solutions of stochastic differential equations with delay, *Ukr. Mat. Zh.* **27: 1** (1975), pp. 94–97. (In Russian.) 12) Control and estimation in systems with aftereffect, in: *Math. Systems Theory*, Nauka, Moscow, 1986, pp.

45–88. (In Russian.) 13) Periodic solutions of stochastic equations with delay, *Diff. Uravn.* **24: 6** (1988), pp. 965–972. (In Russian.)

[282] V.B. Kolmanovskii, R.Z. Hasminskii, Stability of linear systems with retardation, *Ivz. Vyssh. Uchebn. Zav. Mat.* **53: 4** (1966), pp. 58–65. (In Russian.)

[283] V.B. Kolmanovskii, N.I. Koroleva, Optimal control of bilinear hereditary systems and their applications, *Appl. Math. Letters* **4: 2** (1991), pp. 1–3.

[284] V.B. Kolmanovskii, T.L. Maizenberg, Optimal estimation of system states and problems of control of systems with delay, *Prikl. Mat. & Mekh.* **41: 3** (1977), pp. 446–456. (In Russian.)

[285] V.B. Kolmanovskii, A.D. Myshkis, V.R. Nosov, Modern theory of equations with delay from the point of view of its application, *Sovr. Probl. Mat. Fiz., Tbilisi* (1987), pp. 280–290.

[286] V.B. Kolmanovskii, V.R. Nosov, 1) *Stability and Periodic Modes of Regulated Systems with Delay*, Nauka, Moscow, 1981. 2) Stability of neutral-type functional differential equations, *Nonlinear Anal., Theory, Methods & Appl.* **6: 9** (1982), pp. 873–910. 3) *Stability of Functional Differential Equations*, Acad. Press, New York–London, 1986.

[287] V.B. Kolmanovskii, L.E. Shaikhet, 1) On the estimation of solutions of linear stochastic integral equations, *Prikl. Mat. & Mekh.* **51: 5** (1987), pp. 775–782. (In Russian.) 2) State estimate of hereditary stochastic systems, *Stoch. Anal. & Appl.* **7: 4** (1989), pp. 387–411.

[288] M. Konstantinov, D.D. Bainov, Nonlinear vibrations of viscous elastic bodies under the action of periodical forces, *Teor. i Primen. Mekh.* **3** (1977), pp. 23–28. (In Russian.)

[289] M. Kot, Delay differential models and the effects of economic delays on the stability of open-access fisheries, *M.S. Thesis, Cornell Univ.* (1979).

[290] Yu.D. Kozlov, Stability of quasilinear parabolic equations with delay, *Stability & Nonlinear Vibr.* (1980), pp. 41-54. (In Russian.)

[291] N.N. Krasovskii, *Stability of Motion*, Stanford Univ. Press, 1963.

[292] L. Krivy, 1) Some models of predator-prey with time deviation, *Sb. Pr. Ped. Ostrave* **A-17: 77** (1982), pp. 15–24. (In Russian.) 2) Retardation in models of isolated population reproduction, *Sb. Pr. Ped. Ostrave* **A-18: 82** (1983), pp. 3–12. (In Russian.) 3) The Maya's model of predator-prey system with time delay, *Sb. Pr. Ped. Ostrave* **A-20: 92** (1985), pp. 3–15. (In Russian.)

[293] I. Krolik, M. Eizenman, S. Pasupathy, Time delay estimate of signals with uncertain spectra, *IEEE Trans. Acoustics, Speech & Signal Proc.* **36: 12** (1988).

[294] Y. Kuang, H.L. Smith, Global stability in diffusive delay Lotka–Volterra systems, *Diff. & Integral Eq.* **4: 1** (1991), pp. 117–128.

[295] E.P. Kubyshkin, Local methods in the investigation of systems of differential-difference equations, modelling the dynamics of numeral insects population change, *Nonlinear Vibr. in Ecological Probl., Yaroslavl* (1985), pp. 70–92. (In Russian.)

[296] M.R.S. Kulenovic, Harmless and harmful retarded and advanced actions in

mathematical models in biology, *4th Conf. Appl. Math., Split, May 28-30 1984* (1985), pp. 11–18.

[297] M.R.S. Kulenovic, G. Ladas, Oscillations of the sunflower equation, *Quart. Appl. Math.* **46: 2** (1988), pp. 23–29.

[298] V.G. Kurbatov, *Linear Differential-Difference Equations*, Voronezh Univ. Press, 1990.

[299] V.I. Kuznetzova, Stability of the neutral type systems with a small delay, *Autom. & Remote Control* **2** (1986), pp. 31–38. (In Russian.)

[300] W.H. Kwon, C.W. Lee, S.W. Kim, Performance improvement using time delays in multivariable controller design, *Int. J. Control* **52: 6** (1990), pp. 1455–1473.

[301] G. Ladas, I.P. Stavroulakis, On delay differential inequalities of first order, *Funkcialaj* **25** (1982), pp. 105-111.

[302] G.S. Ladde, 1) Differential inequalities and stochastic functional differential equations, *J. Math. Phys.* **1: 6** (1974), pp. 738–743. 2) Stochastic stability analysis of model ecosystems with time-delay, *Link. Sci. & Appl. Autom. Control. Proc. 7th World Congr. Int. Fed. Autom. Control, Helsinki 1978* **2** (1979), pp. 1149–1154.

[303] V. Lakshmikantham, S. Leela, A technique in stability theory delay-differential equations, *Nonlin. Anal., Th., Methods & Appl.* **3: 3** (1979), pp. 317–323.

[304] V. Lakshmikantham, S. Leela, A. Martynyuk, *Stability Analysis of Nonlinear Systems*, M. Dekker, New York, 1989.

[305] K.A. Landmen, Bifurcation and stability theory of periodic solutions for integrodifferential equations, *Stud. Appl. Math.* **62: 3** (1980), pp. 217–248.

[306] M.R. Langlais, Stabilization in nonlinear age dependent population model, *Biomath. & Relat. Comput. Probl. Dordrecht* (1988), pp. 337–344.

[307] G.S. Larionov, Averaging method in 'beast of prey-sacrifice' system, *Diff. Uravn.* **16: 12** (1980), pp. 2247–2254. (In Russian.)

[308] A.B. Larson, The hog cycle as harmonic motion, *J. Farm. Econ.* **46** (1946), pp. 375–386.

[309] A. Lasota, Ergodic problems in biology, *Asterisque* **50** (1977), pp. 239–250.

[310] R. Laugalis, D. Shvitra, 1) Red blood production mathematical modelling, *Diff. Eqs. & Appl. Vilnius* **34** (1983), pp. 29–39. (In Russian.) 2) Control of white blood cells production problem, *Lit. Mat. Sb.* **26: 1** (1986), pp. 70–80. (In Russian.)

[311] H.A. Lauwerier, Mathemathical models of epidemics, *Math. Centre Tracts* **138** (1981).

[312] A.Y. Lee, Hereditary optimal control problems: numerical method based upon Pade approximation, *J. Optim. Th. & Appl.* **56: 1** (1988), pp. 157–166.

[313] C.S. Lee, G. Leitmann, Continuous feedback guaranteeing uniform ultimate boundedness for uncertain linear delay systems: an application to river polution control, *Comp. Math. Appl.* **16: 10/11** (1988), pp. 929–938.

[314] E.B. Lee, Variational problems for systems having delay in the control action, *IEEE Trans. Autom. Control* **AC-13** (1968), pp. 387–399.

[315] E.B. Lee, Y. You, Optimal synthesis for infinite-dimensional linear delayed state-output systems: A semicausality approach, *Appl. Math. Optim.* **19** (1989), pp. 113–136.

[316] S.M. Lenhart, C.C. Travis, Global stability of a biological model with time delay, *Proc. Amer. Math. Soc.* **96: 1** (1986), pp. 75–78.

[317] G.A. Leonov, V.B. Smirnova, Asymptotic of one class of phase synchronization, *Appl. Math.* **2** (1979), pp. 144–153.

[318] Le Suan Kan, Calculating solutions for second order equation with delay describing vibrational processes with large grad force in resonance case, *Ukr. Mat. Zh.* **24: 6** (1972), pp. 838–843. (In Russian.)

[319] A. Leung, 1) Conditions for global stability concerning a prey-predator model with delay effects, *SIAM J. Appl. Math.* **36: 2** (1979), pp. 281–286. 2) , Periodic solutions for a prey-predator differential delay equation *J. Diff. Eq.* **26: 3** (1977), pp. 391–403.

[320] D. Levi, *Nonlinear differential-difference equations as Backlund transformations J. Phys. A. Math. & Gen.* **14: 5** (1981), pp. 1083–1098.

[321] D. Levi, O. Ragnisso, Nonlinear differential-difference equations with N-dependent coefficients 1-2, *J. Phys* **A12: 7** (1979), pp. L157–162; L163–167.

[322] J.J. Levin, Boundeness and oscillation of some Volterra and delay equations, *Diff. Uravn.* **5: 2** (1969), pp. 369–398. (In Russian.)

[323] S. Levin, L. Segel, Pattern generation in space and aspects, *SIAM Review* **17** (1985), pp. 45–67.

[324] Lê Xuân Cân, 1) Influence of delay on the parametric simulation process in multidimensional systems, *Tap chi khoa hoc ky thuat* **8** (1977), pp. 31–32. (In Vietnamese.) 2) Influence of delay on the parametric simulation process in multidimensional systems, *Tap chi khoa hoc ky thuat* **9** (1977), pp. 26–32.(In Vietnamese.)

[325] X.Y. Li, S.N. Chow, Maximum principle of optimal control for functional differential systems, *J. Optim. Theory & Appl.* **54: 2** (1987).

[326] Juan Lin, P.B. Kahn, Averaging methods in the delayed-logistic equation, *J. Math. Biol.* **10: 1** (1980), pp. 89–96.

[327] Laifn Lin, J.E. Cihen, Equilibrium and local stability in a logistic matrix model for age-structured populations, *J. Math. Biol.* **25: 1** (1987), pp. 73–88.

[328] D.A. Linkins, R.I. Vitney, Mode analysis of physiological oscillations intercoupled via pure time delays, *Bull. Math. Biol.* **44: 1** (1982), pp. 57–74.

[329] Ching-Tien Liou, Yi. Shyong Chou, Maximum energy control of time delay systems via piecewise linear polynomial functions, *Int. J. Control* **46: 5** (1987).

[330] L.W. Liou, A. Ray, Integrated communication and control systems, *J. Dynam. Systems, Measurement & Control* **112** (1990), pp. 357–364.

[331] R.Sh. Liptser, A.N. Shiryaev, *Statistics of Random Processes*, Springer, 1977.

[332] Y. Liu, B. Anderson, Model reduction with time delay, *Proc. IEEE* **134: 6** (1987), pp. 349–367.

[333] M. Lizana, Asymptotic expansion for the initial value problem of the sunflower equation, *Proc. 11th Int. Conf. Nonlin. Oscill., Budapest Aug. 1987* **17-23**

(1987), pp. 297–300.

[334] A.S. Lodge, J.B. McLeod, J.A. Nohel, A nonlinear singular perturbed Volterra integrodifferential equation occurring in polymer rheology, *Proc. R. Soc. Edinburgh* **A80** (1978), pp. 99–137.

[335] S.O. Londen, I.A. Nohel, Nonlinear Volterra integrodifferential equation occurring in heat blow, *J. Int. Eqs.* **6** (1984), pp. 11–50.

[336] O. Lopes, Stability and forced oscillations, *J. Math. Anal. & Appl.* **55: 3** (1976), pp. 686–698.

[337] K.A. Lukin, Yu.L. Maistrenko, A.N. Sharkovskiy, V.P. Shestopalov, Difference equations method in resonator problem with nonlinear reverberation, *Dokl. Akad. Nauk USSR* **309: 2** (1989), pp. 321–331. (In Russian.)

[338] S.T. Lunt, G. Tolman, Some effects of delays in control systems, *Meas. & Control* **13: 5** (1980), pp. 184–185.

[339] N. Macdonald, 1) Time delay in prey-predator models, *Math. Biosc.* **28: 3/4** (1976), pp. 321–330. 2) Tone lag in systems described by difference or differential equations, *Int. J. Systems Sc.* **8: 4** (1977), pp. 467–479. 3) Time lags in biological models, *Lect. Notes Biomath.* **27** (1978). 112pp. 4) An activation-inhibition model of cyclic granulopoiesis in chronic granulocytic leukomia, *Math. Biosc.* **54: 1/2** (1981), pp. 61–70.

[340] M.C. Mackey, U. an der Heiden, The dynamics of recurrent inhibition, *J. Math. Biol.* **19: 2** (1984), pp. 211–225.

[341] J.M. Mahaffy, 1) Periodic solutions for certain protein synthesis models, *J. Math. Anal. & Appl.* **74: 1** (1980), pp. 72–105. 2) Stability of periodic solutions for a model of genetic repression with delays, *J. Math. Biol.* **22: 2** (1985), pp. 137–144.

[342] C.M. Mahaffy, C.V. Pao, Models of genetic control by repression with time delays and spatial effects, *J. Math. Biol.* **20: 1** (1984), pp. 39–57.

[343] Yu.L. Maistrenko, E.Yu. Romanenko, A.N. Sharkovskii, Integrodifferential equations and numerical population dynamics, *Diff. Eqs. & Appl., Vilnius* **29** (1981), pp. 69–78. (In Russian.)

[344] Zavarei M. Malek, M. Jamshidi, *Time-Delay Systems; Analysis, Optimization & Application*, North-Holland, Amsterdam, 1987.

[345] J. Mallet-Paret, Morse decomposition for delay-differential equations, *J. Diff. Eqs.* **62** (1986), pp. 266-274.

[346] J. Mallet-Paret, R.D. Nussbaum, Global continuation and complicated trajectories for periodic solutions of a differential delay equation, *Proc. Symp. Pure Math.* **45: 2** (1986), pp. 155–167.

[347] M. Mangel, C.S. Peters, Computation of the fitness and functional response of Holling's 'hungary mantid' by the WKB method, *Appl. Math. Letters* **2: 4** (1989), pp. 361–365.

[348] A. Manitius, R. Triggiani, Function space controllability of linaer retarded systems, a derivation from abstract operator conditions, *SIAM J. Control & Optim.* **16: 4** (1978), pp. 599–645.

[349] P. Marcati, Some considerations on the mathematical approach to nonlinear

age dependent population dynamics, *Comput. & Math.* **9: 3** (1983), pp. 361–370.

[350] G.I. Marchuk, *Mathematical Models in Immunology*, Nauka, Moscow, 1985. Second ed. (In Russian.)

[351] G.I. Marchuk, L.N. Belykh, On the treatment of chronic forms of disease according to a mathematical model, *Lecture Notes & Int. Sci.* **38** (1982), pp. 77–87.

[352] M. Marcus, V.J. Mizel, 1) Semilinear hereditary hyperbolic systems with nonlocal boundary conditions, *J. Math. Anal. & Appl.* **77: 1** (1980), pp. 1–19. 2) Stochastic functional differential equations modelling materials with selective recall, *Stochastics* **25** (1988), pp. 195–230.

[353] I.E. Marshall, Control of time delay systems, *IEEE Control Eng. Series* **10** (1978).

[354] I.E. Marshall, S.V. Salehi, Improvement of system performance by the use of time-delay elements, *IEEE Proc. D. Control Th. & Appl.* **129: 5** (1982), pp. 177–181.

[355] M. Martelli, K. Schmitt, H. Smith, Periodic solutions of some nonlinear delay-differential equations, *J. Math. Anal. & Appl.* **74: 2** (1980), pp. 494–503.

[356] R.H. Martin, H.L. Smith, Convergence in Lotka–Volterra systems with diffusion and delay, *Proc. Workshop Diff. Eqs. & Appl., Retzhof Austria* , M. Dekker, 1989.

[357] P. Martiner-Amores, Una applicacion del principio de invarianze en ecuaciones differenciales functionales de tipo neutro, *Rev. Mat. Hisp.-Amer.* **38: 1** (1978), pp. 23–37.

[358] V.M. Matrosov, N.A. Karatueva, Vector Lyapunov function method in the analysis of functions with delay and its applications to immunology, *Math. Modell. Immunol. & Med., Proc. 7th IFIP TC Workconf. Moscow, July 5–11, 1982* (1983), pp. 175–186.

[359] A. Mazanov, On the differential-difference growth equation, *Search* **4: 6** (1973), pp. 199–201.

[360] V.M. Mazurov, D.I. Malov, V.I. Salomyrov, Systems of PH value control in absorbing column with recycle, *Himich. Promysh.* **4** (1974), pp. 63–65. (In Russian.)

[361] M.C. Memory, Stable and unstable manifolds for partial functional differential equations, *Nonlinear Anal., Th., Methods & Appl.* **16: 2** (1991), pp. 131–142.

[362] V.N. Michelkevich, Yu.A. Chabanov, Identification and synthesis of control system of infeed grinding process, *Elektromash. & Elektrooborudov.* **31** (1980), pp. 42–52. (In Russian.)

[363] M.M. Milman, 1) Representation for the optimal control law for stable hereditary systems, *Lecture Notes in Control & Inform. Sc.* **102** (1987), pp. 253–271. 2) Appromimation of the linear quadratic optimal control law for hereditary systems with delays in the control, *SIAM J. Control* **26: 2** (1988), pp. 291–320.

[364] J. Milota, Stability in models with long memories, *Biomath. & Relat. Comput. Probl., Dordrecht* (1988), pp. 523–527.

[365] Masayasu Mimura, Some convection-diffusion equations arising in population dynamics, *Contemp. Math.* **17** (1983), pp. 343–350.

[366] N. Minorsky, Control problems, *J. Franklin Inst.* **232: 6** (1941), pp. 519–551.

[367] Yu.A. Mitropolsky, A.Yu. Shvets, 1) Delay effect influence on stability of a pendulum with vibrating central point, *Anal. Meth. of Nonlin. Vibrations Investigation, Kiev* (1980), pp. 115–120. (In Russian.) 2) Vibration of a pendulum with vibrating central point regarding delay effect, *Anal. Meth. of Nonlin. Vibrations Investigation, Kiev* (1980), pp. 120–128. (In Russian.)

[368] S.E.A. Mohammed, 1) Retarded functional differential equations: A global point of view, *Pitman Research Notes in Math.* **21** (1980). 2) Stachastic functional differentila equations, *Pitman Research Notes in Math.* **99** (1983). 3) Unstable invariant distributions for a class of stochastic delay equations, *Proc. Edinburgh Math. Soc.* **31: 1** (1988), pp. 1–23.

[369] R.R. Mohler, Z. Farooqi, T. Heilig, Lymphocyte distribution and lymphatic dynamics, *Immun. Volume Ded. to G.I. Marchuk on the Occas. of his 60th Birthday* , New York, 1986, pp. 317–333.

[370] T. Mori, S. Barnett, On stability tests for some classes of dynamical systems with perturbed coefficients, *IMA J. Math. Control & Inform.* **5** (1988), pp. 117–123.

[371] H.C. Morris, Approximate periodic solutions of a delay differential model of interacting populations, *Int. J. Systems Sc.* **9: 1** (1978), pp. 111–119.

[372] H.C. Morris, R.K. Dodd, Pattern formation in a delay equation with diffusion, *Int. J. Systems Sc.* **11: 6** (1980), pp. 787–792.

[373] H.C. Morris, E.E. Ryan, R.K. Dodd, Periodic solutions and chaos in a delay-differential equation modelling haematopoesis, *Nonlin. Anal., Th., Methods & Appl.* **7: 6** (1983), pp. 623–660.

[374] I.H. Mufti, On the stability of a single-species population model with time lag, *Int. J. Systems Sc.* **10: 10** (1979), pp. 1149–1154.

[375] J.D. Murray, *Mathematical Biology*, Springer, New York, 1989.

[376] J. Mycielski, S. Swierczkowski, A model of the Neocortex, *Adv. Appl. Math.* **9: 4** (1988), pp. 465–480.

[377] A.D. Myshkis, 1) General theory of differential equations with delay, *Usp. Mat. Nauk* **4: 5** (1949), pp. 99–141. (In Russian.) (Engl. transl.: *Transl. AMS* **55** (1951), pp. 1–62. Also: *Transl. AMS* **4** (1962), pp. 207–267.) 2) General theorems of the theory of ordinary differential equations in nonclassical cases, *1st Math. Summer School, Kiev* , Nauk. Dumka, 1964, pp.45–116. (In Russian.) 3) *Linear Differential Equations with Delayed Argument*, Nauka, Moscow, 1972. (In Russian.) (First ed.: 1951; German transl.: *Lineare Differentialgleichungen mit nacheilendem Argument*, VEB Deutsch. Verlag, Berlin, 1955.) 4) On some problems of the theory of differential equations with deviating argument, *Usp. Mat. Nauk* **32: 2** (1977), pp. 174–202. (In Russian.) (Engl. transl.: *Russian Math. Surveys* **32** (1977), pp. 181–210.)

[378] A.D. Myshkis, L.E. El'sgol'tz, The status and problems of the theory of differentail equations with deviating arguments, *Usp. Mat. Nauk* **22: 2** (1967),

pp. 21–57.

[379] A.D. Myshkis, S.N. Shimanov, L.E. El'sgol'tz, Stability and oscillations of systems with time lag, *Tr. ICNO-1, Kiev* **2** (1963), pp. 241–267.

[380] W. Nababan, K.L. Teo, Necessary conditions for optimal control for systems governed by parabolic partial delay-differential equations in divergence form with boundary conditions, *J. Optim. Theory & Appl.* **36** (1982), pp. 565–513.

[381] B. Neta, J.O. Igwe, Finite differences versus finite elements for solving nonlinear integro-differential equations, *J. Math. Anal. & Appl.* **112: 2** (1985), pp. 607–618.

[382] V.W. Noonlurg, Competing species model with behavioral adaptation, *J. Math. Biol.* **24: 5** (1986), pp. 543–555.

[383] F.U. Nosprev, Averaging in some gyroscopical systems with delay, *Ukr. Mat. Zh.* **35: 2** (1983), pp. 245–248. (In Russian.)

[384] L.I. Novakovskaya, Construction of asymptotic solutions for first order differential equations with random delay, *Ukr. Mat. Zh.* **41: 11** (1989), pp. 1569–1573. (In Russian.)

[385] V.Yu. Novokshenov, I.T. Habibullin, Nonlinear differential difference schemes integrated by methods of reversed problem of diffusion. Asymptotic of solution with $t \to \infty$, *Dokl. Akad. Nauk SSSR* **257: 3** (1981), pp. 543–547. (In Russian.)

[386] C. Nowak, A maximum principle for optimal control problems with neutral functional differential systems, In: *Cybernetics and Systems* (Ed.: K. Trappl), Kluwer Acad. Publ., 1988, pp. 23–30.

[387] L. Nunney, The effect of long time delays in predator-prey systems, *Theoret. Popul. Biol.* **27: 2** (1985), pp. 202–221.

[388] R.D. Nussbaum, 1) Periodic solutions of analytic functional differential equations are analytic, *Michigan Math. J.* **20** (1973), pp. 249–255. 2) Differential-delay equations with two time lags, *Memoirs AMS* **6: 205** (1978). 62pp. 3) Asymptotic analysis of functional differential equations and solutions of long period, *Arch. Rat. Mech. & Anal.* **81: 4** (1983), pp. 373–397.

[389] M.N. Oguztoreli, 1) Activities in a one-dimensional continuous neural network, *Arch. Math.* **14: 3** (1978), pp. 161–169. 2) *Time-Lag Control Systems*, Acad. Press, New York, 1966.

[390] M.N. Oguztoreli, T.M. Caelli, An inverse problem in neutral processing, *Biol. Cybern.* **53: 4** (1986), pp. 239–245.

[391] A.W. Olbrot, L. Pandlofi, Null controllability of a class of functional differential systems, *Int. J. Control* **47: 1** (1988), pp. 193–208.

[392] J.G. Ollason, Learning to forage in a regenerating patchy environment: can it fail to be optimal?, *Theoret. Popul. Biol.* **31: 1** (1987), pp. 13–32.

[393] A.P. Oskolkov, M.M. Ahmatov, Asymptotic difference scheme for filtering liquid equation with delay, *Dokl. Akad. Nauk SSSR* **163** (1987), pp. 138–142. (In Russian.)

[394] K.R. Palanisamy, G.P. Rao, Minimum energy control of time-delayed systems via Walsh functions, *Opt. Control Appl. Methods* **4** (1983).

[395] C.V. Pao, J.M. Mahaffy, Qualitative analysis of a coupled reaction-diffusion model in biology with time delays, *J. Math. Anal. & Appl.* **109: 2** (1985), pp. 355–371.

[396] W. Pauwels, Optimal dynamic advertising policies in the presence of continuously distributed time lags, *JOTA* **22** (1977), pp. 79–89.

[397] M.L. Pena, Asymptotic expansion for the initial value problem of the sunflower equation, *J. Math. Anal. & Appl.* **143: 2** (1989), pp. 471–479.

[398] M. Peschel, W. Mende, *The Predator-Prey Model: Do We Live in a Volterra World?*, Akad. Verlag, Berlin, 1986.

[399] R.V. Petrov, R.N. Stepanenko, Yu.I. Stalko, The mechanism of immune response amplification by stimulator of antibody prediction (SAP) (studies by means ofa mathematical model), *Vistas Appl. Math. Numer. Anal. Atmos. Sc., Immun. (dedicated to G.I. Marchuk on the occas. of his 60th Birthday)* (1986), pp. 355–365.

[400] E. Pinney, *Ordinary Difference-Differential Equations*, Univ. Calif. Press, Berkeley, 1958.

[401] V.G. Pisarenko, *Problems of Relativistic Dynamics of Many Bodies and Nonlinear Field Theory*, Naukova Dumka, Kiev, 1974.

[402] R.E. Plant, A. Fitzhugh differential-difference equation modelling recurrent neutral feedback, *SIAM J. Appl. Math.* **40: 1** (1981), pp. 150–162.

[403] S.R. Pliska, On a functional differential equation that arises in Markov control problems, *J. Diff. Eqs.* **28: 1** (1978), pp. 390–405.

[404] M. Podovski, Analysis of a class of differential equations with delay effect and its application to nuclear reactor kinetics, *God. Vissh. Uchebn. Zav. Pril. Mat.* **11: 3** (1977), pp. 83–93. (In Russian.)

[405] L.S. Pontryagin, V.G. Boltyanskii, R.V. Gamkrelidze, E.F. Mishchenko, *The Mathematical Theory of Optimal Processes*, Interscience, 1962.

[406] R.F. Poppi, J.E. Bonillet, The steady-state equations of a crystallizer with secondary nucleation and changing sign growth kinetics, *Mat. Apl. e Comp.* **4: 2** (1985), pp. 125–137.

[407] H.M. Portella, F.M.O. Castro, N. Arata, Exact solution of the diffusion equation of cosmic-ray nucleons with rising cross section, *J. Phys. G.* **14: 8** (1988), pp. 1157–1165.

[408] W.M. Post, C.C. Travis, Global stability in ecological models with continuous time delays, In: *Integral and Functional Diff. Eqs.* (ed.: T. Herdman), M. Dekkker, New York, 1981.

[409] M.A. Pozio, Behavior of solutions of some abstract functional differential equations and applicatons to predator-prey dynamics, *Nonlinear Anal., Th., Methods & Appl.* **4: 5** (1980), pp. 917–938.

[410] Prajneshu, A stochastic Gomperz model with hereditary effect, *J. Math. Biol.* **8: 2** (1979), pp. 189–196.

[411] J. Prub, Stability analysis for equilibria in age-specific population dynamics, *Nonlinear Anal., Th., Methods & Appl.* **7: 12** (1983), pp. 1291–1313.

[412] R. Pyrz, Dynamical stability of mechanical systems with hereditary properties,

Stability in the Mechanics of Continuous Media, IUTAM Symp. (1981), pp. 273–281.

[413] G.R.W. Quispel, F.W. Nijhoff, H.W. Capel, J. van der Linden, Linear integral equations and nonlinear difference-differential equations, *Physica* **A125: 2/3** (1984), pp. 344–380.

[414] J. Radcliffe, L. Rass, The spatial spread and final size of the deterministic non-reducible n-type epidemic, *J. Math. Biol.* **19: 3** (1984), pp. 309–327.

[415] V. Rasvan, Approximation of distributed systems by lumped parameter systems, *Proc. Int. Symp. Appl. Math. Syst. Theory, Brasdov, 1978* **2** (1979), pp. 365–373.

[416] B.S. Razumihin, *Stability of Hereditary Systems*, Nauka, Moscow, 1988. (In Russian.)

[417] J. Real, Stochastic partial differential equations with delay, *Stochastics* **8: 2** (1982), pp. 81–102.

[418] D.C. Reber, A finite difference technique for solving optimization problems governed by linear functional differential equations, *J. Diff. Eqs.* **32: 2** (1979), pp. 193–232.

[419] M. Renardy, 1) A class of quasilinear parabolic equations with infinite delay and application to aproblem of viscoelasticity, *J. Diff. Eqs.* **48: 2** (1983), pp. 280–292. 2) Singular perturbed hyperbolic evolution problems with infinite delay and an application to a polymer rheology, *SIAM J. Math. Anal.* **15: 2** (1984), pp. 333–349.

[420] P. Renno, On some viscoelastic models, *Atti Acad. Naz. Lincei. Rend. Cl. Sc. Fis. Mat. e Natur.* **75: 6** (1984), pp. 339–348.

[421] E.Y. Rodin, K.F.W. Lee, J. Rosenbluth, A species intercation model, *Math. & Comput. Model.* **10: 3** (1988), pp. 229–234.

[422] G. Rosen, Time delays produced by essentail nonlinearity in population growth models, *Bull. Math. Biol.* **49** (1987), pp. 253–255.

[423] A.B. Rubin, G.Yu. Riznichenko, Mathematical models in ecology, *Itogi Nauk. i Tekhn.* **2** (1988), pp. 113–172. (In Russian.)

[424] O. Rudin, Delay lines in analog computer studies, *IEEE Trans. Industrial Electr. & Control Instrumentation* **11** (1976), pp. 39–46.

[425] I.S. Sadek, Optimal conrol of time-delay systems with distributed parameters, *J. Optim. Th. & Appl.* **67: 3** (1990), pp. 567–585.

[426] D. Salamon, Control and observation of neutral systems, *Pitman Research Notes in Math.* **91** (1982).

[427] M. Saleem, S.U. Siddiqui, V. Gupta, A mathematical model with young predation, *J. Math. Biol.* **25: 1** (1987), pp. 89–101.

[428] N.J. Sancho, C. Mitchell, Economic optimization in controlled fisheries, *Math. Biosc.* **27: 1/2** (1975), pp. 1–7.

[429] T.J. Santiago, Las equaciones difero-diferenciale. Una nota introductoria, *Rev. Math.* **9/10** (1971), pp. 30–33.

[430] T. Sasagava, Optimal control of stochastic systems dependent on the past, *Trans. Soc. Instrum. & Control Eng.* **12: 3** (1976), pp. 257–263.

[431] M. Schatzman, Self organising mathematical models: nonlinear evolution equations with convolution term, *Disord. Systems & Biol. Org. Proc. NATO Adv., Res. Workshop Les Houches, Febr. 25–March 8, 1985* (1986), pp. 385–388.

[432] A. Schiaffino, A. Tesei, Time periodic solutions for Volterra population equations, *Mem. Inst. Ital. Idrobiol. Dott. M. Marchi* **37** (1979), pp. 193–199 (Supplement).

[433] W.A. Schlup, The retarded Josephson equation, *J. Math. Phys.* **19: 12** (1978), pp. 2469–2475.

[434] K. Schmidt (ed.), *Delay and Functional Differential Equations and their Applications*, Acad. Press, New York, 1972.

[435] L.S. Schulman, Some differential-difference equations containing both advance and retardation, *J. Math. Phys.* **15: 3** (1974), pp. 295–298.

[436] K. Schumacher, Traveling-front solutions for integro-differential equations, *J. Reine u. Angew. Math.* **316** (1980), pp. 54–70.

[437] I. Schutte, Über einige Wachstummodelle, die Nachwirkingen berücksichtigen, *Wiss. Z. Humboldt Univ. Berlin, Math.-Nat. R.* **28: 4** (1979), pp. 587–596.

[438] G. Seifert, 1) A temperature equation for a right heat conductor with memory, *Quart. Appl. Math.* **38: 2** (1980), pp. 246–252. 2) Uniform stability of almost periodic solutions of delay-differential equations, *Lecture Notes in Math.* **1017** (1983), pp. 563–567. 3) On a delay-differential equation for single species population variations, *Nonlin. Anal., Th., Methods & Appl.* **11: 9** (1987), pp. 1051–1059.

[439] S.P. Sethi, T.W. McGuire, Optimal skill mix: An application of the maximum principle for systems with retarded controls, *J. Optim. Th. & Apl.* **23: 2** (1977), pp. 245–275.

[440] N.C. Severo, A recursion theorem on solving differential-difference equations and applications to some stochastic processes, *J. App. Probab.* **6: 3** (1969), pp. 673–681.

[441] D. Sforza, Infinite delay parabolic integro-differentila equations with singular kernels, *Nonlinear Anal., Th., Methods & Appl.* **16: 6** (1990), pp. 559–596.

[442] K.P. Shadursky, Vibrations near the statistical equilibrium for quasilinear differential-functional equations, *Probl. Sluch. Poiska* **11** (1988), pp. 184–201. (In Russian.)

[443] L.E. Shaikhet, On an optimal control problem of partly observable stochastic Volterra's process, *Probl. Control & Inform. Th.* **16: 6** (1987), pp. 439–448.

[444] N. Shanmugathasan, R.D. Johnson, Exploitation of time delays for improved process control, *Int. J. Control* **48: 3** (1988), pp. 1137–1152.

[445] A.N. Sharkovsky, Yu.L. Maistrenko, E.Yu. Romanenko, *Difference Equations and their Applications*, Naukova Dumka, Kiev, 1986.

[446] R. Shigui, Connective stability for large scale systems described by functional differential equations, *IEEE Trans. Autom. Control* **33: 2** (1988), pp. 198–200.

[447] Akira Shilata, Nobuhiko Saito, Time delays and chaos in two competing species, *Math. Biosc.* **51: 3/4** (1980), pp. 199–211.

[448] S.V. Shilman, *Method of Generating Functions in the Theory of Dynamical Systems*, Nauka, Moscow, 1978. (In Russian.)

[449] A. Shimbell, Contributions to the mathematical biophysics of the control nervous system with special reference to learning, *Bull. Math. Biophys.* **12** (1950), pp. 241–275.

[450] W.M. Shong, T.J. Tarn, On the spectral controllability of delay differential equations, *IEEE Trans. Autom. Control* **AC26: 2** (1981), pp. 527–528.

[451] Hu Shouchuan, V. Lakshmikantham, M.M. Rao, Nonlinear variation of parameters formula for integro-differential equations of Volterra type, *J. Math. Anal. & Appl.* **129: 1** (1986), pp. 223–230.

[452] R.E. Showalter, N.J. Walkington, A hyperbolic Stefan problem, *Quart. Appl. Math.* **45: 4** (1987), pp. 768–781.

[453] K. Shrestha, The lag relationship between producer and consumer prices. An unobservable variable approach, *Econ. Letters* **22: 2/3** (1986), pp. 173–179.

[454] V.P. Shukla, Conditions for global stability of two-species population models with discrete time delay, *Bull. Math. Biol.* **45: 5** (1983), pp. 793–805.

[455] A.Yu. Shvets, 1) Influence of delay on vibrational-rotational movements of double pendulum, *Methods of Nonlin. Mech. & Appl., Kiev* (1982), pp. 141–149. (In Russian.) 2) Resonance pendulum vibrations regarding nonconstant delay effects, *Asympt. Integration of Diff. Eqs., Kiev* (1985), pp. 173–179. (In Russian.)

[456] D.I. Shvitra, *Physiology Systems Dynamics*, Mikslas, Vilnius, 1989. (In Russian.)

[457] C. Simo, Strange attractors for differential delay equations, *Lect. Notes in Phys.* **179** (1983), pp. 188–191.

[458] C.L. Simonescu-Badea, Controlled return to equilibrium of a dynamical system, *Cybernet. & Systems* **17: 1** (1986), pp. 43–66.

[459] E. Sinestrari, G.F. Webb, Nonlinear hyperbolic systems with nonlocal boundary conditions, *J. Math. Anal. & Appl.* **121: 21** (1987), pp. 449–464.

[460] A.S.C. Sinha, 1) Further stability results of delay-differential systems for digital networks, *Int. J. Systems Sc.* **5: 4** (1974), pp. 317–321. 2) Solution of a certain nonlinear functional differential equation arising from articular cartilage, *Int. J. Systems Sc.* **15: 10** (1984), pp. 1049–1053.

[461] G.M. Smith, *Models in Ecology*, Cambridge Univ. Press, Cambridge, 1974.

[462] H.L. Smith, 1) On periodic solutions of a delay integral equation modelling epidemics, *J. Math. Biol.* **4: 1** (1977), pp. 69–80. 2) Periodic solutions for a class of epidemic equations, *J. Math. Anal. & Appl.* **64: 2** (1978), pp. 467–479. 3) Oscillations and multiply steady states in a cyclic gene model with repression, *J. Math. Biol.* **25: 2** (1987), pp. 169–190.

[463] M.A. Soliman, W. Ray, The optimal control of processes containing pure time delays. Transfer function models, *Proc. 13th Joint Autom. Control Conf. Amer. Autom. Control Councel, Stanford, Calif., 1972* (1972), pp. 294–298.

[464] F. Solimano, E. Bereta, Existence of a globally asymptotically stable equilibrium in Volterra models with continuous time delay, *J. Math. Biol.* **18: 2**

(1983), pp. 93–102.

[465] A. Solominos, 1) A generaliztaion of a problem of Lurie to functional equations, *Dyn. Systems Proc. Univ. Fla. Internat. Symp., Gainesville, 1976* (1977), pp. 489–492. 2) Periodic solutions of the sunflower equation, *Quart. Appl. Math.* **35: 4** (1978), pp. 465–478.

[466] O.J. Staffans, A neutral FDE with stable D-operator is retarded, *J. Diff. Eqs.* **49: 2** (1983), pp. 208–217.

[467] H. Stats, G.A. de Mars, D.T. Wilson, C.L. Tang, Problem of spike elimination in lasers, *J. Appl. Physics* **36: 1** (1965), p. 1510.

[468] H.W. Stech, Hopf bifurcation calculations for functional differential equations, *J. Math. Anal. & Appl.* **109: 2** (1985), pp. 472–491.

[469] H. Stech, M. Williams, Stability in a class of cyclic epidemic models with delay, *J. Math. Biol.* **11: 1** (1981), pp. 95–103.

[470] G. Stepan, 1) A stability criterion for retarded dynamcal systems, *Z. Angew. Math. & Mech.* **64: 5** (1984), pp. 345–346. 2) Great delay in a predator-prey model, *Nonlinear Anal., Th., Methods & Appl.* **10: 4** (1986), pp. 913–929. 3) *Retarded Dynamical Systems: Stability and Characteristic Functions*, Pitman, New York, 1990.

[471] I.H. Suh, Z. Bien, Use of time delay action in the controller design, *IEEE Trans. Autom. Control* **25** (1980), pp. 600–603.

[472] H.H. Swart, Viable controls in age-dependent population dynamics, *J. Math. Biol.* **27: 3** (1989), pp. 297–308.

[473] G.M. Swisher, S. Tenqchen, Design of proportional-minus delay action feedback controller for second and third order systems, *Proc. Amer. Control Conf.* (1988), pp. 254–260.

[474] G. Szabo, A remark of M. Farkas: stable oscilations in a predator-prey model with time lag, *J. Math. Anal. & Appl.* **128: 1** (1987), pp. 205–206.

[475] L. Szekelyhidi, Stability properties of FDE describing the scientific laws, *J. Math. Anal. & Appl.* **150: 1** (1990), pp. 151–158.

[476] G. Tadmor, Optimal controls and their discontinuities in quadratic problems of delay systems, *IEEE Trans. Autom. Control* **AC-30: 7** (1985).

[477] H. Tamura, A discrete dynamical model with distributed transport delays and its hierarchical optimization to preserve stream quality, *IEEE Trans SMC* **41** (1974), pp. 424–429.

[478] V.H. Tetrode, Über den Wirkungszusammenhang der Welt. Eine Erweiterung der klassischen Dynamik, *Z. f. Physik* **10** pp. 317–328.

[479] H.R. Thieme, 1) Asymptotic proportionality (weak ergodicity) and conditional asymptotic equality of solutions to time-heterogenous sublinear difference and differential equations, *J. Diff. Eqs.* **73: 2** (1988), pp. 237–268. 2) Global asymptotic stability in epidemic model, *Lect. Notes in Math.* **1017** (1983), pp. 608–615.

[480] Michio Tokuyama, I. Oppenheim, Statistical-mechanical theory of Brownian motion. Translational motion in an equilibrium fluid, *Physics* **A94: 3/4** (1978), pp. 501–520.

[481] A. Tomaras, 1) Oscillations of an equation relevant to an industrial problem, *Bull. Austral. Math. Soc.* **12: 3** (1975), pp. 425–431. 2) Oscillatory behaviour of an equation arising from an industrial problem, *Bull. Austral. Math. Soc.* **13: 2** (1985), pp. 255–260.

[482] R. Torrejon, J. Young, On a quasilinear wave equation with memory, *Nonlinear Anal., Th., Methods & Appl.* **16: 1** (1991), pp. 61–78.

[483] T. Trombetti, A model approach to nonlinear stability in nuclear reactor space dynamics, *Int. J. Engin. Sc.* **21: 9** (1983), pp. 1069–1079.

[484] S. Tuljapurkar, Population dynamics in variable environments, *Lecture Notes in Math.* **85** (1980).

[485] M. Tylikowski, On stability of the stochastic partial differential equation with deviated argument describing transversal oscillation of a beam squeezed by the axial force, *Zes. Nauk. PSI* **465** (1979), pp. 217–224. (In Polish.)

[486] E.F. Tzar'kov, *Stochastic Disturbances of Functional Differential Equations*, Zinantne, Riga, 1989. (In Russian.)

[487] Ya.Z. Tzypkin, Optimal adaptive systems of objects with delay control, *Dokl. Akad. Nauk SSSR* **287: 5** (1986), pp. 1092–1096. (In Russian.)

[488] R.G. Underwood, A functional differential equation approach to solving infinite games, *SIAM J. Control & Optim.* **14; 2** (1976), pp. 251–267.

[489] H.R. van der Vaart, A comparative investigation of certain difference equations and related differential equations. Implications for model building, *Bull. Math. Biol.* **35: 1/2** (1973), pp. 195–201.

[490] R. Vallee, C. Delisle, 1) Mode description of the dynamical evolution of an acousto-optics bistable device, *IEEE J. Quantum Electronics* **QE-21: 8** (1985), pp. 1423–1427. 2) Periodicity windows in a dynamical system with a delayed feedback, *Phys. Rev.* **A 34: 1** (1986), pp. 309–318.

[491] R. Vallee, C. Delisle, J. Chrostowski, Noise versus chaos in acousto-optic bistability, *Phys. Rev.* **A 30: 1** (1984), pp. 336–342.

[492] J.M.E. van de Vegte, P. Milgram, R.H. Kwong, Teleoperator control models: effect of time delay and imperfect system knowledge, *IEEE Trans. Systems, Man, & Cybernet.* **20: 6** (1990), pp. 1258–1272.

[493] V.A. Vasiliev, V.M. Romanovskii, V.G. Yachno, *Autowave Processes*, Nauka, Moscow, 1987. (In Russian.)

[494] W. Voigt, On finite-difference methods for parabolic functional differential equations on unbounded domains, *Num. Methods & Appl. Proc. 3rd Internat. Conf. Sofia, Aug. 22–27 1989* (1989), pp. 559–567.

[495] V. Volterra, *Théorie mathématique de la lutte pour la vie*, Gauthier-Villars, Paris, 1931.

[496] I.N. Voznesenskiy, *To Choose Control of Power and Heat Supply Turbines*, Soviet Energooborudov., Moscow, 1934. (In Russian.)

[497] H.O. Walther, 1) Asymptotic stability for some functional differential equations, *Proc. R. Soc. Edinburgh* **A74** (1976), pp. 253–255. 2) One transcendental equation in the stability analysis of a population growth model, *J. Math. Biol.* **3: 2** (1976), pp. 187–195. 3) Hyperbolic periodic solutions, heteroclinic connections

and transversal homoclinic points in autonomous differential delay equations, *Mem. AMS* **79: 402** (1989). 104pp.

[498] Xue-Ping Wang, Phase-space description of time delay in scattering theory, *Comm. Partial Diff. Eqs.* **13: 2** (1988), pp. 236–259.

[499] V. Watanabe, Finite spectrum assignment and observer or multivariable systems with commensurable delays, *IEEE Trans. Autom. Control* **31: 6** (1986), pp. 543–550.

[500] M. Wazewska-Czyzevsia, A. Lasota, Mathematical problems of dynamics of the red blood cells system, *Ann. Polish Math. Soc. Ser. 3. Appl. Math.* **17** (1988), pp. 23–40.

[501] G.F. Webb, 1) Nonlinear age-dependent population dynamics in L1, *J. Integral Eqs.* **5: 4** (1983), pp. 309–328. 2) Dynamics of populations structured by internal variables, *Math. Z.* **185: 3** (1985), pp. 319–335.

[502] W. Wendt, Über eine Klasse von parameterabhangingen Funktionaldifferentialgleichungen und ihre diskrete Approximation mit einer Anwendung auf stationäre Boltzmanngleichungen der Halbleiterphysik, *Prepr. Humboldt Univ. Berlin, Sekt. Math.* **52** (1983), pp. 1–31; 51.

[503] J. Weyer, A mathematical model for chemical mass recruitment of ants, *J. Math. Biol.* **21: 3** (1985), pp. 307–315.

[504] F.S. Wheeler, Two differenial difference equations arising in number theory, *Trans. AMS* **318: 2** (1989), pp. 491–523.

[505] J.A. Wheeler, R.P. Feyman, Classical electrodynamics in terms of direct interparticle action, *Rev. Modern Phys.* **21** (1949), pp. 425–433.

[506] B.S. White, The effect of a rapidly-fluctuating random environment on systems of interacting species, *SIAM J. Appl. Math.* **32: 3** (1977), pp. 666–693.

[507] H. Wilhelmsson, Solutions of reaction-diffusion integral equations describing explosive evolution of densities for localized structures, *J. Math. Phys.* **29: 8** (1988), pp. 1776–1779.

[508] D.L. William, Multivariable singularly perturbed feedback systems with time delay, *IEEE Trans. Autom. Control* **32: 11** (1987), pp. 990–994.

[509] L.R. Williams, R.W. Leggetti, Nonzero solutions of nonlinear integral equations modelling infections disease, *SIAM J. Math. Anal.* **13: 1** (1982), pp. 112–121.

[510] A.N. Willson, *Stability and instability criteria for nonlinear distributed networks*, *IEEE Trans. Circuits Th.* **C19: 6** (1972), pp. 615–622.

[511] S.I. Wilson, K.H. Wong, Convergence of a feasible direction algorithm for relaxed controls in time lag systems, *J. Optim. Th. & Appl.* **53: 3** (1987), pp. 461–479.

[512] A. Winfree, 24 hard problems about the mathematics of 24 hours rythms, *Nonlin. Oscillations Biol.* (1979), pp. 93–126.

[513] B. Wittenmark, Sampling of a system with a time delay, *IEEE Trans. Autom. Control* **30: 5** (1985), pp. 507–510.

[514] K.H. Wong, Optimal control computation for parabolic systems with boundary conditions involving time delays, *J. Optim. Th. & Appl.* **53** (1987), pp. 475–507.

[515] K.H. Wong, D. Clements, K.L. Teo, 1) Optimal control computation for linear time lag systems with linear terminal constraints, *J. Optim. Th. & Appl.* **44** (1984), pp. 509–526. 2) Optimal control computation for nonlinear time lag systems, *J. Optim. Th. & Appl.* **47** (1985), pp. 91–107.
[516] A. Worz-Busekros, Global stability in ecological systems with continous time delay, *SIAM J. Appl. Math.* **35: 1** (1978), pp. 123–134.
[517] E.M. Wright, A functional equation in the heuristic theory of primes, *Math. Gazette* **45** (1961), pp. 15–16.
[518] Jiong-yu Wu, On a class of prey-predator population models with time lag, *Acta Math. Appl. Sin.* **11: 1** (1988), pp. 12–21.
[519] Yodhio Yamada, On acertain class of semilinear Volterra diffusion equations, *J. Math. Anal. & Appl.* **88: 2** (1982), pp. 433–451.
[520] R.T. Yanushevsky, *Control of Objects with Delay*, Nauka, Moscow, 1979. (In Russian.)
[521] Yu.P. Yatsenko, Modelling of some vibrational bio-processes, *Cybernetics* **5** (1978), pp. 108–113.
[522] J. Yong, A. Arapostathis, Stabilization of discrete-time linear systems with a time delay in the feedback loop, *Int. J. Control* **48: 4** (1988), pp. 1475–1485.
[523] Kiyoshi Yoshida, The Hopf bifurcation and its stability for semilinear diffusion equations with time delay arising in ecology, *Hiroshima Math. J.* **12: 2** (1982), pp. 321–348.
[524] Toshinobu Yoshida, On periodic responses of a mathematical neuron model, *Biol. Cybernetics* **52: 6** (1985), pp. 383–390.
[525] K. Youcef-Toumi, O. Ito, A time delay controller for systems with unknown dynamics, *J. Dyn. Systems, Measurem. & Control* **112: 3** (1990), pp. 133–142.
[526] R.L. Zandamela, Dortman on economic interpretation of optimal control theory: an extension, *Riv. Mat. Sc. Econ. e Soc.* **8: 2** (1985), pp. 115–122.
[527] B.G. Zhang, K. Gopalsamy, Global attractivity and oscillation in a periodic delay logistic equation, *J. Math. Anal. & Appl.* **150: 1** (1990), pp. 274–283.
[528] V.I. Zhdanov, 1) About doublesided solutions of charged particles movement equation with delay, *Izv. Vuz. Ser. Mat.* **5** (1983), pp. 20–25. (In Russian.) 2) Finite parameter decision's family of the problem of two bodies in dynamics with delay, *Dokl. Akad. Nauk SSSR* **A: 7** (1983), pp. 13–16. (In Russian.) 3) On the one dimensional two-body problem of clasical electrodynamics, *Int. J. Theoret. Phys.* **15** (1976), pp. 157–167.
[529] Meike Zhou, A class of evolution equations for the dynamics of nonlinear age dependent populations, *J. Beljing Normal Univ. (Natur. Sc.)* **1** (1987), pp. 18–24.
[530] W.X.l. Zheng, C.B. Feng, A novel approach to simplification of large scale systems, *IEEE Proc. D. Control Th. & Appl.* (1988), pp. 730–732.
[531] P. Zitek, Stability criterion for anisochronic dynamic systems, *Acta Techn. CSAV* **4** (1984), pp. 399–407.
[532] S.L. Zolotarev, To the investigation of numerical species dynamics on diffusional

models regarding delay effect, *Dyn. Biological Populations, Gorki* (1987), pp. 55–60. (In Russian.)

[533] A.M. Zsykunov, *Adaptive Control of Objects with Delay*, Nauka, Moscow, 1984. (In Russian.)

[534] N.V. Zubov, About existence and exponential stability of periodic and near periodic solutions in systems with delay, *Optimal Control of Mechanical Systems* **L** (1983), pp. 139–144. (In Russian.)

[535] A.M. Zverkin, 1) Dependence of the stability of solutions of linear differential equations with lag upon the choice of the initial moment, *Vestnik Moskov. Univ. Ser. Mat. Mekh. Astr.* **5** (1959), pp. 15–20. (In Russian.) 2) Theorems of existence and unicity for equations with deviating argument in critical case, *Tr. Sem. Teor. Diff. Uravn. Otkl. Argum.* **1** (1962), pp. 37–46.. (In Russian.)

Index

MIX
Papier aus verantwortungsvollen Quellen
Paper from responsible sources
FSC® C105338

If you have any concerns about our products,
you can contact us on
ProductSafety@springernature.com

In case Publisher is established outside the EU,
the EU authorized representative is:
Springer Nature Customer Service Center GmbH
Europaplatz 3, 69115 Heidelberg, Germany

Printed by Libri Plureos GmbH
in Hamburg, Germany